Education and Technology

Education and Technology

An Encyclopedia
Volume 1: A–I

Edited by

Ann Kovalchick and Kara Dawson

A B C CLIO

Santa Barbara, California • Denver, Colorado • Oxford, England

The article entitled Cognitive Flexibility Theory by Rand J. Spiro, Brian P. Collins, Jose Jagadish Thota, and Paul J. Feltovich, on pages 108–115, is a reprint of an article, "Cognitive Flexibility Theory: Hypermedia for Complex Learning, Adaptive Knowledge Application, and Experience Acceleration," that appeared in Educational Technology, 43(5), September-October, 2003, pp. 5–10. This article appears with the permission of Educational Technology Publications.

Library of Congress Cataloging-in-Publication Data

Education and technology : an encyclopedia / edited by Ann Kovalchick and Kara Dawson.
p. cm.
Includes bibliographical references and index.
ISBN 1-57607-351-3 (hardcover : alk. paper) ISBN 1-57607-749-7 (e-book)
1. Educational technology—Encyclopedias. 2. Computer-assisted instruction—Encyclopedias. 3. Internet in education—Encyclopedias. 4. Distance education—Encyclopedias. I. Kovalchick, Ann. II. Dawson, Kara.

LB1028.3.E17 2003
371.33'03—dc22

2003020749

08 07 06 05 04 10 9 8 7 6 5 4 3 2 1

ABC-CLIO, Inc.
130 Cremona Drive, P.O. Box 1911
Santa Barbara, California 93116 1911

This book is printed on acid-free paper ♾
Manufactured in the United States of America

Contents

Volume 2: J–W

Contents by Category

Issues

Leaders

Professional Associations

Projects

Research and Theory

Introduction

It has never been easy to define educational technology or the scope of the profession. Educational technology has roots in the broadly defined disciplines of education, psychology, and communication, as well as more specialized areas of interest such as organizational development, business and management, and computer sciences. As a result, individuals with different professional training experiences often find that they share similar goals and methods, though they may use different words to describe what they do. Educational technology may also refer to specific technological devices and machines. When used in this way, educational technology describes a particular method: the use of a technology or a technique toward achieving an educational outcome. Without an educational focus, the technology in question is only a device. And as if this weren't confusing enough, in some cases the technology may provide educational value only because it is the most convenient and efficient delivery medium, or because only one of its characteristics serves a specific teaching or learning goal.

Although there may be no singular definition for educational technology, any definition is likely to include reference to the use of technology for instruction, training, learning, or teaching. In practice, definitions serve to focus the interest of associations of individuals by emphasizing a particular scope of interest. From 1990 to 1994, Barbara Seels and Rita Richey (1994) led members of the Association for Educational Communications and Technology (AECT) in an effort to update one definition of the field. The AECT definition highlights the scope of activities of member individuals who share common interests in the systematic design and development of instruction and instructional resources using education technologies. These activities are: design, development, utilization, management, and evaluation. More recently, Robert A. Reiser (2001) has suggested a redefinition of the field along with a shift in emphasis, from "instructional technology" to "instructional design and technology." His proposal highlights the numerous ways in which educational technology is regarded as a framework for action. Though not necessarily synonymous, terms such as "instructional technology," "instructional systems design," and "instructional media" have all been used to describe topics relevant to educational technology. To understand this dynamic nomenclature is to understand the historical development of educational technology as a discipline-based effort to establish a consistent and comprehensive set of methods for the use of technology as an educational tool. Yet each new technological invention presents a new set of possibilities as well as challenges for integrating practices with past achievements.

In short, educational technology may describe a process, a product, or a profession defined by a shared knowledge base. Selecting terms that accurately demonstrate the history, achievements, and accomplishments of the field is an effort fraught with ongoing debate and discussion. In this encyclopedia, we have elected to focus on an enduring vision of educational technology. This is a vision that defines educational technology by its service to learning. As such, education provides a framework for selecting and using technology, regardless of whether the technology in question is a machine, a technique, or an innovative idea.

A number of excellent books already exist that offer insight into educational technology as a domain of knowledge. David Jonassen has edited the *Handbook of Research for Educational Communications and Technology: A Project of the Association for Educational Communications and Technology* (1996, 2001), a comprehensive overview of educational technology research and scholarship. Paul Saettler's *The Evolution of American Educational Technology* (1990) provides a historical look at educational technology's evolution as a profession and its supporting institutions and industries. Tjeerd Plomp and Don P. Ely's *International Encyclopedia of Educational Technology* (1996) includes more than 100 in-depth articles that together represent educational technology's global influence as a domain of practice across many academic disciplines. Most recently, Robert A. Reiser and John V. Dempsey have edited *Trends and Issues in Instructional Design and Technology* (2002), an exploration of past and current trends and issues in the field that emphasizes the junction of instructional design, instructional technology, and performance technology. Although these reference books are written for those with a professional or scholarly interest in the field, this encyclopedia is written for a general, nonpractitioner audience.

The broad sphere of influence of educational technology means that this volume should be considered as a complement to the other excellent reference handbooks cited above. In this encyclopedia, we aim to provide an opportunity for those who have had little or no formal introduction to the field of educational technology to learn about its numerous applications and to recognize the relevance of educational technology to many endeavors. We have tried to present concise entries written in layperson's terms that include avenues for further exploration for readers. Where possible we have chosen to highlight the ways that new technologies, enabled by the wide adoption of the personal computer, demonstrate the core principles of educational technology. Though we've tried to ensure a contemporary focus, the history and foundations of the field are critical to understanding its current value and purpose. Therefore, we have included references to established concepts, principles, tools, and individuals that have shaped the field as a professional discipline. For example, we have included such entries as Analysis because the use of educational technologies nearly always requires this professional competency. It is through analysis that we select appropriate technologies and define methods of addressing learning or training needs. We've documented the multiple perspectives found within the field with entries such as Wilbur Schramm, which is a biographical sketch of a pioneer in research on the utilization of television; his work shaped early uses of broadcast technologies as tools for educational and social change. And we have included long-standing approaches to good practice (see Instructional Design and Performance Support). It is on the basis of these foundational concepts and practices that such current applications as Open-ended Learning Environments, Virtual Reality, and Probeware have their significance.

Another challenge that we share with many authors writing on this topic is the never-ending effort to stay current. The multifaceted quality of educational technology means that it is a field on the go. Thus our challenge is compounded not only by the effort to stay current with the pace of technological change but also by the need to stay current with what we know about educational processes—in short, all the activities that we associate with teaching and learning. This became a particularly pressing challenge in the 1990s as the Internet and personal computer became common tools for education and industry; this challenge continues in the first decade of the twenty-first century. Consider, for example, the shifting fortunes of Virtual Universities (see the entry of that same name, as well as the entry Western Governors University), or the impact of changing delivery technologies on the development of educational resources as described in the entry *Who Built America?* Finally, the many entries that reference technologies and strategies that support online communications and collaboration suggest that, even though such topics have been vital to the educational process, we are examining their significance in new ways using new tools.

Take but one example of the pervasiveness of educational technologies: expectations for their use in K-12 schools as evident in the standards developed by professional organizations such as the International Society for Technology in Education (ISTE) and the National Council for Accreditation of Teacher Education (NCATE). These organizations have now established indicators and criteria that define effective uses of technology for instruction, outline adequate levels of access to technology within the schools, consider planning, implementing, and assessing technology usage, and attend to the social, ethical, and human concerns related to technology. Likewise, content-specific associations, such as the National Council for Social Studies and the National Council of Teachers of Mathematics, call for the use of technology to advance content-area learning.

The amount of technology available in schools has increased dramatically in only a few years. Unfortunately, a disproportionate amount of money has been spent on professional development for teachers, technical support, and curricular support. This often results in a lack of technology use, the use of technology in additive, nonessential ways, or the use of technology to replace traditional instructional methods. Many of those traditional methods are grounded in solid pedagogy and research but often do not require the use of expensive hardware and software. This volume thus provides an overview (see the entry Technology in K-12 Schools) of the current status of uses of technology that enhance student learning experiences, encourage higher-level thinking skills, and promote interdisciplinary understandings; another entry on strategies (Curriculum Integration) complements this overview.

As a result of the amount of technology now common in schools, teacher professional development has received increased attention at both the state and national levels. In addition to strengthening teachers' content and pedagogical knowledge, current efforts include expectations for preparing teachers to use information technologies effectively in the classroom. In the entry Teachers: Preparation and Training, the authors review the status of teachers' professional development relative to technology integration. This includes preparing teachers to use technology as an administrative tool, as a productivity tool, and as an instructional tool.

Even with sufficient technical resources and increased attention to teachers' professional development, technology integration may be hindered by a lack of vision (see School Reform); another entry (Technology Planning) contains concrete steps for implementing a school's vision. Part of preparing a vision for technology in K-12 schools includes consideration of social factors. Many of these are addressed in this encyclopedia in entries such as Internet Safety, Digital Divide, and Assistive Technology.

Although there is much progress to be made relative to the integration of educational technology in the K-12 environment, many exemplary projects and activities can serve as models for others. These include the entry Electronic Emissary, describing a resource designed to bring together mentors who are experts in various disciplines with K-12 students; and Learning Circles, an initiative to facilitate collaboration among groups of classrooms around the globe.

Colleges and universities have always valued technology as a research tool to support data analysis and as an administrative tool to support instructional management activities such as student registration and facilities scheduling. Furthermore, within various disciplines, technologies have always been a part of curricula focused on professional and applied sciences such as engineering and broadcast journalism. Within the physical sciences, technologies have long served to model and simulate data analysis and natural phenomena. However, such academic uses of technology often were limited because they required expensive initial investments and their maintenance required highly specialized knowledge; allowing novice learners access presented significant risks. With the introduction of the desktop personal computer and the Internet, access to technology tools and resources has become an expected feature of the postsecondary experience. See the entry on Campus Computing Project for ample evidence of technology's increasing importance in post-secondary education.

The integration of technology into postsecondary environments as an educational support tool has presented three primary challenges for colleges and universities. The first is the organizational change required to ensure the sufficient financial, technical, and human support for educational technology. Traditionally, support for educational technologies has been provided by separate service units; for example, slide projectors and videotape players were distributed by the library, statistical computing support came from the survey research center, videotaping was provided by a media productions unit associated with marketing and publications, and so on. Support was typically device-centered, that is, defined according to the type of media or technology delivery system. As networked computing environments became the norm, the technologies used within these separate units have converged. The management and utilization of educational technology within the postsecondary environment is now less device-centered and more service-centered as the desktop computer has become the predominant gateway to digital content. The ubiquitous nature of networked computers means that less emphasis is placed on managing a particular device and more on providing services to design, create, deliver, and use digital content in an appropriate format. Thus the entries on Learning

Objects and Virtual Library underscore how technology resources and services crisscross the postsecondary environment and present challenges for the effective design of learning experiences and the usage of specific technology tools.

The second challenge that those who teach at colleges and universities face—similar to those who teach in the K-12 level—is defining the role of technology as an educational tool. This is essentially a question of good instructional and curriculum design and often teaching or instructional support units exist within colleges and universities to assist university faculty in developing effective teaching skills. Unlike K-12 teachers, who often learn their profession as students seeking a degree in the field of education, postsecondary faculty typically have little formal exposure to educational theory. Rather, they study to become specialists in a particular discipline. Consequently, they must learn not only to skillfully use technologies but also to use them in ways that enhance and support learning. Teaching and instructional support units are often charged with the responsibility of assisting faculty in gaining the basic teaching skills important to the effective use of technology as a presentation tool and to author and design content for teaching and learning applications. The entries on Computer-Mediated Communication and Web-Based Instruction describe effective principles for designing and delivering content via digital technologies.

Lastly, with the wild popularity of the World Wide Web beginning in 1995, universities and colleges saw technologies as playing a significant role in extending education beyond the physical classroom. The entry on Distance Education offers a detailed consideration of how educational technology has been used to deliver postsecondary learning opportunities in new formats and to new audiences. Numerous entries also address the processes and tools associated with online communication and collaboration. Yet the expectations for using technology for education at a distance have raised many issues for colleges and universities as access to the Internet and levels of technology literacy among potential distance learners are problematic. In addition, the costs of providing quality content to serve distance learner populations while supporting the effective use of technology in the classroom campus can be steep. Effective teaching at a distance often requires a set of skills other than those used to teach in a classroom, where face-to-face student-teacher interaction is the norm.

Educational technologies have also gained a prominent foothold in corporate industry. Employee training and professional development have always been a key concern within corporate environments since they are so crucial to economic productivity. Whether employed for nonformal continuing education or to support ongoing credentialing and skills training, the use of educational technologies within the corporate and industry contexts is most commonly identified with Just-in-Time Training; the entry on Collaborative Technologies also provides an excellent overview of digital tools used to meet not only the communication needs of end-users within corporate environments but also project management and team-building needs within and across organizations. Finally, the value of using technologies to support formal and nonformal communication that enhances the workplace and serves professional and personal growth is aptly described in the entry on Communities of Practice.

We devised an organizational scheme for including entries in this volume, looking at seven overall categories: (1) Foundations; (2) Implementation (e.g., strategies, methods, processes); (3) Issues; (4) Leaders; (5) Professional Associations; (6) Projects; and (7) Research and Theory. These categories, while useful for organizing the study of educational technology, should not be considered exclusive. In fact, there is considerable overlap among them (see Contents by Category). How a term is categorized is a matter of emphasis. Each of the categories is described below.

Foundations are principles that have sustained educational technology as a professional practice over time. These are the building blocks of the field, and any student of educational technology should aim to grasp the significance of these essential terms. Such terms focus on the theories and concepts drawn from a wide range of disciplines and help to provide an orientation toward educational technology as a discipline.

Implementation includes terms that describe the use of technologies to support general educational outcomes. Such terms represent broad applications of technology, typically implemented on principles described within the Foundations category. These entries note examples of the ways that educational technologies may be used, and they are often applicable to a broad cross-section of educational contexts. Thus we have included a sample of entries that describe implementations characterized by specific educational goals, objectives, learners, or learning conditions. We also include terms that define particular technologies existing apart from any educational purpose but that can be defined by their implementation within an educational context (see, e.g., Interactive Television).

Issues relate to the policies and procedures relevant to educational technology. Many terms (see, e.g., Copyright) have always had an impact on the use of educational technologies. Others, such as Acceptable Use Policies and Web Accessibility reflect emerging concerns brought about by new technologies. Many of these terms relate to the way in which teachers, students, and institutions must manage and organize technologies as resources. Others provide insight into the social and cultural changes that result when new technologies are introduced.

Leaders includes biographical sketches of luminaries in the field, individuals who have contributed in numerous ways and whose work can serve as references for further study. Selecting leaders is always a difficult task—particularly in a field as broad and dynamic as educational technology. Therefore, the reader should not assume that the entries provided in this category are exhaustive. Rather, we've aimed to sample the contributions that make educational technology an exciting field by selecting scholars and practitioners whom we view as models of the quality of work worth aspiring toward.

Professional Associations are as varied as the disciplines influenced by educational technology. Those selected for inclusion here represent a wide cross-section of the associations in existence today and should not be considered exhaustive. Rather, we seek to highlight the important roles that professional organizations play in advancing the practice of educational technology principles.

Projects includes terms that are intended to draw attention to best practices and exemplify uses of educational technologies. Like those terms in the Implementation category, the projects described here draw on established strategies and methods specific to the foundations of educational technologies. These terms present work that is widely recognized as a model practice, that have spurred a series of research studies to establish standards for ensuring project success, or that have functioned to demonstrate important proof-of-concept applications of strategies or technologies. The entries selected, however, should in no way be considered inclusive of all the noteworthy projects in the field.

Research and Theory includes terms that summarize important intellectual developments that help to define the significance of educational technology as a field of inquiry. Here we include topics that attempt to explain, predict, or describe learning processes tethered to educational technologies. We also include research that serves as a standard of excellence for the scholarly inquiry that takes place within the field.

Given the range of technologies now available to the general public, we thought it useful to also include a glossary of terms to define many of the technologies now commonly found in the educational technologist's toolbox. These definitions provide a basic description of many technologies and how they function. A broader description of the context for the use of many of these terms can then be located within the alphabetized section of the book. In this way, we seek to give readers a solid grounding in the specific technologies referenced by the contributors who highlight educational technology as a set of shared practices.

References

Jonassen, David., ed. 1996 [2001]. *Handbook of Research for Educational Communications and Technology.* New York: Simon and Schuster Macmillan.

Plomp, Tjeerd, and Donald P. Ely, eds. 1996. *International Encyclopedia of Educational Technology,* 2nd ed. Oxford, UK: Pergamon.

Reiser, Robert A. 2001. "A History of Instructional Design and Technology, Part 1: A History of Instructional Media." *Educational Technology Research and Development* 49(1): 53–64.

Reiser, Robert A., and John V. Dempsey, eds. 2002. *Trends and Issues in Instructional Design and Technology*. Merrill/Prentice-Hall.

Saettler, Paul. 1990. *The Evolution of American Educational Technology.* Englewood, CO: Libraries Unlimited.

Seels, Barbara B., and Rita C. Richey. 1994. *Instructional Technology: The Definitions and Domains of the Field.* Washington, DC: Association for Educational Communications and Technology.

Education and Technology

A

Acceptable Use Policies (AUP)

One of the most powerful trends in education today is the integration of technology throughout the K-12 curriculum. As our schools begin traveling the information superhighway, it is crucial that certain guardrails are put into place to ensure the safest route to effective integration. As schools use the Internet and technology on an everyday basis, it becomes critical that a clear set of guidelines is established to promote a positive and efficient use of these resources. According to the Center for Democracy and Technology, 92 percent of U.S. schools utilize an acceptable use policy as their guidepost on the Internet.

Simply stated, an AUP is a legal document that identifies the parameters of the acceptable use of technology as indicated by a school. Its purpose is to establish a standard of behavior that is expected by those individuals who use the technology (computer-based services, information networks, Internet access, etc.) made available through the school system. These technical resources are intended to serve as educational supplements, as tools for instruction, and as tools for personal productivity. The AUP is a policy statement that focuses on maintaining the intended purpose of such resources in an effort to sustain their use within the educational setting.

A variety of individuals should be involved in the development of a school system's AUP. A legal expert skilled in both school law and state policy development should be a primary participant in the development of an AUP. Others include school administrators, school technology officials, teachers, parents, and community leaders.

Most schools require that these policies be signed by administrators, teachers, students, and parents in an effort to promote awareness on all levels. Although each AUP is unique, most include a similar set of content. Examples of content include: a description of who is authorized to use the resources; a working definition of what type of use is acceptable; and a set of repercussions for the misuse of such resources.

Enforcement and accountability are the key ingredients in maintaining a usable AUP. Without these two features, the AUP simply becomes another form filed away with little to no actual meaning. With the combined efforts of all parties involved, an effective AUP can help to protect the use of technology in the realm of education.

Certain limitations may lead to ineffectiveness. Due to the legal nature of the document, the vocabulary and sentence construction may not easily convey the meaning of the policy to most students and many adults. Due to this lack of clarification, parents and students often sign the policies without deciphering its true purpose. This haphazard approach leads to a student population that simply does not understand the true nature of the policy or the behavior that is expected of them. One solution is to draft a separate memo to accompany the AUP in an effort to more clearly identify the meaning of the school district's definition of "acceptable use." Additionally, it would also be beneficial to rework the AUP into an age-appropriate set of guidelines that would serve to identify student behavior expectations. These guidelines could then be discussed in the classroom and posted as reminders in visible locations.

Even with the most detailed AUP, it is a good idea for teachers to expand upon the school's AUP and establish an additional set of rules (to complement the AUP) specific for their individual class and written in an age-appropriate language. By discussing these rules as a group, the teacher will have an opportunity to reemphasize that students will be held accountable for their actions, to clarify the expectations, and to underscore that it is a privilege for the students to be able to use such resources. It also highlights the fact that the rules will be enforced on a regular basis.

Judith Oates Lewandowski

See also Copyright; Educational Fair Use; Internet Safety; Netiquette; Technology Planning

References

Critiquing Acceptable Use Policies (www.io.com/~kinnaman/aupessay.html). A critique of acceptable use policies with many interesting links.

Crystal, Jerry. 2000. "Empowering Your Acceptable Use Policy." *Technology and Learning* 21(4): 27.

Indiana Department of Education Model AUP (doe.state.in.us/olr/aup/aupmod.html). This is the officially sanctioned state model for policies to be incorporated by the state.

K-12 Acceptable Use Policies (www.erehwon.com/k12aup). This site offers an assortment of reference materials, including example AUPs, student account agreements, and examples of letters to parents. Great resource for the online teacher.

Sample AUP in Spanish (www.rice.edu/armadillo/aupspanish.html). Bilingual forms are necessary for use in some communities. The emphasis should be placed on informing as many parents as possible.

Active Learning

Active learning constitutes learning that helps students to think critically, analyze, synthesize, and evaluate information, work efficiently and effectively in groups, and solve problems within a variety of different disciplines. Active learning is an attempt to counter traditional instructional models that primarily consist of knowledge transmission and development of inert knowledge (Whitehead 1929). Inert knowledge refers to knowledge learned out of context that is not readily transferable to novel situations. For example, students may learn the formula for determining distance but be unable to apply this information within the context of an interdisciplinary scenario.

Assumptions of Active Learning

S. E. Berryman (1991) outlines several assumptions of traditional educational practice that have been with us since the industrial age:

1. Knowledge transfer occurs when students learn decontextualized concepts;
2. Learners are information receivers or knowledge sponges and teachers are information providers;
3. Learning is a behavioristic endeavor;
4. Learners are blank slates waiting to be written upon; and
5. Knowledge is best attained independent of context.

R. S. Grabinger (1996, 666) counters these "erroneous assumptions" with the following assumptions on which active learning is based:

1. Knowledge transfer is difficult and is best accomplished with content and context learning;
2. Learners are active participants in the learning process;
3. Learning involves cognitive functioning and is constantly growing and evolving;

4. Learners' experiences and prior knowledge must be considered in all learning situations; and
5. Skills and knowledge are best acquired and assessed in authentic and holistic forms.

Rich Environments for Active Learning

In order to operationalize these assumptions, Grabinger (1996) proposes the use of instructional systems: real environments for active learning (REALs). These environments have six main characteristics or attributes:

1. *Constructivist underpinnings* guide the development of REALs and build on the notion that learning is an evolutionary process by which students modify their personal representations of knowledge as new knowledge is explored. This process involves social interaction and collaboration.
2. *Authenticity* also guides the development of REALs. Learning tasks should be as realistic as possible in terms of context and task. Authenticity is important because it encourages student ownership in learning. Through authenticity, problems hold more relevance, develop deeper meaning and understanding (thus increasing the likelihood of transfer to other situations), and encourage collaboration, cooperation, and negotiation.
3. *Student-centeredness* is essential in REALs because it encourages intentional, responsible learners and lifelong learning skills such as reflection and metacognition or thinking about and analyzing what one is learning and how well learning is progressing.
4. *Collaboration* is a key feature in REALs because students are able to shape their personal knowledge by learning from others, are willing to take more learning risks in a group setting, are able to learn about individual accountability within a group setting, and are exposed to skills and issues that they will face in an increasingly collaborative workforce.
5. *Generative learning* refers to learning where students generate knowledge through active participation in the learning process. Students may generate knowledge as they attempt to make sense of multiple perspectives or as they compare and contrast their knowledge representation with those of their peers.
6. *Authentic assessment* of the environment and of student learning is required in REALs. Teacher observation, student interviews, focus groups, product analysis, portfolios, journals, and

peer evaluations are among the techniques that could be used during authentic assessment.

Instructional Strategies Associated with Active Learning

Numerous instructional strategies may be considered when implementing the principles of active learning, including: anchored instruction, collaborative learning, problem-based learning, cognitive apprenticeships, and case-based instruction. A brief definition of each follows. While each strategy is presented separately, they are rarely used independently, as there is considerable overlap among them.

1. *Anchored instruction* is grounded in a realistic event or problem that is meaningful and motivating to students, is complex, requires the consideration of multiple perspectives and solutions and the use of multiple processes, and facilitates collaboration, cooperation, and negotiation.
2. *Collaborative learning* is built on the need for students to collaborate with each other to share perspectives, solutions, and plans related to a complex task or scenario. Collaborative learning requires individual accountability within a group situation and parallels expectations in the modern workforce.
3. *Problem-based learning* is grounded in the process students go through to solve a realistic problem and requires self-directed learners, acquisition of content knowledge, and use of metacognitive strategies (Savery and Duffy 1994).
4. *Cognitive apprenticeships* are modeled after traditional apprenticeships. Whereas traditional apprenticeships involved learning a visible activity or skills, cognitive apprenticeships involve using mentors to model processes that are typically invisible, such as problem-solving, comprehension, and computation.
5. *Case-based instruction* involves the use of stories or teaching "cases" to facilitate contextual knowledge and understanding.

Applications of Active Learning

Educational technology provides an excellent medium to facilitate active learning. Technology provides a nonlinear, multimedia context in which to develop teaching cases and anchored scenarios. It also enables easy revisions to these contexts. Likewise, educational technology may facilitate the collaborative process and enable students to communicate with geographically disparate peers and experts. The following three projects exemplify or facilitate the principles of active learning and are detailed in separate entries in this encyclopedia.

The Adventures of Jasper Woodbury

The Adventures of Jasper Woodbury is a series of video-based instructional materials with accompanying printed teacher manuals designed to facilitate higher-level thinking skills via motivating and authentic problem-solving and reasoning, interdisciplinary mathematics instruction, and collaboration. The Jasper Woodbury series is distinct from other video-based instructional efforts because students are presented with complex, believable scenarios and a challenge that must be solved by integrating mathematic concepts and skills with the story details. The series was developed by an interdisciplinary staff at the Learning Technology Center at Peabody College, Vanderbilt University; its goals include improving instructional tools for teachers through the use of technology. The series integrates learning theory such as anchored instruction, generative learning, and constructivism with classroom practice.

Electronic Learning Circles

Learning Circles groups six to ten classes of students attending different schools in diverse locations with the goal of publishing their collective work organized around an interdisciplinary, curriculum-based theme. Each school takes the responsibility to create a project, direct the student responses from the others schools, monitor the progress of their partners in completing the task, and then integrate the responses into a final report or anthology.

Electronic Emissary

The Electronic Emissary is an example of how technology can facilitate active learning. It is a web-based service and resource center that helps preschool and K-12 teachers and students with Internet access locate mentors who are experts in various disciplines, for purposes of setting up curriculum-based, extended electronic exchanges among teachers, students, and experts.

Kara Dawson

See also Anchored Instruction; Cognitive Apprenticeship; Electronic Emissary; Jasper Woodbury; Learning Circles; Problem-Based Learning

References

Berryman, Sue E. 1991. *Designing Effective Learning Environments: Cognitive Apprenticeship Models. ERIC Document 337 689,* 1–5.

Grabinger, R. Scott. 1996. "Rich Environments for Active Learning." In David H. Jonassen (ed.), *Handbook of Research for Educational Communications and Technology* (pp. 665–692). New York: Simon and Schuster Macmillan.

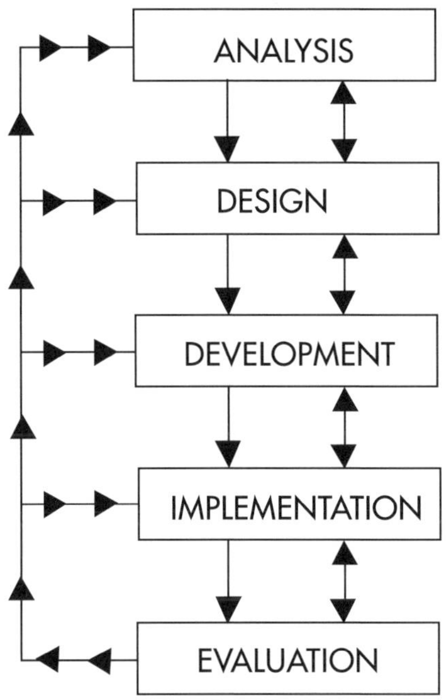

FIGURE 1: An ISD Model Featuring the ADDIE Processes

SOURCE: Deborah J. Grafinger (1988), *Basics of Instructional Systems Development,* INFO-LINE Issue 8803 (Alexandria, VA: American Society for Training and Development).

Savery, John R., and Thomas M. Duffy. 1994. "Problem Based Learning: An Instructional Model and Its Constructivist Framework." *Educational Technology* (August).

Whitehead, Alfred N. 1929. *The Aims of Education.* New York: MacMillan.

ADDIE Model

The ADDIE model describes a systematic approach to instructional development. The term is virtually synonymous with instructional systems development (ISD) and evolved informally through oral tradition. It is not a specific, fully elaborated model in its own right but rather an umbrella term that represents a family of models that share a common underlying structure. The acronym ADDIE refers to the major processes that comprise the generic ISD process: Analysis, Design, Development, Implementation, and Evaluation. When used in ISD models, these processes are considered to be sequential but also iterative, as depicted in Figure 1.

The basic engine of ISD models is the systems approach: viewing human organizations and activities as systems in which inputs, outputs, processes (throughputs), and feedback and control elements are the salient features. Advocates claim that the process of designing instruction can be carried out more efficiently and effectively if the steps are followed in a logical order so that the output of each step provides the input for the next. For example, the output of the Analysis phase is a set of performance deficiencies (such as errors being made by workers), which can be broken down to determine what ought to be taught. This output is converted into statements of performance objectives. In the Design phase the content and objectives are examined to decide on appropriate sequencing, media, and methods—specifications that comprise the blueprint for the instruction. The blueprint created in the Design phase is converted into instructional materials and procedures in the Development phase. The materials and procedures are used by actual learners in the Implementation phase. In the Evaluation phase the learners and the instructional system are probed to decide whether revisions are necessary, in which case the process would be repeated with the next version of instruction.

The iterative aspect of the model is represented by the line and arrows running vertically down the left side of the model and the two-headed arrows between each component, as depicted in Figure 1. Each major phase of the process is accompanied by some sort of formative evaluation to test the adequacy of the decisions made during that phase. After Analysis, for example, are the descriptions of the audience and the learning needs accurate? After Design, are the objectives and methods judged appropriate by experts? After Development, does the prototype work in a small-scale tryout (or how can it be improved)? After Implementation, did the entire intervention achieve its goal (or what remains to be done)? This summative evaluation is what is symbolized by the final Evaluation phase. At each of these phases, the results of the evaluative activity could lead the developers to revisit earlier steps (thus the two-headed arrows).

The origin of the label itself is obscure, but the underlying concepts of ISD can be traced to the model developed for the United States armed forces in the mid-1970s. As Robert Branson (1978) recounts, the Center for Educational Technology at Florida State University worked with a branch of the U.S. Army to develop a model, which evolved into the Interservice Procedures for Instructional Systems Development (IPISD), intended for the Army, Navy, Air Force, and Marine Corps. Branson (1978) provides a graphic overview of the IPISD, which shows five top-level headings: Analyze, Design, Develop, Implement, and Control. This model is referenced in virtually all subsequent historical reviews of instructional development, but notably it is not referred to by the ADDIC acronym, thus it is clearly not the source of the ADDIE acronym either.

The underlying concepts of the IPISD model can be found in an earlier handbook by Leslie Briggs (1970), who also was affiliated with Florida

State University. Briggs's model (1970) incorporates ideas similar to the IPISD model, but without the ADDIC headings.

Although Sivasailam Thiagarajan (1976) is sometimes cited as the originator of the ADDIE label, this is not satisfactory because he only refers to "the basic systems approach A-D-E model" and not "ADDIE," nor does he provide a visual or verbal model as such.

In fact, the term "ADDIE" does not appear at all in the many textbooks on instructional design, the dictionaries or encyclopedias of education, or the several histories of instructional design written in the 1980s and 1990s. The name itself seems to have been disseminated by word-of-mouth, beginning perhaps in the 1980s. The ADDIE processes appear in a figure in a how-to monograph distributed by the American Society for Training and Development on "basics of instructional systems development" (Grafinger 1988), as shown in Figure 1, but nowhere in the monograph is the acronym ADDIE itself given (it is consistently referred to as the "ISD model" in Grafinger [1988]). Similarly, Allison Rossett (1987) includes a figure showing an ISD model in which the superordinate boxes are labeled with the five ADDIE names, but the caption reads "What happens during ISD." ADDIE also appears quite frequently on the World Wide Web in various manifestations. One of the better-known web sources is "Big Dog's ISD Page" (Clark 1995). As with Deborah Grafinger and Rossett, Donald Clark provides a visual model incorporating the ADDIE terms but refers to it as the "ISD model."

One of the few explicit narrative references to the ADDIE model in the academic literature of the field is found in one work (Molenda, Pershing, and Reigeluth 1996) and is used as a major organizing principle in another (Gustafson and Branch 2002). Neither work provides any citation for the references to ADDIE. Thus it is only in the recent literature that the term is beginning to take on a more fully elaborated meaning. However, authors appear to be creating their own interpretations, as there does not appear to be an original, authoritative version of the ADDIE model.

Michael Molenda

See also Analysis; Evaluation; Instructional Design

References

Andrews, Dee H., and Ludwika A. Goodson. 1980. "A Comparative Analysis of Models of Instructional Design." *Journal of Instructional Development* 3(4): 2–16.

Branson, Robert K. 1978. "The Interservice Procedures for Instructional Systems Development." *Educational Technology* (March): 11–14.

Briggs, Leslie J. 1970. *Handbook of Procedures for the Design of Instruction.* Pittsburgh: American Institutes for Research.

Clark, Donald. 1995. Big Dog's ISD Page. Available online at www.nwlink.com/~donclark/hrd/sat1.html#model (December 9, 2002).
Grafinger, Deborah J. 1988. *Basics of Instructional Systems Development.* INFO-LINE Issue 8803. Alexandria, VA: American Society for Training and Development.
Gustafson, Kent L. 1994. "Instructional Design Models." In T. Husen and T. N. Postlethwaite (eds.), *The International Encyclopedia of Education,* 2nd ed. Oxford, UK: Pergamon.
Gustafson, Kent L., and Robert M. Branch. 2002. "What Is Instructional Design?" In Robert A. Reiser and John V. Dempsey (eds.), *Trends and Issues in Instructional Design and Technology.* Columbus, OH: Merrill Prentice-Hall.
Molenda, Michael, James A. Pershing, and Charles M. Reigeluth. 1996. "Designing Instructional Systems." In Robert L. Craig (ed.), *The ASTD Training and Development Handbook,* 4th ed. (pp. 266–293). New York: McGraw-Hill.
Rossett, Allison. 1987. *Training Needs Assessment.* Englewood Cliffs, NJ: Educational Technology Publications.
Thiagarajan, Sivasailam. 1976. "Help, I Am Trapped Inside an ID Model!" *NSPI Journal* (November): 10–11.

Adult Learners

Adult learners are people over the age of eighteen in an instructional situation, whether formal or informal. Lifelong learning is now a common expectation of society, but in the field of educational technology most adult learners undertake their studies in employee education and training programs. This stems from the fact that since the 1980s the preponderance of instructional design practice has occurred within the private sector, primarily in business and industrial settings. This coincides with the steady growth of employee training as an integral part of most organizations. This growth reflects the emphasis on producing a more knowledgeable workforce and, increasingly, on improving employee on-the-job performance and solving organizational problems.

For many, the term "adult learner" is too comprehensive. Unlike children, adults span a wide range of ages and developmental phases. There is little agreement upon what these phases are. Some view adult learners in terms of age groupings with characteristic physical changes, others in terms of a person's social and psychological history, and others in terms of typical life events. Nonetheless, it is common to assume that life changes have implications for adult learning in terms of learning style, the motivation to learn, and the capacity to learn. At the least, many scholars recommend distinguishing younger adult learners from older adult learners.

Theories of adult learning encompass a range of topics, many of which are unlike those addressed in generic learning theory. Most adult-oriented theories speak to issues such as how adults learn, what they learn, and why they learn. Such learning theory has implications for a variety of instructional design factors, including how to encourage participation in learning activities, adult motivation and persistence, how to promote using information learned in one's job, and how to modify adult attitudes.

The interest in adults as learners stems from questions surrounding the extent to which they differ from children in learning situations. Popular views of adult learners present them as being more self-directed, autonomous, and problem-oriented. However, others cite many examples of adults who are not mature and self-directed learners. Most scholars in the field recognize many similarities between the learning patterns of adults and children while acknowledging those characteristics that are unique to adults.

The characteristics of adults that seem to most influence learning pertain to demographic and experiential backgrounds, capacity and competence, and attitudes. One of the most important demographic questions is whether age influences one's ability to learn. Most feel that age per se is not a good indication of who is likely to be the best learner. Although at some point age can suggest the likelihood of declining physical abilities, for most people it is a less powerful predictor of learning success than are the experiences one has had. In employee training situations, experiences that are more likely to be associated with effective learning relate to the nature and extent of a person's previous education and training as well as her work, professional, and cultural and language backgrounds.

In employee training especially, it seems that one's experiences can, to a great extent, counterbalance the debilitating effects of age on one's capacity to learn. Although a person may become less adept in general learning with age, people often continue to grow as learners when the instruction pertains to areas of special interest or areas with which they have considerable experience.

Adult attitudes are often important predictors of successful learning. Of particular importance to success in the classroom are feelings of self-confidence, perceptions of the instruction's relevance, and one's personal priorities. These attitudes are often difficult to change and are typically ingrained in well-established work habits. In colloquial language, this is called being "set in your ways," a characteristic of many older adults.

In employee training, learners' attitudes toward their jobs, the company, their supervisors, and even their coworkers can influence the learning process. These attitudes influence not only how much learning occurs but also whether the information learned will be used on the job. Attitudes are an element effecting motivation in the classroom and in the workplace.

Special emphasis is warranted in relation to adult attitudes toward the new technologies used in employee training. Many training programs, especially in larger organizations, are turning to the use of advanced, technology-based instructional delivery systems. A prime example of this is the preponderance of web-based training. Even though there are many advantages of such training methods, there can be problems. Many adult learners and instructors alike have had little experience with computers as a learning tool. Some persons not only don't like learning with computers, they also feel a sense of computer anxiety. These feelings can constrain the learning process.

Any learning situation is characterized to some extent by the nature of the learners, and employee training is no exception. In what ways do instructional designers build upon the distinctive nature of adult learners and the training environment? Today, predesign analysis includes a consideration of the learner and the work context, in addition to the training content. Of particular importance in this analysis is that adults come to employee training programs with a certain amount of related knowledge, skills, and attitudes gathered from previous experiences they've had, both on the job and off. Instructional strategies and motivation tactics must be attuned to these characteristics. Since participation in training programs is usually not voluntary and often detracts from pressing job responsibilities, instruction needs to be efficient, relevant, and engaging to be considered worthwhile. Instruction should build upon real-life situations that will be encountered and provide sufficient hands-on practice for people to feel comfortable with using their new skills. New technologies should not be used in training without providing sufficient practice and orientation to their use. Instructors of adult learners should often take the role of a facilitator rather than a lecturer. Furthermore, there should be an immediate opportunity for trainees to use the knowledge and skills learned back on the job. If the ultimate goal is to change the employee's work habits, training programs must also include tactics for continuing on-the-job support such as refresher experiences or using job aids and tools. Often supervisors are also trained in an effort to ensure daily support.

Rita Richey

See also Analysis; ARCS: A Model of Motivational Design; Instructional Design; Just-in-Time Training; Performance Support; Web-Based Instruction

References

Knowles, Malcom S., Elwood F. Holton, and Richard A. Swanson. 2000. *The Adult Learner: The Definitive Classic in Adult Education and Human Resource Development,* 5th ed. Houston, TX: Gulf Professional Publishing.

Merriam, Sharan B., and Rosemary S. Caffarella. 1998. *Learning in Adulthood: A Comprehensive Guide,* 2nd ed. San Francisco: Jossey-Bass.

Wlodkowski, Raymond J. 1998. *Enhancing Adult Motivation to Learn: A Comprehensive Guide for Teaching All Adults,* 2nd ed. San Francisco: Jossey-Bass.

Alternative Assessment

Alternative assessment refers to nontraditional assessment methods. Such assessment methods are often applied to complex knowledge and performances and engage students in higher-level thinking and authentic performances while simultaneously permitting reliable scoring and manageable record-keeping. Most frequently, alternative assessment is used as a complement to rather than a replacement for traditional objective tests. Objective tests are an efficient way to assess students' knowledge of rote facts, whereas alternative assessment strategies move beyond such testing and meet the learning styles of more students. Alternative assessment is closely linked to curriculum planning and instructional methods. Thus the push for constructivist learning environments, authentic learning experiences, collaboration, and facilitation of higher-order thinking skills has fueled interest in alternative assessment.

Alternative assessment strategies often involve the use of rubrics. The term "rubric" has multiple meanings, but in the world of education it refers to formal guidelines used to assess student work. Typically, rubrics are presented in a matrix format with performance levels in the top row, performance indicators along the left column, and evaluation criteria in the cells. Teachers often involve students in the creation of rubrics so that they have a vested interest in the learning experience. A properly designed rubric provides enough guidance for students to know what is expected of them but enough generality to encourage creativity and higher-level thinking.

Educational technology has many characteristics that can enhance alternative assessment strategies, including multiple ways of representing knowledge, support of diverse communication strategies, facilitation of collaborative work, and multiple data collection strategies. Four common alternative assessment strategies that can be enhanced by educational technology are (1) performance-based assessment; (2) project assessment; (3) portfolio assessment; and (4) journal assessment.

Performance-Based Assessment

In performance-based assessment, teachers observe students performing a particular task such as using science laboratory equipment or giving an oral presentation. Educational technology can enhance this type of alternative assessment by providing means of record-keeping through spreadsheets and video technology. Video technology also enables students to review their work and for teachers to share this work with parents. Educational technology can also enhance this type of assessment through the use of simulations. Students may manipulate a variety of variables related to maintaining an ecosystem, managing a city, or negotiating with other countries. Such simulations enable students to demonstrate competency through performances that may otherwise be impossible for fiscal, managerial, or safety reasons.

Project Assessment

In project-based assessment, students are given a project to complete related to integrated curriculum goals. Ideal projects simulate real-life situations, are rich in design and long-term, encourage the development of multiple responses, facilitate collaboration and individual accountability, and encourage creativity, knowledge integration, and planning skills. Examples of projects include development of a school-wide election campaign, creation of a product designed to perform a particular task, or composing a persuasive presentation about a controversial issue. Educational technology can enhance this type of assessment by providing a variety of formats for final products or presentations (i.e., text, graphics, multimedia, video, etc.) by enabling communication with subject-matter experts and by allowing research opportunities beyond those available in the school library.

Portfolio Assessment

Portfolio assessment involves students purposefully selecting works that show effort and achievement in specified areas. Portfolios can be used to assess progress over time. For example, some schools ask each student to keep an ongoing writing portfolio throughout their elementary years. Portfolios can also be used to encourage student reflection and provide summative evaluation information. For example, some teacher education programs require graduating students to complete a portfolio documenting the teaching standards they have met throughout the program. Electronic portfolios are one of the most popular ways that educational technology is used to facilitate authentic assessment and an entry is devoted to them in this encyclopedia.

Journal Assessment

This type of assessment involves students' expressions and reflections and is frequently ongoing. Journal assessment can be highly structured, with strict guidelines and directions, or open-ended, with minimal direction. Students are frequently asked to keep writing journals where they document ideas they have for stories or poems, or math journals where they explain how they came to solve particular problems. Students may also be asked to keep journals about personal experiences related to class topics. For example, when studying habitat destruction, students may be asked to document instances they notice throughout their communities; or when studying nutrition, students may be asked to document their eating and exercise habits for a specified period. Educational technology can be used to allow students to share journals with others beyond the four walls of the classroom. For example, when studying Japan and its culture, American students may be able to "journal" with Japanese students via e-mail. Likewise, journals need not be in written form. Educational technology allows for a variety of formats, thus meeting the needs of more students and encouraging multiple approaches in the same assignment.

Kara Dawson

See also Electronic Portfolios; Evaluation

References

Stiggins, Richard. 2000. *Student-Involved Classroom Assessment.* 3rd ed. Upper Saddle River, NJ: Prentice-Hall.

Wiggins, Grant. 1999. *Assessing Student Performance: Exploring the Purpose and Limits of Testing.* San Francisco: Jossey-Bass.

Analysis

Analysis represents the planning phase critical to the effective use of educational technology. It is what educational technology professionals undertake when determining what action should be taken. Whether called "analysis," "assessment," "needs analysis," or "performance analysis," the focus is on making decisions regarding the design and use of technologies toward achieving educational outcomes by examining and considering many viewpoints and data elements rather than by habit, intuition, or fiat. Through analysis, it is possible to cultivate an analytical point of view rather than a view that favors technology or formal classroom instructional strategies. Similar to the diagnostic activities carried out by doctors, architects, and consultants, analysis is a data-driven process that helps educational technologists identify key problems to design effective solutions. During analysis, many sources, questions, instruments, work products, and approaches are used to plan a solution.

Analysis frequently occurs "up front," or "first," especially when influenced by conventional linear models of instructional systems development. However, analysis can be both responsive *and* proactive, immediate *and* perpetual. Examples of when and how an analyst might act include:

- When a client asks, for example, for an online learning program to handle time-management challenges, the educational technologist endeavors to examine the situation before buying or building a program.
- The educational technologist is aware that time is a critical resource in an engineering company and that there's been some concern about its efficient uses in the past. For eight months, she's been gathering data associated with that topic, including

supervisory comments, engineering opinion and references from the literature, and best practices. Now, when the topic comes up with her vice president, she is ready to speak with authority on the subject.

- The educational technologist adds several questions to the annual assessment survey and spends hours poring over the data. She also plays an active role on the organizational intranet, reading listservs, and asking questions. This extant data provides key directions for planning.

American philosopher John Dewey deserves credit as a parent of analysis. He turned away from the European and manufacturing traditions and urged interest in the needs of the learner. While modern practice now perceives soliciting learner perspectives as commonplace, it was revolutionary at the time for educators to be interested in anything other than the job or subject matter.

Ralph Tyler, generally recognized as the originator of behavioral objectives, emphasized a questioning process as a means of determining objectives. Two decades later, Robert Mager made Tyler's work accessible by crystallizing tools and approaches associated with analysis as well as finding and writing instructional objectives. Objectives assist in planning only if they are the "right" objectives, which in this case is linked to a robust definition of where the organization or individual wants to go, what the current state is, and what it will take to narrow that gap. Thus analysis assures fresh, varied, and data-driven roots for objectives.

Several scholars (Mager 1984, 1970; Kaufman and English 1979; and Rossett 1999) urge an expanding view of analysis predicated on a holistic vision of how organizations and people operate. Roger B. Kaufman presses the profession to consider how gaps and interventions affect the group, the organization, and society, not just the individual performer. Analysis activities must occur throughout the organization, resulting in more robust and data-driven human resources development, product development, organizational development, human resources management, and environmental engineering.

Other researchers (Gilbert 1978; Mager and Pipe 1984; Harless 1975; Robinson and Robinson 1995; Rossett 1999; and Hale 1999) would probably agree that:

- Analysis has two overarching purposes: (1) to make good decisions about technology-based and non–technology-based approaches based on data from many sources; and (2) to influence the organization in systematic and systemic directions.

- Analysis is critical to figuring out what to put in any particular lesson or class; larger questions about systems and readiness are addressed in analysis, increasing the likelihood that the class or lesson will make a difference.
- Analysis is for the purpose of divining a tailored array of interventions that can be used to enhance performance, including, but not limited to, instruction. Examples of interventions are e-learning modules, job aids, personnel selection strategies, electronic knowledge bases, compensation and incentive programs, reengineered processes, and job redesign.
- Analysis ensures that many fingerprints are on the effort.
- Analysis requires that the analyst then collaborate with appropriate colleagues and market the aligned approach to the organization.
- Analysis increases the need for a fresh and data-driven look necessary to enabling global and culturally diverse contexts.

Two questions dominate analysis. The first is about the "what." What is the essence of good performance? What are the problems? Where are we attempting to go (e.g., what constitutes good auditing, good sales, or good sportsmanship)? The "what" questions are answered by reviewing the literature, best practices, and observing and querying experts and model performers. The second question that dominates analysis is the "why," which defines the means for the educational technologist.

Analyzing Causes and Identifying Solutions

Several scholars (Mager and Pipe 1984; Harless 1975; Gilbert 1978; and Rossett 1987, 1999) offer templates for ascertaining why—ferreting out the causes and then using them to derive solutions. Tom Gilbert (1978), for example, elegantly distinguished between two kinds of deficient performers: those who cannot do what is expected of them, and those who can but do not for some other reason. The analyst then must ascertain why they do not, then propose what to do about it. The incumbents themselves and their supervisors are key sources here, not surprisingly. Typical questions are: Why are sales down this quarter? Do they know how to sell *x?* Do they believe in that new product and feel they can represent it positively? Table 1 provides one statement of the relationship between kinds of causes and kinds of solutions.

Four common causes of suboptimal achievement are:

- *Employees lack skill or knowledge or information:* Even if they tried, they could not do it. They do not possess the knowledge

TABLE 1: Common Relationships between Causes and Solution Identified through Analysis

Driver	Description	Solutions
Lack of skill, clarity about expectations, knowledge, or information	People don't because they don't know how, or they've forgotten, or there's just too much to know.	• Education/training • Information support (job aids) • Documentation • Coaching and mentoring • Clarity re standards • Communications initiatives
Weak or absent motivation	People don't because they don't care, or see the benefits, or they don't believe they can.	• Education/training • Information support (job aids) • Documentation • Coaching, mentoring • Participatory goal-setting • Communications initiatives
Ineffective environment, tools, or processes	People don't because processes or jobs are poorly designed, or necessary tools are unavailable.	• Reengineered work processes • New or improved tools or technologies or work spaces • Job design or redesign • Job enrichment • Participatory decisionmaking
Ineffective or absent incentives	People don't because doing it isn't recognized, doing it is a hassle, or not doing it is ignored.	• Improved appraisal/ recognition programs • Management development • New policies

SOURCE: Created by the author.

necessary to add the memory to the computer that explains the reasons for the new operating system.

- *The environment is in the way:* Individuals do not have access to the equipment, tools, forms, or work space necessary to perform. Software, for example, has not been installed.
- *There are no, few, or improper incentives:* What are the consequences of doing the job badly or not doing it at all? Are supervisors paying attention to desired outcomes? Does the compensation program recognize excellence and extraordinary effort? If managers are expected to implement a program, will their performance appraisals reflect that priority? What happens if they provide only halting support?

- *Employees are unmotivated:* Here we direct attention to the internal state of the individuals. What is going on within employees as they contemplate a new product or system?

Tools for Gathering Information and Perspectives

Whether conducted in person or on the phone, interviews are used to gather information about optimal and actual performance, feelings, causes, and solutions. When the information is technical, detailed, or emotionally charged, an interview is appropriate. The interview, especially when conducted in person, is also an effective device for creating relationships, another key purpose of analysis. Another advantage of the interview is that the analyst is able to generate follow-up questions. For example, when an employee says, "I'm not going to say negative things in a performance appraisal," the analyst can follow with, "Why not? What keeps you from doing that?"

Observations are used to determine what is going on in the workplace. They are a powerful tool for capturing information about current skills and knowledge, as well as for examining the context surrounding the individual. Historically associated with task analysis, many have used interviews to observe model deck-swabbing or motherboard repair in order to generate the details of excellent performance.

According to one study (Zemke and Kramlinger 1982), there are two levels of observation. The first level provides a general "take" on the situation—providing, say, an "establishing shot" of what is going on during the initial stages of an audit. The second level of observation captures the details of the interactions.

Observations often focus subsequent inquiries, and the analysts will often choose to conduct interviews or distribute surveys after conducting observations at the work site. Surveys solicit the thoughts and concerns of many people at lower cost than observations or interviews. The survey can capture mundane information, such as which classes have been taken in the past, and emotionally loaded information, such as the employees' confidence surrounding it or views about the causes of the problem. Because the survey can be anonymous, the analyst can be hopeful that people will report honestly.

Successful surveys are clear about their purposes, and even the most experienced survey-writer benefits from piloting. This involves distributing the survey to colleagues and selected members of the target population prior to widespread mail or electronic distribution.

Structured focus-group meetings provide an effective means for obtaining and dispensing information and for generating relationships among people and across units. It is essential to carefully construct an agenda based on the purpose of the analysis and to anticipate differences in opinion. The analyst must know ahead of time whether the experts drawn from across the globe to talk about maintenance of turbine engines in desert climates disagree. If they do, it is better to work with them alone initially, to put them in smaller work groups, and ask upper management to designate the expert who is "first among equals." Participants should be thoroughly briefed before they appear for a focus group.

Technology Tools

Technology tools can serve the analysis process in numerous ways. For example, technology may be used to collect data from many at a reasonable cost. Interviews can be conducted via e-mail, and with instant messaging it's not necessary to lose the benefits of follow-up questions. Time and distance become irrelevant because respondents can answer at their own convenience. Using prior notification with substantive explanations encourages participation.

Likewise, e-mail surveys are an excellent way to reach out for individual, anonymous opinions. Data collection is ongoing, and results are omnipresent. Listservs can be used to conduct focus groups by querying participants in a structured way. Participants respond to those questions, and to other participants' responses. Two-way videoconferencing presents a way to talk and see each other during analysis at remote sites by providing a means for conducting observations and examining visual cues. Finally, technology is useful for making sense of the data and to communicate it to others. Quantitative tools (such as SPSS and SAS) and qualitative tools (such as NUD*IST, the Ethnograph, and HyperResearch) can contribute to analysis of data.

Building Support for Analysis

According to one study (Rossett and Czech 1996), while analysis is judged to be critical to most learning professionals, leadership within an organization does not allow them to do as much of it as they believe is critical. Five strategies for increasing support for analysis include:

1. Conduct effective analyses, then document what has been done and how it has contributed to the bottom line. It is important to collect examples of how analysis has unearthed information and opinions that altered subsequent decisions and impacts.
2. Make a case for analysis through analogies to other professions. Would an engineer launch a project without serious scoping activities? Would a physician prescribe treatment without diagnostic tests? Would an urban planner develop low-income housing without constant interaction with community leaders?

3. Avoid professional jargon, such as "performance analysis" or "needs assessment" or "task analysis," words with little meaning beyond familiar professional boundaries. Until there is clarity and a successful history behind these phrases, consider using others that are more familiar in the customer's world: "planning" or "auditing" or "customization."
4. Step into the shoes of the customers, clients, or colleagues who are blocking analytical activities. Ask them about their hesitancies and respond accordingly. If analyses have delayed progress and failed to deliver useful information in the past, it's not surprising to find halting acceptance. Why would a colleague in information technology or organizational effectiveness hesitate? Again, cultural and historical factors deserve attention.
5. Do whatever you do speedily and virtually. One of the best ways to encourage support for analysis is to speed up the process. Use technology. Gather data even before it is needed. Anticipate and repurpose existing data.

If every interaction with clients and customers involves tailoring solutions to particular needs and circumstances, then analysis provides the defining energy and direction. Trainers, instructional designers, educational technologists, and web-learning managers are examples of professionals who approach their work with an analytical and performance perspective. No solution is assumed until analysis is completed.

Allison Rossett

See also ADDIE Model; Evaluation; Instructional Design; Performance Support

References

Gilbert, Thomas. 1978. *Human Competence: Engineering Worthy Performance.* New York: McGraw Hill.

Hale, Judith. 1998. *Performance Consultant's Fieldbook.* San Francisco: Jossey Bass.

Harless, J. H. 1975. *An Ounce of Analysis Is Worth a Pound of Objectives.* Newnan, GA: Harless Performance Guild.

Kaufman, R., and F. W. English. 1979. *Needs Assessment Concept and Applications.* Englewood Cliffs, NJ: Educational Technology Publications.

Mager, R. M. 1970. *Goal Analysis.* Belmont, CA: Pitman Learning.

Mager, Robert M., and Peter Pipe. 1984. *Analyzing Performance Problems.* Belmont, CA: Pitman Learning.

Robinson, Dana G., and James C. Robinson. 1995. *Performance Consulting.* San Francisco: Berrett.

Rossett, Allison. 1987. *Training Needs Assessment.* Englewood Cliffs, NJ: Educational Technology Publications.

———. 1999. *First Things Fast: A Handbook for Performance Analysis.* San Francisco: Jossey-Bass/Pfeiffer.

Rossett, Allison, and Kendra Sheldon. 2001. *Beyond the Podium: Delivering Training and Performance to a Digital World.* San Francisco: Jossey-Bass.

Rummler, Geary A. 1986. "Organization Redesign." In National Society for Performance and Instruction, *Introduction to Performance Technology,* vol. 1. Washington, DC: National Society for Performance and Instruction.

Zemke, Ronald, and Thomas Kramlinger. 1982. *Figuring Things Out: A Trainer's Guide to Needs and Task Analysis.* Boston, MA: Addison Wesley.

Anchored Instruction

Anchored instruction can be defined as an attempt to help students become actively engaged in learning by situating, or anchoring, instruction in interesting and realistic problem-solving environments. A primary goal of anchored instruction is to help students develop the confidence, skills, and knowledge necessary to solve problems and become independent thinkers and learners (Cognition and Technology Group at Vanderbilt [CTGV] 1990). Most of the education reform literature centers on the failure of traditional instruction to accomplish this goal. The concerns about traditional approaches to instruction have been influenced in part by A. Whitehead's (1929) discussion of what he termed "inert knowledge." Inert knowledge is the knowledge that can usually be recalled when people are explicitly asked to do so but is not used spontaneously in problem-solving even though it is relevant. Meaningful, problem-oriented approaches to learning are more likely than fact-oriented approaches to overcome inert knowledge problems (CTGV 1990).

Anchored instruction helps to overcome the problem of inert knowledge. Creating environments that permit sustained exploration by students and teachers enables them to understand the kinds of problems and opportunities that experts in various areas encounter and the knowledge that these experts use as tools (CTGV 1990). Anchored instruction also helps students develop representations or mental models of their experiences in order to set the stage for positive transfer (CTGV 1993).

Educational technology provides an excellent medium for creating and implementing effective anchored instruction because: visual formats allow students to develop pattern recognition skills; technology allows a more vertical representation of events than text; it is dynamic, visual, and spatial; students can more easily form rich mental models of the problem-solving situations; and technology has random access capabilities. All this allows teachers to almost instantly access information for discussion.

Stages of Anchored Instruction

When using an anchor, the following steps or phases of instruction are distinct and sequential, each contributing to the process: (1) introduce the anchor; (2) develop shared experience around the anchor; (3) expand the anchor; (4) use knowledge as tools for problem-solving; (5) work on projects related to the anchor; and (6) share what was learned.

The students are introduced to the anchor in phase one. The anchor might be a video segment, which contains a complex problem with embedded data to help solve the problem. In another class, the video content might be rich with information that supports sustained thinking about target concepts or that is needed to comprehend related text and for class discussions. In another class, the anchor might be a videoconference with a subject-matter expert who poses a problem to the students. By using such anchors, students and the teacher have a shared learning context (McLarty et al. 1990).

In phase two, students develop shared expertise around the anchor. Multiple visits to specific scenes in the anchor will allow students to develop expertise on particular aspects. In this phase, the teacher might lead a discussion of the anchor. However, as their knowledge of the anchor increases, students might assume more responsibility for their own learning. Once the teacher and students have developed expertise on the anchor, the links across the curriculum and to their prior experiences become a common occurrence within the classroom.

The students expand the anchor by conducting their own research in phase three. Gaps in information provided by the anchor might require students to research related materials in the library or via the Internet. Students might learn new technologies using the anchor for content material. For example, the students might create a HyperStudio stack about one of the topics in the anchor.

In phase four, students use their knowledge as tools for problem-solving. They might use this knowledge to solve problems posed in the anchor itself or relate the information to problems in other content areas. In this phase, teachers might provide scaffolds to help students solve the problems, or students may collaborate with geographically disparate peers via e-mail, videoconference, or other electronic means. For example, teachers who are using the Jasper Woodbury series to teach problem-solving and math skills might encourage the students to determine how to approach the problem and then provide them with the resources necessary to make progress, or students may discuss their plans with students from another school via a newsgroup.

Students work on projects related to the anchor in phase five. In this phase, students are given the opportunity to extend their knowledge and relate it to other areas. Some examples of this phase might include reading more about the subject, writing a report or an essay, or creating a multimedia report.

In phase six, students share what they learned from the project. The process of sharing not only creates pride in their own work but also gives them valuable insight into how their classmates solved the problem. At this point the students are encouraged to compare their solutions with those developed by their classmates and by experts, as well as to evaluate the strengths and weaknesses of each approach.

Advantages of Anchored Instruction

Advantages of anchored instruction include student ownership of their learning because problems hold more relevance; development of deeper meaning and understanding, thus increasing the likelihood of transfer to other situations; and use of collaboration, cooperation, and negotiation skills.

There are several advantages to organizing instruction around an anchor and then moving to hands-on activities. First, it provides everyone involved with a common background about the subject. Because it is visual, it is easier for students who are not good readers to participate in class discussions. Teachers often find this approach more manageable than finding all the resources necessary to accomplish a community-based project. Students often focus on an issue from a macro context that was not noticed as a potential issue by other members of the class. Once this issue is noticed, further research can be done on it.

Challenges of Anchored Instruction

The current emphasis on student-centered instruction means that teachers need to change their role from a provider of information to coach and, often, fellow-learner. Anchored instruction provides the means to make the shift from a teacher-dominated to a learner-centered classroom. With anchored instruction, the teacher can no longer follow a fully scripted lesson plan. Students are encouraged to identify their own questions, goals, and issues that arise as they explore the anchors. Since the students construct their own learning, teachers struggle with how to help the students reconceptualize problems without being overly directive. Another challenge for teachers is how and where to fit anchored instruction into their existing curricula and make sure that it meets their needs with respect to mandated achievement testing common to K-12 environments (CTGV 1993).

Applications of Anchored Instruction

Anchored instruction has been used with students ranging from fifth grade through college. It has been used in a variety of disciplines: language arts, social studies, math, science, and educational technology. Some examples of anchored instruction have used the World Wide Web as a resource by anchoring students in the principles of aerodynamics to learn about science and math concepts, presenting multiple perspectives of the study of global warming and its effects on our environment, and putting students in the role of determining whether to defend or oppose nuclear research.

The Cognition and Technology Group at Vanderbilt has implemented several successful anchored instruction projects. The Young Sherlock Holmes project, organized around a movie on videodisc, was implemented in two fifth-grade classrooms. The students were below average and average in academic ability. The project was designed to help the students learn language arts and social studies content by helping them to observe relevant historical information in movie settings and use their observations to make inferences (Risko et al. 1990). This study found that the video motivated students to form well-defined goals when reading to learn. Students in the anchored group were more likely to spontaneously use new vocabulary terms than were those in the comparison group (CTGV 1993).

A second CTGC project, the Jasper Woodbury series, focuses on mathematical problem formulation and problem-solving. It also involves the development of applications to enable students to learn science, history, and literary concepts. Although the series was designed for fifth- and sixth-graders, it could be used with fourth-graders through college freshmen (CTGV 1989). An important feature of this series is that information needed to solve the problems is embedded in the story. The embedded data design allows teachers to help students try to generate what they need to know, attempt to retrieve this information from memory, and then review segments of the disc to see if they were accurate (CTGV 1989). The Jasper Woodbury series provides examples of problems that occur in everyday life and how they might be solved. The Cognition and Technology Group at Vanderbilt has also explored the anchored approach to science instruction in its Scientists in Action series.

Donna Baumbach
Mary Bird
Sally Brewer

See also Active Learning; Constructivism; Jasper Woodbury; Learner-Centered Environment; Problem-Based Learning

References

Cognition and Technology Group at Vanderbilt. 1993. "Anchored Instruction and Situated Cognition Revisited." *Educational Technology* 33(3): 52–70.

———. 1990. "Anchored Instruction and Its Relationship to Situated Cognition." *Educational Researcher* 19(6): 2–10.

———. 1989. "Enhancing Learning in At-Risk Students: Applications of Video Technology." ERIC Document Reproduction Service No. ED 318 464.

McLarty, K., et al. 1990. "Implementing Anchored Instruction: Guiding Principles for Curriculum Development." Paper presented at the annual meeting of the National Reading Conference. ERIC Document Reproduction Service No. ED 315 736.

Risko, V., C. Kinzer, N. Vye, and D. Rowe. (1990). "Effects of Videodisc Macrocontexts on Comprehension and Composition of Causally Coherent Stories." Paper presented at the annual meeting of the American Educational Research Association. ERIC Document Reproduction Service No. ED 318 996.

Whitehead, A. N. 1929. *The Aims of Education.* New York: MacMillan.

Apple Classrooms of Tomorrow Project (ACOT)

The Apple Classrooms of Tomorrow Project, a ten-year, longitudinal research and development collaboration among public schools, universities, research agencies, and Apple Computer, began in 1985. The project set out to investigate how routine use of technology by teachers and students would affect teaching and learning. ACOT worked independently of the company's product, sales, and marketing divisions. The results of this project have influenced staff development and technology integration plans throughout the country. This project is also one of the most widely cited longitudinal studies in the field.

After soliciting and reviewing proposals from interested school districts, ACOT began work in five schools in four states. The initial sites represented a cross-section of America's K-12 schools in terms of grade level, socioeconomic status, and community setting. Each of the ACOT sites began with one classroom, then added classrooms, staff, and students in subsequent years. ACOT staff asked that the gender and ethnic composition of the classes mirror the school as a whole; all other decisions about student selection were left up to school personnel.

ACOT equipped project classrooms with computers, printers, scanners, laser-disc and videotape players, modems, CD-ROM drives, and a variety of software packages. ACOT provided participating teachers and students with two computers, one for the classroom and one for the home. Since hardware in 1986 was big and heavy, the two-computer formula was the only way to provide teachers and students with constant access to technology. This option offered a way to simulate a future time when technology would be more accessible because of its smaller size, portability, and lower cost.

Project teachers—all volunteers selected by the individual school districts—ranged from novices with one or two years of experience to veterans with more than twenty years in the classroom. Few had worked closely with technology before joining the project. ACOT staff provided training for teachers on telecommunications, basic troubleshooting, and tools software such as spreadsheets, databases, and graphics programs. ACOT also helped fund a full-time coordinator at each site to provide technical and instructional assistance.

Technology was viewed as a tool to support learning across the curriculum. No attempt was made to replace existing instructional technologies with computers. Instead, project classrooms included—along with computers—multiple instructional resources such as textbooks, workbooks, and manipulative math materials. The operating principle in ACOT classrooms was to use the tool that best supported the learning goal.

When computers were first introduced to ACOT classrooms, the technology added another layer of complexity to the classroom, a whole new set of things for already overworked and stressed teachers to learn and manage. Yet as the project continued, teachers found strategic ways to use the technology. Its use in instruction and learning changed as teachers themselves changed. Researchers described these changes in a model of instructional evolution that includes five stages: entry, adoption, adaptation, appropriation, and invention. In this model, instruction is first strengthened through the use of technology and then gradually replaced by far more dynamic learning experiences for students.

Entry

Teachers who were beginning with the project had little or no experience with computer technology and were in various stages of trepidation and excitement. They spent the first weeks unpacking boxes, running extension cords, untangling cables, inserting cards, formatting disks, checking out home computers, and generally trying to establish order in radically transformed physical environments. During this unavoidable initiation, even experienced teachers found themselves facing problems typical of first-year teachers: discipline, resource management, and personal frustration. They expressed serious reservations about students' access to computers and whether the new technology would ever fit in. Overwhelmed by other issues, the teachers showed little inclination to change instruction.

From the initial arrival of the computers, the students showed a high level of interest in and curiosity about technology. Students enthusiastically worked on the computers, sometimes becoming so intrigued that they resisted moving on to other classroom activities. Some students became so enamored with technology that they became unwilling to do work with pencil and paper. The children's inquisitive nature often seemed like misbehavior to the teachers. For instance, students took apart the mice to see how they worked, experimented with magnets to see if they really would erase a disk, figured out how to get into the teacher's management system, or unintentionally caused problems with the network. In this stage, students demonstrated a steady fascination with the technology, but teachers didn't yet know how to use that excitement to benefit teaching and learning.

Adoption

Although teachers continued to deal with technology issues in this stage, they began to show more concern about how technology could be integrated into daily instructional plans. They began to incorporate computer-based activities aimed primarily at teaching students how to use technology: keyboarding, word-processing, and saving, storing, and organizing work. Given their lack of experience with technology, teachers adopted it to replicate traditional instructional strategies and attempted to blend its use into the most familiar form of classroom practice: direct instruction. They searched for software they could adapt to established curricular and pedagogical preferences. Although much had changed physically in the classrooms, more remained the same.

Given frequent disruptions to normal classroom operation that came with beginning attempts to use the computers, researchers anticipated short-term declines in student performance. Traditional measures of achievement, however, showed no significant decline or improvement in student performance aggregated at the classroom level, and teachers reported that individual students were performing better. Site coordinators noted improvements in student attendance, self-esteem, and discipline. Teachers and researchers observed changes in student motivation and engagement. Students became more excited about learning, and their enthusiasm aided their own progress while also reinforcing teachers' efforts.

Adaptation

In this phase, technology became thoroughly integrated into traditional classroom practice. Lecture, recitation, and seat work remained the dominant form of student tasks; but students used word processors, databases, some graphic programs, and many computer-assisted instructional packages for approximately 30–40 percent of the school day. More frequent and purposeful use of technology began to return dividends.

The most significant change for students occurred in productivity. Students produced more and at a faster rate. In a self-paced computational math program, for example, sixth-grade students completed the year's curriculum by the beginning of April. Rather than beginning the seventh-grade curriculum, teachers decided to use the remainder of the year focusing the students on application and problem-solving. Students who were not usually enamored of math and rarely performed well became engaged in the hands-on, problem-solving approach. These students then became recognized as creative math problem-solvers. In subsequent years, teachers at this site assigned more of the computational math activities as homework to be completed on the home computers and used class time for application and problem-solving activities.

At an elementary site, teachers focused on basic math and language arts skills and used their computers purposefully to raise student test scores. For two years in a row, the district reported that ACOT students scored significantly higher on the California Achievement Test than non-ACOT students in vocabulary, reading comprehension, language mechanics, math computation, and math concepts and application. At other sites, which were less focused on basic skills development, teachers worried that time spent on developing technology skills (rather than on covering standard curriculum) might erode student test scores. Less time spent on the basics, however, did not have a negative impact on student performance on tests at these sites.

Increased productivity emerged at the high school level as well. In chemistry, for example, students learned to use a simple graphics program to illustrate molecules and the exchange of atoms in chemical reactions. The teacher reported that students learned how to write and balance chemical formulas faster and more accurately than in his previous experience. The instructional process also became more efficient, as students could access assignments on the network, work on assigned tasks, and send their work to the printer to be picked up by the teacher.

Teachers and researchers also noted improvement in students' writing during this stage. Many students could type faster than they could write, thus preferring to prepare assignments with the word processor. Researchers who examined writing in one third-grade ACOT classroom determined that children maintained a high level of enthusiasm for and interest in writing; that computers made compositions more presentable to others, thus encouraging sharing and writing; and that students wrote more and better as a function of the accessibility of computers. The opportunity to write regularly on computers made a substantial difference for low-achieving students who demonstrated significant improvement in the quantity and elaboration of their writing. With increased productivity in writing, teachers could work with even young students on narrative skills. Students willingly reworked their papers because the process was much easier than with pencil and paper.

During the adaptation phase, teachers also noted changes in the quality of student engagement in classroom tasks. They found that students, during computer activities, were highly involved in their assignments and were frequently able to work with little assistance. Students spent more time on assignments and projects, and they chose to use the computers during recesses, lunch periods, and after-school hours. In addition, teachers reported that students were increasingly more curious and assistive learners, taking on new challenges far beyond the normal assignments. Some students developed technology-related skills that led to jobs in their communities.

Students started to take on new roles in the classroom as they led classes, became peer tutors, and spontaneously organized collaborative work groups. Teachers also discovered that students who did not do well in a typical setting frequently excelled when working with technology. For example, a first-grade student who was low to average on academics turned out to be a whiz at word-processing and finished all twenty-one lessons of a program in one day. Low achievers had a chance to experience success and began concentrating and applying themselves to their projects. Teachers saw less advanced students blossom, unpopular students gain peer approval, and unmotivated students stay in to work at recess.

Appropriation

Appropriation was less a phase in instructional evolution and more a turning point. It was evidenced less by changes in classroom practices and more by changes in personal attitudes toward technology. Appropriation was the point at which an individual came to understand technology and use it effortlessly as a tool to accomplish real work. As a turning point, appropriation marked the end of efforts simply to computerize their traditional practice and led to the next stage, where new teaching approaches promoted the basics yet opened the possibility of a new set of student competencies.

Invention

In the invention stage, teachers experimented with new instructional patterns and ways of relating to students and to other teachers. As more teachers reached this stage, the whole tenor of the sites began to change. Interdisciplinary project-based instruction, team teaching, and individually paced instruction became common. Students were busier, more active learners. ACOT teachers tended to see learning as an active, creative, and socially interactive process. In this stage, knowledge came to be viewed more as something that children must construct for themselves and less like something that can be transferred intact.

Teachers reported on students' highly evolved skills with technology, their ability to learn on their own, and their movement away from competitive work patterns toward collaborative ones. Students helped other students over hurdles with the technology, and as teachers adapted to their students' growing expertise, students helped their teachers as well. Students not only coached each other but also shared their expertise with people beyond the ACOT classrooms. In addition, some teachers began to allow students to teach one another subject-matter content in addition to technological information.

A researcher studying one of the elementary sites observed changes in communications patterns and the extent of collaborative work among even the youngest students. She reported that the children interacted differently at the computers. They talked to each other more, frequently asked for assistance from their neighbors, quickly interrupted their own work to help someone else, and displayed tremendous curiosity about what others were doing.

Another perspective on the benefits for students comes from examining the record of students at one high school who spent four years in the program. Their collective record compared to the entire 216-student, non-ACOT graduating class showed a marked difference. Although the ACOT graduating class did not constitute a technical random sample of the high school students, they were representative of the school as a whole. ACOT students' absentee rate was 50 percent less, and they had no dropouts, compared to the school's 30 percent rate. Although half of the students who joined ACOT as freshmen had not planned to go to college, 90 percent of them graduated and went on to college, compared to 15 percent for the non-ACOT graduates. Moreover, this ACOT graduating class amassed twenty-seven academic awards, including inductees into the National Honor Society and Who's Who among American High School Students, and recognition for outstanding accomplishments in history, calculus, foreign language, and writing. A four-year longitudinal study of these students showed the greatest difference was the manner in which they organized and accomplished their work. They routinely employed inquiry, collaborative, technological, and problem-solving skills uncommon to graduates of traditional high school programs. These skills are remarkably similar to competencies argued for by the U.S. Department of Labor.

Teachers who changed to more integrated and problem-solving teaching approaches found themselves hampered by traditional forms of assessment that emphasize basic skills, memorization of facts, and direct instruction. When ACOT students demonstrated new learning outcomes such as creative problem-solving strategies or greater abilities to collaborate in performing tasks, teachers struggled with how to translate those types of student gains into quantitative measures that could be entered into grade books. Rewarding students for their successes with new competencies proved difficult. In addition, some teachers interrupted the natural flow of project-based activities to "demonstrate" whole-class direct instruction for district evaluators using instruments too inflexible to accommodate more active classroom environments. Frequently the larger context of school, district, and state policies inhibited rather than encouraged change.

ACOT Teacher Development Centers

In 1992 ACOT, the National Science Foundation, and three school districts joined to create teacher development centers at three of ACOT's original sites. A primary goal of the program was to prepare teams of teachers who could become technology leaders at their own schools. More than 600 teachers—representing thirty districts, fifteen states, and two foreign countries—participated in the program during its first three years of operation. Participants represented a variety of grade levels and disciplines and ranged in teaching experience and technological expertise. At the centers, visiting teachers observed and worked in ACOT classrooms for one-week practicums during the school year or for four-week institutes during the summer. This model of staff development allowed participants to see expert teachers modeling instructional use of technology. Participants learned about integrating specific hardware and software into their instruction, and they explored issues such as interdisciplinary instruction, alternative assessment, project-based teaching, and team teaching. ACOT coordinators also provided follow-up support for participants for one year after visiting the centers.

Researchers investigating the impact of the teacher development center program on participants found that teachers' abilities to implement what they had learned in their classroom varied significantly, depending on access to equipment, administrative support, technical support, and collegial support. First, although ACOT considered the importance of access to equipment in designing their program, the project had little control once participants returned to their schools, and consequently teachers' classroom equipment ranged widely. Second, participants experienced a broad range of administrative support when they returned to their sites. Principals varied dramatically in their attitudes toward technology and consequently in what actions they took to help teachers. Third, unlike the original ACOT teachers, most participants returned to schools that did not offer on-site technical support. When teachers could not obtain sufficient technical assistance, they frequently altered or abandoned plans to use technology in their classroom instruction. Finally, although ACOT required participants to attend the center in teams to establish a source of collegial support for teachers once they returned to their schools, participants sometimes encountered resentment, rather than support, from their other colleagues. Given the limited enrollment in the ACOT program and the limited access to technology in many of the schools, faculties sometimes became fragmented and competitive. Clearly, the variations among ACOT participants in terms of computer access and administrative, technical, and collegial support naturally led to variation in their ability to integrate technology into instruction in their classrooms. When barriers such as limited access to technology and lack of support could be overcome or minimized, teachers made significant changes in their instructional practices and technology use.

Key Conditions for Effective Use of Technology

The experience of the ACOT project suggests that technology has the potential to improve education, but only under certain conditions. First, the successful use of technology requires teachers to confront their beliefs about learning and the efficacy of different instructional activities. Teachers experimented with new ideas until old habits gave way to new. As they successfully attempted new methods of instruction, they saw for themselves the value of different strategies and began to reevaluate their beliefs about learning and teaching. Second, technology should be viewed as one tool among many. As such, it should be used only when it is the most appropriate means of reaching the learning goal. Third, technology has little influence unless it is integrated into a meaningful curricular and instructional framework. The benefits of technology are best realized when the goal is to empower students as thinkers and problem-solvers. Technology provides an excellent platform for learning environments where students can collect information in multiple formats and then organize, visualize, link, and discover relationships among facts and events.

Fourth, the benefits of technology integration are not realized unless students and teachers have adequate access. Like any other tool for learning, technology should be in classrooms. The potential of technology goes unrealized if the goal is to teach technology, as frequently occurs in a lab setting; technology is best learned within the context of meaningful tasks tied to the curriculum. Fifth, teachers need contexts that support risk-taking and experimentation and that provide opportunities for collegial sharing and ongoing professional growth. Sixth, technology can be a catalyst for change, but the process of technology integration should be viewed as a long-term, challenging enterprise. Change is slow, even when teachers are willing. In addition, contextual supports are rarely in place when technology is added to schools. Teacher commitment will come only after they see positive benefits for themselves and their students.

Judith Haymore Sandholtz
Cathy Ringstaff

See also Curriculum Integration; Diffusion of Innovations; Teachers: Preparation and Training; Technology in K-12 Schools

References

ACOT Research Reports are available online at www.apple.com/education/k12/leadership/acot/library.html.

Fisher, Charles, David Dwyer, and Keith Yocam, eds. 1996. *Education and Technology: Reflections on Computing in Classrooms.* San Francisco: Jossey-Bass.

Sandholtz, Judith H., Cathy Ringstaff, and D. Dwyer. 1997. *Teaching with Technology: Creating Student-Centered Classrooms.* New York: Teachers College Press.

ARCS: A Model of Motivational Design

Motivation refers to a person's desire to pursue a goal or perform a task (Keller and Litchfield 2002) and is typically used to examine work-related rather than play-related goals and behaviors (Weiner 1992). Motivation is influenced by internal factors, such as perceptions and personal goals, as well as external factors, such as opportunities and rewards. Because motivation affects learning and performance outcomes, it is important for instructional designers to understand how to use specific design strategies to increase learners' motivation. In 1979, John Keller proposed the ARCS model of motivational design as a framework for addressing four primary factors that influenced learners' motivation for learning: attention, relevance, confidence, and satisfaction.

The ARCS model was based on the premise that people's learning efforts can be influenced by changes in the learning environment, specifically those instructional events that define learners' perceptions of the relevance, or value, of the instruction and/or their expectations for success (Peters 2001). Prior to the 1970s, design models were based on the assumption that instruction would be effective if learners wanted to learn. However, little was written to describe how to make the instruction more appealing, especially for learners who were not motivated to learn. The ARCS model provided designers with an easy to remember and simple to use method for gaining and sustaining learner motivation, particularly when this did not occur naturally during the course of instruction. Currently, as we search for new and effective ways to promote students' learning through the use of new educational technologies (e.g., web-based instruction, sophisticated modeling tools), the ARCS model of motivational design will provide a useful foundation.

Components of the ARCS Model

In general, the *attention* and *relevance* components of the ARCS model relate to the need to make instruction exciting, to meet learners' needs, and to build on learners' previous experiences. The *confidence* and *satisfaction* components relate to the need to create positive expectancies of success for students, based on the amount and quality of the effort they expend (Peters 2001). Below, each category is defined further, and specific examples are included in Table 1.

Attention refers to students' overt curiosity or interest in the topic at hand and can be addressed by increasing perceptual arousal (heightened awareness or interest), inquiry arousal (an attitude of curiosity or inquis-

TABLE 1: Strategies for Increasing Student Motivation

To increase	The teacher/designer can
Attention	• Vary the methods, media, activities, and materials used during the instruction. • Use challenging questions, realistic problems, or novel situations to stimulate students' curiosity.
Relevance	• Include examples that are familiar to the students. • Show students how the content relates to some previous experience. • Show students how the content relates to some personal need or goal.
Confidence	• Provide students with clear requirements for course assignments. • Provide students with opportunities to exercise control over what and how they learn. • Provide students with exercises or assignments that gradually increase in difficulty.
Satisfaction	• Encourage students' enjoyment of learning for its own sake. • Provide students with positive feedback. • Provide students with verbal praise and other forms of tangible reward.

SOURCE: Created by the author.

itiveness), or variability of materials (inclusion of a variety of presentation tactics). Attention strategies typically include methods for capturing and sustaining learners' interest.

Relevance refers to students' perceptions that learning about the subject will meet personal needs or goals. Relevance can be established by considering learners' goal orientations, motives, and previous experiences and then linking instructional objectives, content, or examples to these needs and experiences.

Confidence refers to students' expectations for success. By providing opportunities for students to build positive expectations for their performances, as well as to actually experience success while learning, greater confidence can be built. Moreover, students must be able to see how their efforts and actions have led to their successes. Successful results that can be attributed to luck or other uncontrollable factors tend not to build students' confidence.

Satisfaction refers to the tangible and intangible rewards that students receive for completing the instruction. Satisfaction can be achieved by providing intrinsic reinforcement (a sense of accomplishment) or extrinsic rewards (payment or other symbolic awards, or verbal reinforcement like "Good job!"), and by ensuring consistent application of success criteria.

According to Keller (1987), instructors and instructional designers implement or incorporate these strategies into their instruction in somewhat of a sequential manner. First, the attention of the learners is gained as they begin to engage in the learning activity. Before proceeding very far, however, learners must be convinced of the relevance of the instruction to their own needs and goals. They must also be fairly confident that they will be able to succeed in the instruction. If learners believe that they have little to no chance of being successful, they are not likely to participate for long. During the instruction, it may be necessary to recapture learners' attention, as well as to indicate to them how well they are progressing toward their learning goals. This will keep both interest and confidence high. At the end of the instruction, learners should have gained a sense of accomplishment that, hopefully, will then fuel their desire to learn more about the topic.

Application of the Model

The model is typically applied in four phases that integrate well with the processes of instructional planning and instructional design. These phases include (1) perform an audience analysis; (2) develop motivation objectives; (3) design a motivational strategy; and (4) try out and revise as needed. The selection of motivational strategies and tactics, including number as well as type, comprises a systematic design process, beginning with an analysis of audience motivation.

Analyze the Audience

The first step involves developing an audience profile, by using the ARCS model, to determine existing motivational gaps. Are the learners interested in the content? Do they have adequate levels of confidence to complete the instruction successfully? To complete this step, designers or instructors need to know the goals of the instruction so that they can identify learners' levels of interest, perceptions of relevance, levels of confidence, and effective intrinsic and extrinsic rewards, specifically related to the content to be learned. Data on each of these categories, then, need to be collected and analyzed. If the designer should discover, however, that motivation is not a problem for this particular audience, it is neither necessary nor desirable to add motivation strategies to the instruction.

Develop Motivational Objectives

Based on the results of the audience analysis, motivational objectives should be defined, specific to the motivational needs of the audience. For example, if perceptions of relevance were determined to be low for the audience members, then objectives should indicate that learners would be able to state the relevance of the instruction to their needs. As is true for instructional objectives, motivational objectives should be written to convey what *learners* will know, or be able to do, by the end of the instruction.

Design a Motivational Strategy

Motivational strategies should be selected or developed based on the specific needs of the audience. Consider a variety of different ideas and then select those that best fit the specific situation. F. Peters (2001) recommended selecting strategies that: do not take up too much time; complement, rather than overshadow, the learning objectives; are feasible given the time, money, and implementation constraints of the instruction; are acceptable to the audience; and are compatible with the instructor's personal teaching style.

Try Out and Revise as Needed

Designers or instructors need to pilot the selected strategies to determine their overall effectiveness with the specific audience. This typically occurs during the normal course of implementing the instruction. Instructors should also consider including items specifically related to motivational factors in their course evaluations, including questions related to attention, relevance, confidence, and satisfaction. Based on the results, instruction and/or the motivational strategies may need to be modified prior to subsequent use.

General Guidelines for Use

Although motivational strategies can be added to any instructional materials, they are most effective when learners' initial motivation is low or nonexistent. Peters (2001) recommended the following guidelines for implementing motivational strategies: maintain a balance between motivational and instructional strategies; maintain a balance among the kinds of motivational strategies used; and match strategies to learner needs.

According to Peters, too few strategies will result in learners feeling bored, whereas too many strategies may cause learners to become annoyed or anxious. Consider your learners' initial levels of motivation and the inherent motivational appeal of the topic prior to selecting the appropriate number and types of strategies.

Research Support

A number of studies have been conducted that support the application of the ARCS model in instructional situations. For example, T. Newby (1991) found a strong positive correlation between elementary teachers' use of relevance strategies and on-task student behaviors. This was true despite the fact that teachers used relevance strategies only 7 percent of the time. Furthermore, a negative correlation was found between teachers' use of satisfaction strategies and students' on-task behaviors, even though teachers used this strategy 58 percent of the time. This suggests that relevance strategies, even when applied at relatively low levels, are related to higher levels of students' on-task behaviors. By contrast, satisfaction strategies are associated with higher levels of students' off-task behaviors. Based on these findings, instructors and designers might consider increasing their use of relevance strategies while decreasing their use of satisfaction strategies within their instruction.

The ARCS model has also been used to determine the motivational needs of adult learners. In one study (Small, Dodge, and Jiang 1996) college students were more likely to describe instruction as being interesting if attention and relevance strategies were used. Results from another study (Gluck and Small, cited in Peters 2001) suggested that adults perceive the value factors of attention and relevance to be significantly different from the expectancy factors of confidence and satisfaction, supporting the contention of Keller (1979) that different strategies should be used to address different motivational needs.

Peggy A. Ertmer

See also Adult Learners; Instructional Design; Learning Styles

References

Keller, John M. 1979. "Motivation and Instructional Design: A Theoretical Perspective." *Journal of Instructional Development* 2(4): 26–34.

———. 1987. "Strategies for Stimulating the Motivation to Learn." *Performance and Instruction* (October): 1–7.

Keller, John M., and B. C. Litchfield. 2002. "Motivation and Performance." In Robert A. Reiser and John V. Dempsey (eds.), *Trends and Issues in Instructional Design and Technology* (pp. 83–98). Upper Saddle River, NJ: Merrill/Prentice-Hall.

Newby, Timothy J. 1991. "Classroom Motivation: Strategies of First-Year Teachers." *Journal of Educational Psychology* 83: 195–200.

Peters, F. 2001. "ARCS Motivational Design." In K. L. Medsker and K. M. Holdsworth (eds.), *Models and Strategies for Training Design* (pp. 297–317). Silver Spring, MD: International Society for Performance Improvement.

Small, Ruth V., B. M. Dodge, and X. Jiang. 1996. "Dimensions of Interest and Boredom in Instructional Situations." In *Proceedings of Annual Conference of Association for Educational Communications and Technology.* Indianapolis, IN: Association for Educational Communications and Technology.

Weiner, Bernard. 1992. *Human Motivation: Metaphors, Theories, and Research.* Newbury Park, CA: Sage.

Assistive Technology

For the estimated 15 percent of the U.S. population with a disability or chronic health condition, assistive technology offers the potential to ameliorate problems associated with mobility, communication, and learning. Assistive technology encompasses the tools that are used to increase an individual's functional capabilities and the services that assist with evaluation, selection, and use of technology by special populations. Students with disabilities in preschool through postsecondary institutions benefit from assistive technology that promotes independence and allows them to participate more fully in the general education curriculum and access educational technology products along with their peers. A vast array of technology solutions exist, ranging from high-tech devices such as a power wheelchair or talking computer to everyday, low-tech products such as a cane or a large-print book. Assessment of an individual's need for assistive technology presents the challenge of matching the most appropriate technology devices and services with the user's profile and the tasks or goals to be accomplished within a particular environment. Funding, training, availability of technical expertise, and strategies to reduce potential abandonment or failure to use a selected device are all critical to the effective provision and implementation of assistive technology.

Assistive technology plays a key role in allowing students with disabilities to interact with instructional technology products and bypass barriers associated with physical, sensory, or cognitive limitations. Beyond simply enabling equal access to educational tools, assistive technology contributes to the individual's full participation in the learning environment and enhances potential vocational and continuing education opportunities. In some circumstances, educational technology that increases or improves learning outcomes can serve the dual purposes of assistive technology that increases, maintains, or improves functional capabilities. A typical example would be the use of a word-processing program by a student who cannot complete handwritten work due to paralysis of the dominant hand. The word-processing program represents assistive technology because it offers a functional way for written work to be completed.

Assistive Technology Devices

Thousands of assistive technology applications span across all disability types; however, some generalizations can be drawn for specific groups based on characteristic functional limitations. The following sections highlight typical possibilities.

Vision Impairment

Vision impairment technically encompasses all degrees of vision loss, including total blindness. Technology that enhances access to print and other information sources has made the most significant impact on educational and career options for individuals who are blind or visually impaired. Some low-tech solutions for difficulties with glare, contrast, or size include use of a small tensor lamp to increase illumination, color transparencies, a magnifying glass or hand-held telescope, and reversed text (white on black). Braille documents or books and tape recorders are familiar items used by the blind. A closed-circuit television unit allows magnification of any printed item. A refreshable Braille display replaces the embossed paper format with a display of vibrating pins controlled by an electronic circuit that can be interfaced with a computer for electronic storage and retrieval of information. Braille note-takers are similar to portable hand-held organizers and offer a convenient way for the blind individual to keep track of phone numbers, assignments, appointments, and other short pieces of information. Computer-based adaptations include screen readers with a speech synthesizer and headphone, screen magnification software, Braille or large print keyboard markers, and Braille printers. Special hardware and software make it possible to print tactile displays from a graphing calculator, map, or illustration. For those with low vision, computer use is enhanced by easily altered background and text color and font and built-in options for large icons, cursors modifications, and magnification tools. Scanners with optical character reading (OCR) capabilities store printed text electronically on computers where it can be read via speech synthesis or printed in large print or Braille.

Hearing Impairment

A hearing impairment is any type or degree of auditory impairment, whereas deafness is the inability to use hearing as a means of communication. Acoustic listening devices amplify and filter sounds, allowing for increased participation and comprehension for the user during classes and lectures. Most assistive listening devices consist of a receiver and a transmitter. The receiver is worn by the person with a hearing impairment, and the transmitter is worn by a speaker or installed as a stationary fixture in the environment. Amplification systems are available for listening to television, and a closed-captioning decoder is now a standard component of any TV set with a 13-inch or larger screen. Signaling systems convert sound to visual, tactile, or vibrating signals and alert the hearing-impaired or deaf individual to smoke alarms, telephones, doorbells, and alarm clocks. Computer use has been particularly helpful in providing the deaf community with a text-based form of communication.

Learning and Cognitive Disabilities

According to the National Joint Committee on Learning Disabilities (1990) "learning disabilities" is a general term that refers to a heterogeneous group of disorders manifested by significant difficulties in the acquisition and use of listening, speaking, reading, writing, reasoning, or mathematical abilities. For individuals with learning and other cognitive disabilities, assistive technology offers strategies or compensatory tools that play to an individual's strengths. In some instances, the tools that assist students with learning disabilities are the same devices used by other individuals with similar functional limitations. Like students with visual impairments, individuals with a specific learning disability affecting reading may use an OCR/speech synthesis system to input text and have it read back with screen-reading software. A four-track variable-speed tape player used with talking books or access to electronic texts provides a way for students to listen to or review books. Students with auditory processing difficulties may use a personal FM listening system to help them focus on a speaker's voice. A personal data manager or hand-held computer could compensate for difficulties with memory and organization. Computer software programs with outlining and graphic organizer features support students with organizational difficulties. Written language disorders represent a common problem for students with learning disabilities. In addition to the use of a tape recorder as a backup for note-taking, various computer software applications provide a way to bypass the motor demands of writing and support written expression, spelling, and grammar usage.

Thus the availability of a word-processing program or portable word processor provides a way for students to prepare a more legible document that can be easily changed and edited. Word prediction can augment the features of a word processor by using rules of grammar, syntax, and vocabulary frequency to predict words the user is entering into the computer. Speech recognition systems have gained popularity with improved technology and more affordable prices. The user dictates into the computer via a microphone, and oral language is converted to written text. Speech recognition requires the user to have sustained patience and persistence in training the system for voice and vocabulary, adequate literacy skills, and good self-monitoring strategies.

Orthopedic and Mobility Impairment

Some orthopedic impairments are the result of a congenital condition such as cerebral palsy, whereas others may be acquired from an accident that results in a spinal-cord injury or other permanent or temporary physical problems. Some orthopedic/mobility impairments are due to neurodegenerative disease, such as muscular dystrophy, that may be compounded by scoliosis or other secondary problems as the disease progresses. Access and physical accommodations are the primary issue for students with orthopedic/mobility disabilities. Many of these individuals are nonambulatory and require power or manual wheelchairs or scooters with adapted seating. Assisted ambulation with canes, crutches, or walkers is adequate for mobilizing some students. Common dilemmas in assistive technology service delivery for this population involve orchestrating transportation to and from school, maintaining equipment, confirming compatibility with existing institutional resources, and ensuring that the student has appropriate positioning for accomplishing required tasks throughout the school day. An array of low-tech devices and adaptations are available to assist with activities of daily living (dressing, eating, bathing) and environmental control of appliances and equipment. Simple tools such as a universal cuff or splint can hold a pencil or typing stick in place for an individual with little or no hand function. Individuals who experience a tremor or uncontrollable movement may use a weighted wristband or pen. A book holder or slanted writing surface with a clip for holding papers can help to position books and writing materials. Students with orthopedic impairments, like those with visual impairments, have benefited immensely from computer technology. Depending on the individual's limitations and physical abilities, various devices positioned at the most functional body site can help to make input more efficient and accurate. Special keyboards or keyboard modifications, mouse alternatives such as a trackball, hands-free access achieved with a head-mounted device, speech recognition, or eye gaze, and switch interfaces with scanning or Morse code entry represent some of the possibilities.

Communication Disorders

Stuttering or fluency problems, poor articulation or unintelligible speech (particularly with unfamiliar communication partners), voice disorders, and aphasia (inability to comprehend words) are among the most familiar expressive language problems. Technology available to assist with some of these difficulties may include a voice amplifier to enhance weak or strident vocal quality or a device that records, analyzes, and then repeats back speech in a more intelligible format. Vocalizations, gestures, eye gaze, and written language, or a combination of these strategies, may be part of a communications system for an individual with more severe limitations who cannot speak. Adaptive assistance may be required as part of an augmentative and alternative communication (AAC) system. Low-tech options include a simple communication board with pictures, symbols, or words; a conversation book; or a simple device that speaks a limited number of messages. Computer-based electronic devices with high-quality digitized speech output, rate enhancement features, and access to a large vocabulary are at the other end of the AAC device continuum and are priced in the $5,000–6,000 range. Low-tech and high-tech devices satisfy interactive needs but may fulfill different purposes depending on the environment and situation.

Other Health Impairment

This category includes individuals who have disabilities that do not fall into any of the other major groups. Some conditions are progressive, whereas others are stable. A partial list would include AIDS, arthritis, asthma, burns, cancer, cardiovascular disorders, diabetes mellitus, epilepsy, psychological disorders, traumatic brain injury, and chronic pain. Depending on the nature of the disability and the expected activity, individuals with chronic or other health impairments could benefit from any of the assistive technologies described above. For instance, problems with memory or concentration may be addressed by suggested technology solutions for students with learning disabilities. Individuals experiencing physical fatigue or weakness may be assisted by tools discussed in the section on mobility/orthopedic impairments.

Assistive Technology Policy and Practice

Although canes and other adaptive tools have been used for centuries, it was not until recently that assistive technology gained legal recognition. The United States Congress first defined assistive technology devices and services in the Technology Related Assistance for Individuals with Disabilities Act of 1988 (P.L. 100–407). The following definition was adapted and used in the Individuals with Disabilities Act of 1990 (IDEA; P.L. 101–476) as amended in 1997:

> An assistive technology device is any item, piece of equipment or product system whether acquired commercially or off the shelf, modified, or customized that is used to increase or improve functional capabilities of a child with a disability.
>
> An assistive technology *service* is any service that directly assists an individual with a disability in the selection, acquisition, or use of an assistive technology device. These services include:
>
> 1. The evaluation of the needs of the child with a disability including a functional evaluation of the child in the child's customary environment;
> 2. Purchasing, leasing, or otherwise providing for the acquisition of assistive technology devices by children with disabilities;
> 3. Selecting, designing, fitting, customizing, adapting, applying, retaining, repairing, or replacing of assistive technology devices;
> 4. Coordinating and using other therapies, interventions, or services with assistive technology devices such as those associated with existing education and rehabilitation plans and programs;
> 5. Training or technical assistance for a child with a disability and when appropriate, the child's family, and;

6. Training or technical assistance for professionals (including individuals providing education or rehabilitative services), employers, or other individuals who provide services to, employ, or are otherwise substantially involved in the major life functions of a child with a disability. (20 U.S.C. Chapter 33 §1401)

The phrase "auxiliary aids and services" is another term used to reference assistive technology in legal statutes. Section 504 of the Rehabilitation Act of 1973 (P.L. 93–112) and subsequent reauthorizations require that public schools and postsecondary institutions be physically accessible. As a civil rights statute, section 504 protects the rights of individuals with a physical or mental impairment that substantially limits one or more life activities (e.g., learning or hearing). Reasonable accommodations and academic adjustments must be provided for those who qualify under section 504's broad definition of "disability." The availability of materials, printed in large type or Braille, is an example of an accommodation. A tape recorder used to tape a lecture, by a student who cannot take notes, represents an academic adjustment. Section 504 mandates that organizations that receive public funding must provide auxiliary aids and services that offer the individual with a disability an equal opportunity to access the full range of programs and activities. In addition to requiring that institutions provide adaptive and compensatory tools, the Rehabilitation Act and the Americans with Disabilities Act (ADA) of 1990 (P.L. 101–336) require that all students, including those with disabilities, must be able to access telecommunications and the available educational technology. The legal and ethical ramifications of these laws suggest that schools and postsecondary institutions should be proactive in reviewing the accessibility of educational technology products as part of the organization's technology plan.

The reauthorization and amendments to IDEA in 1997 strengthened access to assistive technology aids and services for children who are eligible for special education services in early childhood programs through secondary school. According to IDEA, school systems are required to determine if the individual needs assistive technology, and the appropriate devices or aids and services must be provided free of charge as part of the child's "individualized education program." The process of considering a student's assistive technology needs is ongoing since academic demands, requirements, and expectations change as the student progresses through school.

Assistive technology decisions should be informed by an evaluation that is completed by a collaborative interdisciplinary team. A team that includes the user, the family, professionals (i.e., teachers or vocational counselors and therapists), and at least one individual with expertise in assistive technology is considered best practice and implicit in the law. Maintaining a functional approach in conducting the assessment is pivotal to the ultimate goal of using technology to perform a task or compensate for a barrier imposed by an impairment. In striving to identify a device that matches the individual's profile, an evaluation should include analysis of abilities and needs, the activities to be performed, the environment, characteristics of the devices being considered, and the user's preferences. For youngsters of school age, the evaluation may be completed by a school-based team along with input from the family. Various assessment models and instruments have been developed to assist teams in arriving at a consensus, contribute to the development of an action or implementation plan, and facilitate ongoing review (see, e.g., Scherer 1991; Zabala 1996; University of Kentucky Assistive Technology Project 2000; Wisconsin Assistive Technology Initiative 2000).

Efforts to match an individual's needs and a technology solution can be foiled by various organizational obstacles and human factors. A device may be unobtainable due to costs, fail to be useful, or abandoned altogether. Effective assistive technology implementation is enhanced by the availability of expertise; training for the user (as well as caregivers or teachers); administrative support; communication and coordination of service delivery; identification and maintenance of appropriate devices; and including the user and other stakeholders in the decisionmaking process. In a comprehensive study related to funding issues, the National Council on Disability (1993) found that the lack of funding and the intricacies of identifying financial resources to support assistive technology represent the most formidable barriers to delivery of assistive technology. Costs associated with the provision of assistive technology go beyond the purchase of a device and may encompass evaluation, setup of the device in the user's environment, training, follow-up, adjustments, and maintenance or repairs. Payment for devices and services is further confounded by the fact that there is no single strategy or system for financing assistive technology in the United States. Funding may come from public, private, and community-based sources, with each agency having its own eligibility criteria, requirements for documentation of need, and contexts where assistive technology might be provided. Personal aids such as artificial limbs and orthotic splints or braces, adaptive seating, mobility aids, glasses, and hearing aids are usually prescribed by a medical professional or rehabilitation team and acquired through insurance, personal, or community-based funds. Tools provided by an educational institution allow access to the existing environment or provide the student with a learning or compensatory aid to accomplish a required activity. A school would be likely to provide a visual alert system in a college dormitory room, a special keyboard at a computer workstation, or a wheelchair lift on a school bus.

Until recently, there have been few resources to guide consumers and professionals in seeking quality and expertise in obtaining assistive technology. The Rehabilitation Engineering and Assistive Technology Society of North America has actively worked with other organizations to promote standards of quality assurance, outcome measures of assistive technology interventions, and certification of individuals with preparation and qualifications in assistive technology. A credentialing process has been established for assistive technology practitioners, assistive technology suppliers, and rehabilitation engineers (known as ATPs, ATSs, and CRE/CRETs, respectively).

Christine L. Appert

See also Technology Planning; Usability; Web Accessibility

References

ABLEDATA (www.abledata.com).

Beukelman, David, and Pat Mirenda. 1998. *Augmentative and Alternative Communication.* Baltimore: Paul H Brookes.

CAST (www.cast.org).

Cook, Albert, and Susan Hussey. 1995. *Assistive Technologies: Principles and Practices.* St. Louis, MO: Mosby.

Galvin, Jan, and Marcia Scherer, eds. 1996. *Evaluating, Selecting, and Using Appropriate Assistive Technologies.* Gaithersburg, MD: Aspen.

Golden, Diane. 1998. *Assistive Technology in Special Education: Policy and Practices.* Albuquerque, NM: CASE/TAM, Assistive Technology and Practice Group.

National Council on Disability. 1993. *Study on the Financing of Assistive Technology Devices and Services for Individuals with Disabilities.* Washington, DC: National Council on Disability.

National Joint Committee on Learning Disabilities. 1990. "Learning Disabilities: Issues on Definitions." In *Collective Perspectives on Issues Affecting Learning Disabilities: Position Papers, Statements, and Reports,* 2nd ed. (pp. 27–32). Austin, TX: Pro Ed, Inc.

Rehabilitation Engineering and Assistive Technology Society of North America (RESNA) (www.resna.org).

Scherer, Marcia. 1991. *Matching Person and Technology.* Rochester, NY: By the author.

Technology and Media Division of the Council for Exceptional Children (www.tamcec.org/index.html).

TRACE Center (www.trace.wisc.edu).

University of Kentucky Assistive Technology Research Institute (serc.gws.uky.edu/www/ukatri/sitemap.html).

University of Kentucky Assistive Technology Project. 2000. UKAT Toolkit. Unpublished manuscript, University of Kentucky–Lexington.

Wisconsin Assistive Technology Initiative (www.wati.org).

———. 2000. *Assessing Students' Needs for Assistive Technology (ASNAT),* 3rd ed. Oshkosh, WI: Wisconsin Assistive Technology Initiative.

Zabala, Joy. 1996. "SETTing the Stage for Success." In *Proceedings of the Seventeenth Annual Southeast Augmentative Communication Conference* (pp. 129–157). Birmingham, AL: United Cerebral Palsy of Birmingham.

Association for the Advancement of Computing in Education (AACE)

The Association for the Advancement of Computing in Education, founded in 1981, is an international, educational, and professional nonprofit organization dedicated to the advancement of the knowledge, theory, and quality of learning and teaching at all levels with information technology. This is accomplished through the encouragement of scholarly inquiry related to information technology in education and the dissemination of research results and their applications through: publications, conferences, societies and chapters, and interorganizational projects. AACE's membership includes researchers, developers, and practitioners in schools, colleges, and universities, as well as administrators, policymakers and decisionmakers, trainers, adult educators, and other specialists in education, industry, and the government with an interest in advancing knowledge and learning with information technology in education.

Publications

Visit the AACE website (www.aace.org; link to Publications) for more information on the journals listed below.

International Journal on E-Learning (*IJEL*); **Corporate, Government, Healthcare, and Higher Education (ISSN# 1537–2456; quarterly)**

IJEL provides educators and trainers with unique opportunities to enhance learning and teaching in corporate, government, health care, and higher education. *IJEL* serves as a forum to facilitate the international exchange of information on the current research, development, and practice of e-learning in these sectors.

Journal of Computers in Mathematics and Science Teaching (*JCMST*) **(ISSN# 0731–9258; quarterly)**

JCMST is the only periodical devoted specifically to using information technology in the teaching of mathematics and science. The journal offers an in-depth forum for the exchange of information in the fields of science, mathematics, and computer science.

Journal of Interactive Learning Research (*JILR*) **(ISSN# 1093–023X; quarterly)**

The published papers in the *JILR* relate to the underlying theory, design, implementation, effectiveness, and impact on education and training of the following interactive learning environments: authoring systems, CALL, assessment systems, CBT, computer-mediated communications, collaborative learning, distributed learning environments, performance support systems, multimedia systems, simulations and games, intelligent agents on the Internet, intelligent tutoring systems, microworlds, and virtual reality–based learning systems.

Journal of Educational Multimedia and Hypermedia (*JEMH*) **(ISSN# 1055–8896; quarterly)**
Designed to provide a multidisciplinary forum to present and discuss research, development, and applications of multimedia and hypermedia in education. The main goal of the *JEMH* is to contribute to the advancement of the theory and practice of learning and teaching using these powerful and promising technological tools that allow the integration of images, sound, text, and data.

Journal of Technology and Teacher Education (*JTATE*) **(ISSN# 1059–7069; quarterly)**
A forum for the exchange of knowledge about the use of information technology in teacher education, the *JTATE* content covers preservice and in-service teacher education, graduate programs in areas such as curriculum and instruction, educational administration, staff development, instructional technology, and educational computing.

Information Technology in Childhood Education Annual (*ITCE*) **(ISSN# 1522–8185; annual)**
A primary information source and forum to report the research and applications for using information technology in the education of children—early childhood, preschool, and elementary. This annual is a valuable resource for all educators who use computers with children.

Educational Technology Review (*ETR*) **(an electronic journal; ISSN# 1065–6901; quarterly)**
This member journal is the focal point for AACE members to exchange information among disciplines, educational levels, and information technologies. Its purpose is to stimulate the growth of ideas and practical solutions that can contribute to the improvement of education through information technology.

Contemporary Issues in Technology and Teacher Education (*CITE*) **(an electronic journal; ISSN# 1528–5804; quarterly)**
An electronic publication of the Society for Information Technology and Teacher Education (SITE), established as a multimedia, interactive

counterpart of the *Journal of Technology and Teacher Education.* Funded by the U.S. Department of Education's Preparing Tomorrow's Teacher to Use Technology catalyst grant, *CITE* makes possible the inclusion of sound, animated images, and simulation, as well as allowing for ongoing, immediate dialog about theoretical issues.

Conferences

- SITE: Society for Information Technology and Teacher Education International Conference (held in March)
- ED-MEDIA: World Conference on Educational Multimedia, Hypermedia, and Telecommunications (held in June)
- E-Learn: World Conference on E-Learning in Corporate, Government, Healthcare, and Higher Education (held in October or November)
- ICCE: International Conference on Computers in Education (Asia-Pacific Chapter) (held in December)

Membership Benefits

Annual individual membership in the AACE includes:

- subscriptions to AACE's internationally respected journals
- discounts on journal subscriptions, conference registrations, proceedings books and CD-ROMs, and all AACE products
- access to AACE's two electronic journals (*Educational Technology Review* and *Contemporary Issues in Technology and Teacher Education).*

Resources for Members

AACE's Journal Table of Contents Alert Service (www.aace.org/dl/Search/TOC) will automatically e-mail the table of contents of any chosen journal issues whenever the latest issue is available in the AACE Digital Library. These notifications will include links to the issue abstracts weeks in advance of regular publication.

The AACE Digital Library (www.aace.org/DL) has collected thousands of journal articles and conference proceedings and will assist members in their efforts to conduct scholarly research; keep current on the latest research and publications in their field; have access to publications in an easy, searchable, and efficient manner; and maintain their academic interests in the area of information technology in education.

The goal of the Career Center (www.aace.org/careers) is to assist members and others in their efforts to search for new positions and advance their careers within the educational technology and e-learning communities. Included is a job board and career book resources.

The new AACE Store (www.aace.org/store) initially offers conference CD-ROMs at a discount to members. Many new products will be added, all offering member discounts.

Societies and Chapters

The Society for Information Technology and Teacher Education is an international association of individual teacher educators, and affiliated organizations of teacher educators in all disciplines, who are interested in the creation and dissemination of knowledge about the use of information technology in teacher education and faculty/staff development. The society seeks to promote research, scholarship, collaboration, exchange, and support among its membership and to actively foster the development of new national organizations where a need emerges. This is the only organization that has as its sole focus the integration of instructional technologies into teacher education programs.

In 1995 a group of professionals from several Central European countries met at ED-MEDIA 95 (the World Conference on Educational Multimedia and Hypermedia in Austria) and discussed the establishment of a Central European Chapter (CEC) of AACE. It was clear that this chapter was needed and should be formed to promote educational technology research and development in the region. Thus the CEC was established. Member countries are: Bulgaria, Czech Republic, Croatia, Estonia, Hungary, Latvia, Lithuania, Poland, Slovakia, Slovenia, Romania, Russia, Ukraine, and all other republics of the former Soviet Union and Yugoslavia. CEC's objectives are to:

- enhance awareness and promote the conduct and dissemination of research and development related to computing technologies in education both within and outside the Central European region
- encourage collaborative activities within the CEC countries and with other countries throughout the world
- organize and hold an international-level Central European conference series within the CEC countries
- encourage and support academic activities of AACE subchapters of CEC member countries.
- obtain greater representation of active researchers from the Central European region in the committees of related leading professional organizations and the editorial boards of reputable journals

Interorganizational Partnership Projects

The *IMEJ of Computer-Enhanced Learning* is an electronic journal focused on ideas and innovations in education technology. *IMEJ* presents articles in a dynamic format including pictures, videos, sound, discussion forums, and interactive tutorials and exercises. This journal is published biannually at Wake Forest University in association with AACE.

Computers in the Social Studies Journal is an educational electronic publication dedicated to the encouragement of the use of personal computers and related technology in social studies classrooms (K-12 and above).

Marianne Williams

See also Society for Information Technology and Teacher Education

Association for Computing Machinery (ACM)

The Association for Computing Machinery is a major force in advancing the skills of information technology professionals and students. Founded in 1947, ACM is the world's first educational and scientific computing society. ACM serves its global membership of more than 71,000 by delivering cutting-edge technical information and transferring ideas from theory into practice. ACM hosts the computing industry's leading portal to computing literature. With its journals and magazines, special interest groups, conferences, workshops, electronic forums, career resource center, and professional development center, ACM is a primary resource to the information technology field. For more information, visit www.acm.org.

ACM was founded, provisionally, as the Eastern Association for Computing Machinery at a meeting at Columbia University in New York City on September 15, 1947. This was the logical outgrowth of increasing interest in computers as evidenced by several events, including a January 1947 symposium at Harvard University on large-scale digital calculating machinery; the six-meeting series in 1946–1947 on digital and analog computing machinery conducted by the New York chapter of the American Institute of Electrical Engineers; and the six-meeting series in March and April 1947 on electronic computing machinery conducted by the Department of Electrical Engineering at Massachusetts Institute of Technology. In January 1948 the word "Eastern" was dropped from the name; in September 1949 a constitution was instituted by membership approval.

The original notice for the September 15, 1947, organization meeting stated in part: "The purpose of this organization would be to advance the science, development, construction, and application of the new machinery for computing, reasoning, and other handling of information." The first and subsequent constitutions for the association have elaborated on this statement, although the essential content remains. The current constitution states that the association "is an international scientific and educational organization dedicated to advancing the art, science, engineering, and application of information technology, serving both professional and public interests by fostering the open interchange of information and by promoting the highest professional and ethical standards."

ACM membership is drawn from all sectors of the computing sciences and their applications, from the design and construction of computers to the development of appropriate programming theory and languages and the utilization of computers in scientific investigation, industrial control, management data processing, and the humanities. Originally, membership in ACM was open to all interested in the purposes of the association.

In 1966, ACM adopted grades of membership, which currently include "professional member" and "student member." Eligibility for these grades is defined as follows: professional members are those who subscribe to the purposes of the ACM and satisfy one of the following qualifications: bachelor's degree; equivalent level of education; or two years' full-time employment in the information technology field. Student members are those who subscribe to the purposes of the association. Institutional memberships for companies and universities were introduced in 1960. As of 2002 some 750 companies and universities had become institutional members. Professional membership is approximately 54,000, and student membership totals some 17,000, for a total of more than 71,000 members.

ACM is governed by a council consisting of sixteen members that is the highest governing authority in ACM. The council is composed of the president, vice president, secretary-treasurer, the immediate past president, the Special Interest Group Governing Board (SGB) chair, three SGB council representatives, the Publications Board chair, and seven at-large members. The president, vice president, and secretary-treasurer are elected to two-year terms by the members; the chair of the Publications Board is elected to a three-year term by the council; and at-large members are elected to two-year terms by members. The council meets two times per year, and the executive committee meets as necessary.

The headquarters office houses a staff of approximately seventy-five under the supervision of the chief executive officer. The staff performs necessary organization functions (membership, accounting, subscription fulfillment, etc.); coordinates and supports the activities of ACM chapters and committees; provides membership services (professional development center, career resource center, job center, etc.); acts as a liaison for meetings sponsored by the association; and produces ACM periodicals. It serves as an information center for members, news media, and the general public on a diversity of subjects in the general area of computers and their applications.

Four boards, comprising numerous volunteer committees and subgroups, work together with the headquarters staff to manage ACM products and services. These boards are the Publications Board, the Special Interest Group Governing Board, the Education Board, and the Membership Activities Board.

ACM and the ACM Special Interest Groups (SIGs) sponsor, cosponsor, and cooperate with more than 150 technical meetings annually. Because ACM provides an objective arena for the discussion of novel and often competing ideas, many of these conferences have become premier world events.

ACM SIGs, in thirty-three distinct areas of information technology, address varied interests: programming languages, graphics, computer-human interaction, and mobile communications, to name a few. Each SIG organizes itself around those specific activities that best serve both its practitioner- and research-based constituencies. Many SIGs sponsor conferences and workshops and offer members reduced rates for registration and proceedings. SIGs also produce newsletters and other publications or support lively e-mail forums for information exchange.

ACM Special Interest Groups

SIGACT	Algorithms and Computation Theory
SIGAda	Ada Programming Language
SIGAPL	APL Programming Language
SIGAPP	Applied Computing
SIGARCH	Computer Architecture
SIGART	Artificial Intelligence
SIGCAPH	Computers and the Physically Handicapped
SIGCAS	Computers and Society
SIGCHI	Computer-Human Interaction
SIGCOMM	Data Communication
SIGCSE	Computer Science Education
SIGDA	Design Automation
SIGDOC	Systems Documentation
SIGECOM	Electronic Commerce
SIGGRAPH	Computer Graphics
SIGGROUP	Groupware
SIGIR	Information Retrieval
SIGKDD	Knowledge Discovery in Data
SIGMETRICS	Measurement and Evaluation
SIGMICRO	Microprogramming/Microarchitecture
SIGMIS	Management Information Systems
SIGMOBILE	Mobility of Systems, Users, Data, and Computing
SIGMOD	Management of Data
SIGMULTI-MEDIA	Multimedia
SIGOPS	Operating Systems

SIGPLAN	Programming Languages
SIGSAC	Security, Audit, and Control
SIGSAM	Symbolic and Algebraic Manipulation
SIGSIM	Simulation and Modeling
SIGSOFT	Software Engineering
SIGSOUND	Electronic Forum on Sound Technology
SIGUCCS	University and College Computing Services
SIGWEB	Hypertext, Hypermedia, and Web

Publications

ACM publishes, distributes, and archives original research and firsthand perspectives from the world's leading thinkers in computing and information technologies that help computing professionals negotiate the strategic challenges and operating problems of the day. ACM publishes twenty-five journals, more than thirty newsletters, and eighty-five conference proceedings annually. ACM is also recognized worldwide for its published curricula recommendations, for colleges and universities as well as for secondary schools that are increasingly concerned with preparing students for advanced education in the information sciences and technologies.

ACM's flagship publication, *Communications of the ACM,* keeps information technology professionals up to date with articles spanning the full spectrum of information technologies in all fields of interest. Examples include object-oriented technology, multimedia, internetworking, and hypermedia. *Communications* also carries case studies, practitioner-oriented articles, and regular columns, the ACM Forum, and technical correspondence. The monthly magazine is distributed to all ACM members. For a complete list of publications, visit www.acm.org/dl.

The ACM Press books program is a collaborative effort between ACM and Addison Wesley Longman Publishing Company to develop, publish, and distribute a broad range of new books in computer science and engineering. The program also includes a book series with ACM SIGGRAPH, the ACM SIG in computer graphics. ACM also publishes the Portal to Computer Literature, which is the gateway to the ACM Digital Library and the ACM Online Guide to Computing Literature. The ACM portal includes a sophisticated search capability, which is free to all browsers.

The ACM Digital Library contains the citations and full text of 100,000 articles, representing all of ACM's journals, newsletters, and proceedings. Each citation contains links to other works by the same author; clickable references to their original sources; links to similar articles and critical reviews, if available; and digital object identifiers to easily manage electronic linkages to vendors.

The ACM Online Guide to Computing Literature consists of a bibliographic database of more than 500,000 citations, dating back to 1985.

These citations come from a broad range of information technology publications and publishers. Many of these citations contain abstracts and/or reference sections as well. The books database, for example, contains citations to nearly 50,000 volumes, with links to commercial vendors that facilitate online purchasing.

Chapters

ACM professional chapters and local SIG chapters unite colleagues in particular geographical areas, offer the opportunity to gain immediate access to technological advances, and establish a personal networking system in the locale. There are currently more than 160 ACM professional and local SIG chapters worldwide, 25 percent of which are outside the United States. The chapters host lectures by internationally known computer professionals, sponsor state-of-the-art seminars on the most pressing issues in information technology, conduct volunteer training workshops, and publish informal newsletters.

ACM has established student chapters to provide an opportunity for students to play a more active role in the association and its professional activities. More than 600 colleges and universities throughout the world participate in the ACM Student Chapter Program, whose aims are to enhance learning through exchange of ideas among students and among established professionals and students. By encouraging organization of student chapters on college and university campuses, the association is able to introduce students to the benefits of a professional organization. These benefits include periodic meetings.

Student chapter members may take advantage of the activities and services provided by the association, such as the Distinguished Lectureship Program, the International Collegiate Programming Contest, the Student Research Competition, and the Publications Program. Student chapters provide an obvious setting to develop and demonstrate leadership capabilities—an important factor both to students in career development and professional growth, as well as to the future of the association.

Awards

ACM recognizes excellence through its eminent series of awards for outstanding technical and professional achievements and contributions in computer science and information technology. ACM sponsors eight major awards, named for the foremost luminaries in the computing field, as well as several other awards that honor distinguished service in information technology. It also names as fellows each year those outstanding members who have demonstrated achievements in computer science and information technology and who have made significant contributions to the ACM's mission. A prize of $100,000 accompanies ACM's most prestigious technical award, named for A. M. Turing, a pioneer in the computing field. It is awarded to an individual selected for contributions of a technical nature made to the computing community. The contributions should be of lasting and major technical importance to the computer field.

In addition to the A. M. Turing Award, other ACM awards include:

- ACM Software System Award
- ACM/AAAI Allen Newel Award
- Grace Murray Hopper Award
- Karl V. Karlstrom Outstanding Educator Award
- Paris Kanellakis Theory and Practice Award
- Eugene Lawler Award for Humanitarian Contributions within Computer Science and Informatics
- Distinguished Service Award
- Outstanding Contribution to ACM Award
- Doctoral Dissertation Award

Professional and Public-Service Activities

Many of ACM's professional and public-service activities are conducted by standing committees. Examples include Computers and Public Policy, Constitution and Bylaws, Nominating, Elections, USACM Public Policy, Committee on Professional Ethics, and the ACM International Collegiate Programming Contest.

ACM is committed to bringing potentially significant technical and public policy issues to the attention of the ACM membership and community. It sponsors several committees to address these issues, including: the Committee on the Status of Women in Computing; the Committee on Computers and Public Policy; and the U.S. Public Policy Committee (USACM). The ACM Office of Public Policy works with the USACM to assist policymakers and the public in understanding information technology issues and to advance a policy framework that supports innovations in computing and related disciplines.

As the first society in computing, the Association for Computing Machinery continues to provide quality content and information, community-building, reliability, and global vision to its members. With its unique role in advancing the art, science, education, and application of computing, ACM is a leading resource for advancing the skills of information technology professionals and for interpreting the impact of information technology on society.

Virginia Gold

See also Association for the Advancement of Computing in Education; Association for Educational Communications and Technology

References

Cochran, Anita, ed. 1987. "ACM: The Past 15 Years, 1972–1987." *Communications of the ACM* 30(10): 866–872.

Cringely, Robert X. 1992. *Accidental Empires*. Readng, MA: Addison-Wesley.

Fishman, Katharine Davis. 1982. *The Computer Establishment*. New York: McGraw-Hill.

Freiberger, Paul, and Michael Swaine. 2000. *Fire in the Valley: The Making of the Personal Computer*. New York: McGraw-Hill.

Goldstine, Herman H. 1972. *The Computer from Pascal to Von Neumann*. Princeton, NJ: Princeton University Press.

Hodges, Andrew. 1983. *Alan Turing: The Enigma*. New York: Simon and Schuster.

McCartney, Scott. 1999. *ENIAC: The Triumphs and Tragedies of the World's First Computer*. New York: Walker.

Palfreman, Jon, and Doron Swade. 1991. *The Dream Machine: Exploring the Computer Age*. London: BBC Books.

Shurkin, Joel. 1984. *Engines of the Mind: A History of the Computer*. W. W. Norton.

Revens, Lee, ed. 1972. "The First 25 Years: ACM, 1947–1972." *Communications of the ACM* 15(7): 485–490.

Association for Educational Communications and Technology (AECT)

The Association for Educational Communications and Technology is an international association representing professionals in a range of occupations who have an interest in improving learning through the use of media and technology. AECT is the oldest professional home for this field of interest, having been founded in 1923, and it has continuously maintained a central position in its field, promoting high standards in scholarship as well as in practice.

The history of the association is summarized by Paul Saettler (1998) in the seventy-fifth anniversary issue of AECT's monthly magazine. The association was formed in 1923 as the Department of Visual Instruction of the National Education Association (NEA), and it remained a unit of the NEA, located within its Washington, D.C., headquarters for forty-eight years. The immediate charge of the organization was to help teachers and school administrators make more effective use of motion pictures in schools. Scholars of visual instruction of that period were motivated by the belief that students were hindered in their learning by teaching that was abstract, sterile, and limited to verbal representation. The antidote was to infuse teaching with more active experiences and more richly visualized presentations, as indicated by the title of the leading textbook of that era, *Visualizing the Curriculum* (Hoban, Hoban, and Zisman 1937). In the early years participation was open to all NEA members, and there was not a separate membership list; however, attendance at the national convention was in the range of 100–200 people, predominantly school administrators.

In 1947 the name changed to the Department of Audiovisual Instruction to reflect the emerging importance of sound recording. In the post–World War II period a new paradigm was gaining influence: the communications movement. Gradually, people in this field began to see themselves not just as visualizers but also as designers of communications systems and the messages that flowed through them. This orientation is reflected in the name of the first scholarly journal, *AV Communication Review,* founded in 1953.

Membership rose steadily throughout the 1950s and 1960s, spurred by the passage of major federal legislation supporting research and infrastructure development in audiovisual media (the National Defense Education Act of 1958 and the Elementary-Secondary Education Act of 1965). Also in the 1960s, the emergence of the programmed instruction movement stimulated another paradigm shift in the field, incorporating the mission of designing and utilizing interactive self-instructional systems. The concept of technology of teaching was popularized by B. F. Skinner (1968) to describe his view of programmed instruction as an application of the science of learning. Thereafter, technology had the dual meanings of application of scientific thinking and the various communications media and devices. Stimulated by these new ideas and by the funding of research and development in these areas, membership rose dramatically, peaking in 1970 with about 10,000 members.

As the focus of the association shifted away from helping teachers use media to designing self-instructional systems, the rationale for staying with the NEA, which was increasingly functioning as a teacher's union, waned. In 1971 the association became independent, moved to separate quarters in Washington, D.C., and changed its name to the current Association for Educational Communications and Technology—with the key words reflecting the two dominant paradigms: improving communications, and applying technology to education. The scholarly journal was also renamed to *Educational Communication and Technology: A Journal of Theory, Research, and Development,* consistent with the association's new name. Since its inception, the association has also published a more practice-oriented monthly journal, originally named *Educational Screen,* later *Audiovisual Instruction,* then *Instructional Innovator,* and currently *TechTrends.*

For the middle fifty years of its existence AECT predominantly represented professionals working at the elementary-secondary school levels as administrators of audiovisual services, with college professors as the next largest group. In the mid-1970s the balance shifted toward higher education as audiovisual directors began to disappear as a separate job classification at the school building level (Molenda and Cambre 1977). By 2002 professors and graduate students represented about 60 percent of the membership, with school media specialists—building, district, and regional—representing about 30 percent.

Membership numbers declined through the 1970s, 1980s, and 1990s as new telecommunications and information technologies rose to prominence, each stimulating its own professional organization. AECT continued as a generalist organization, retaining 2,000–3,000 core members attracted to its broad scope, bringing together teachers, school and district media specialists, professors and graduate students of instructional technology, corporate instructional designers, military training designers, multimedia developers, and others. The current interests of the membership are indicated by the special-interest groups of the association: instructional design and development; distance learning; information and technology management; school media and technology; research and theory; teacher education; training and performance; systemic change; and international.

AECT has long served as an umbrella organization for more specialized organizations with overlapping concerns. Affiliated organizations include the International Council for Educational Media, the International Visual Literacy Association, and the National Association of Regional Media Centers, among others. In addition, AECT supports seven local chapters and several state affiliates.

The major means by which the association and its members exert influence in the field of educational technology is its program of periodic and nonperiodic publications. In addition to the quarterly scholarly journal, now known as *Educational Technology Research and Development,* AECT publishes *TechTrends,* a practice-oriented magazine (six issues per year), as well as the online journal *Interpersonal Computing and Technology Journal.*

AECT also offers a wide range of nonperiodic publications on instructional technology research and theory, school media programs, distance learning, educational uses of computers and the Internet, copyright law, and administering media programs. Professional development and communications are encouraged by means of an annual convention featuring hundreds of educational sessions, numerous workshops, and the International Student Media Festival. AECT also sponsors a summer institute devoted to leadership development and professional updating. It also conducts an awards program, recognizing outstanding contributions to educational technology.

As befits an organization that advocates electronic communications, AECT offers a wide range of electronic services, including: a website (www.aect.org); a roster of listservs, most linking members of special-interest divisions; online publications (periodic and nonperiodic), including

Handbook of Research for Educational Communications and Technology; and mailing lists, databases, and other electronic services.

The association is governed by a board of directors of seventeen members, with an executive committee composed of the president, president-elect, past president, and secretary-treasurer. There is a full-time professional staff (four people in 2003) at the permanent headquarters (located since 1999 at 1800 N. Stonelake Drive, Bloomington, IN 47404).

Michael Molenda

See also Instructional Design; International Society for Performance Improvement; Research on Media and Learning

References

Hoban, Charles F., Charles F. Hoban Jr., and Samuel B. Zisman. 1937. *Visualizing the Curriculum.* New York: Cordon Company.

Lembo, Diane L. 1970. *A History of the Growth and Development of the DAVI/NEA from 1923 to 1968.* Unpublished doctoral dissertation. New York: New York University.

Molenda, Michael, and Marjorie Cambre. 1977. "The 1976 Member Opinion Survey." *Audiovisual Instruction* (March): 65–69.

Saettler, Paul. 1998. "Antecedents, Origins, and Theoretical Evolution of AECT." *TechTrends* (January/February): 51–57.

Skinner, Burrhus F. 1968. *The Technology of Teaching.* New York: Appleton-Century-Crofts.

B

Behaviorism

Behaviorism is the study of the observable, or outward, aspects of behavior in relation to changes in the environment. From the behaviorist's perspective, the processes that go on inside the head (planning, deciding, thinking, creating, etc.) are merely internal behaviors and, as such, are governed by the same mechanisms that can be seen externally. Behaviorism is the belief that behavior itself is the appropriate object of the study of learning and teaching. That is, behavior is not to be studied because it is an indirect means of studying something else (such as cognition, or the mind, or constructions); it is the end, not the means. Studying the cause and effect of behavior, therefore, is studying the cause and effect of learning. Behaviorism focuses on the connections among contexts, acts, and consequences. Changing context or outcome can change behavior.

Behaviorism has been a force in the psychology of learning beginning around 1900. Its distinguishing precept has always been the belief that people are biologically based systems as opposed to intersections between biology and spirit. In essence, behaviorists believe that it is not necessary to divide mind and body simply because humans are conscious beings. Consciousness—indeed, all mental processes—are simply extensions of the same biology that we see throughout the animal kingdom.

Behaviorism often distinguishes three types of learning: respondent, operant, and observational. Each type relies on the human ability to discriminate among different situations (contexts) and therefore respond differently. Additionally, each relies on our ability to generalize (extend) our behavior to situations or contexts that are similar to ones we have experienced in the past.

Respondent learning (i.e., classical conditioning) involves pairing a cue that produces a behavior or emotion or attitude with a new cue or stimulus. Over time, the new stimulus is associated with the old one and produces a version of the original response. Respondent learning is the basis of advertising. It is why attractive young women appear in advertising targeted toward young males and why distinguished couples in evening dress appear in luxury automobile advertisements targeted to older, more affluent men and women. It is also the basis of political advertisements (and is particularly evident in negative political advertisements) and propaganda. Respondent conditioning has been popularized in the public mind by Ivan Pavlov's famous studies to condition dogs to salivate to a light or bell.

Operant learning is based on the notion that behaviors (the ones we can see and the ones inside our heads) are selected just like genes: because certain physical traits or "instinctive" behaviors will work in certain environmental conditions. Some conditions might favor long necks, warm blood, and so on. Behaviorists believe that, given a certain environmental situation, certain behaviors will "work." Thus in operant learning, environmental cues "tell" us what to do, what will be "functional," what will "work." Behaviors that work in a situation are said to be "reinforced" and are more likely to be used again. Behaviors that don't work in a context will eventually not be used in that situation (they will be "extinguished"). If the behavior not only doesn't work but actually costs the individual (i.e., losing something valuable or "gaining" something painful), then we say the individual has been "punished" for behaving that way in that context and is less likely to behave that way again. Operant learning is most associated with the writings of B. F. Skinner.

Observational learning is the notion that we can learn new behaviors by watching others. In general, it is assumed to be an extension of operant learning in that we watch what people do in a situation; if what they did "worked," then we are likely to do it ourselves in a similar situation. We know that the model can be live, filmed, or even cartoons. Albert Bandura's social learning theory is generally considered to be the most influential work in this area.

Several major assumptions of behaviorism are directly relevant to educational technology. These focus on the role of the learner, the nature of learning and the generality of the learning, and instructional procedures and processes. First, the learner is not passive but must play an active role in learning. Learners learn by doing—by behaving—often through trial and error with feedback from the environment (including *people* in the environment). Similarly, the learner has to behave in some way (even by responding to a test item) to validate that learning has occurred.

Second, and of particular importance to educational technology, is the requirement of behaviorism to focus on the individual in the learning process because an individual's genetic endowment and reinforcement history are unique. Learner goals are developed in terms of what the learner will accomplish based upon the instructional event *and* must consider where the individual "is" relative to that behavior. Learning complex skills often involves learning a series of incremental behaviors that are ordered in small steps and learned through the use of reinforcement based upon successful achievement at each step.

Third, behaviorists place a great deal of emphasis on the context or cues present during learning because cues "cause" behavior in the sense that cues tell us what behaviors will work. Naturally, the more closely the cues mirror the actual conditions under which the behavior should occur, the better. Just as obviously, if the behavior has a "real-world" context, it should be taught in the presence of those cues. Naturally, technology can help bring such "reality" into the classroom.

Fourth, in order for learning to take place in educational settings there has to be feedback; this can be something typically thought of as a reward, or it could just be knowledge of whether the behavior was "right" or "wrong." In an educational technology sense, this means that the learner must be given lots of opportunities to behave (or respond) with feedback. For human beings, social feedback is very powerful.

Finally, behaviorists believe that learners obey universal laws of learning. This means that all learning in all species on the planet respond to the same laws. From the perspective of educational technology it means that if the program or lesson doesn't work, then it was designed wrong. Behaviorism is never getting to say one is finished.

The practical aspects of the theoretical notions of behaviorism were demonstrated in the 1920s with the development of "teaching machines." These machines are designed to be a single device capable of testing, scoring, and informing students of their errors and finding correct solutions all in one step. These machines did not become popular, however, until the 1950s, when Skinner used them to test and develop his operant conditioning principles. As a result, the teaching machine (and programmed instruction) became a primary research emphasis in the 1960s in both instructional psychology and instructional technology. These early efforts involved taking complex content and breaking it down into component parts that were arrayed in a linear fashion. Each step required students to behave and provided feedback for their responses, then comparing responses and taking small steps to learn content. The teaching machines and programmed instruction actually took the theoretical constructs of operant learning and created a working instructional device.

Closely akin to teaching machine technology, and undergirding theoretical concepts, was *programmed instruction,* which usually took the form of teaching texts or programmed books. These materials essentially were carefully arranged sequences of stimuli that required overt interaction and feedback for a particular instructional objective. These materials were generally found in two formats, linear and branching. Although both used the tenets of behavioral theory, branching was more flexible and less rigid in terms of reinforcement, the size of the steps, and type of feedback.

The impact of the military, during and immediately after World War II, on the research of learning, learning materials, and instructional approaches was driven by the need to train millions of military and nonmilitary individuals for the war effort and the technology-oriented future that was to follow. Much of this research stressed the stimulus, response, and reinforcement characteristics of various instructional strategies and devices. These studies looked at such behavioral components as the role of active responding in learning, the optimal size of an informational step in procedural learning, and the use of prompts (cues) in the instructional setting. During this era, major research programs looked at the use of instructional films and, later, television programs. Stimulus-centered techniques stressed the meaning, structure, and organization of stimulus materials while response-centered techniques dealt with the design techniques to ensure that materials produced adequate responses. The techniques of programmed instruction that were combined with film and television depended on the effective design of the stimulus materials (and the theoretical basis for this) as well as the design of the appropriate response practice. Thus research on stimuli, and on response, became major components of educational technology research during and following World War II.

One of the more prominent impacts of behavioral theory on educational technology and on education in general is its influence on *instructional design,* usually defined as the systematic development of instructional sequence. As such, it is a clear combination of the use of a learning theory to guide the development of instruction. In fact, many of the current instructional design models use major components of methodological behaviorism such as the specification of behavioral objectives, the importance of student behavior changes, and the emphasis on the learning environment (stimulus). The behavioral influence is also evident in the association between the stimulus and the student response. Instruction, in this view, must be evaluated by its ability to change the behavior of an individual student. In most instructional design models the following components are found: Objectives of the instruction are stated; all objectives are measurable and observable and meet standards of reliability and validity; and the concept centers on the changes of the student (learner).

At least three modified instructional models have been developed around behavioral concepts and used in recent years: the personalized system of instruction (PSI), precision teaching, and direct instruction. All three share the basic tenets of the behavioral approach. Each instructional approach involves scripted lessons, the requirement of overt student response, and the provision of immediate feedback. PSI, developed by Fred Keller, contained features such as interlocking instruction, highly sequenced progressive tests, individualized learning activity, and self-pacing with reinforcement through the sequence of instruction.

O. R. Lindsley's concept of precision teaching emphasized the counting of specific behaviors with a special emphasis on analyzing both correct and incorrect responses. This program made "second nature" responding plus speed of performance a key feature of each student's progress. The third behaviorally inspired instructional system of note is direct instruction design. This system stressed three types of analysis: learner behavior, communication, and knowledge systems. This system looked at how to prompt and reinforce responses and how to correct errors that are made. Communication was viewed in terms of logical design and sequence of presentation. The analysis of knowledge stressed logical organization and progression of content, from simple to complex.

John K. Burton
David M. Moore

See also Cognitive Psychology; Constructivism; Instructional Design

References

Burton, John K., David M. Moore, and Susan G. Magliaro. 1996. "Behaviorism and Instructional Technology." In D. H. Jonassen (ed.), *Handbook of Research for Educational Communications and Technology* (pp. 46–73). New York: Simon and Schuster Macmillan.

Keller, Fred S., and J. Gilmour Sherman. 1974. *The Keller Plan Handbook.* Menlo Park, CA: Benjamin.

Skinner, Burrhus F. 1938. *The Behavior of Organisms.* New York: Appleton.

———. 1953. *Science and Human Behavior.* New York: Macmillan.

———. 1958. "Teaching Machines." *Science* 128: 969–977.

———. 1964. "Behaviorism at Fifty." In T. W. Wann (ed.), *Behaviorism and Phenomenology.* Chicago: University of Chicago Press.

———. 1968. *The Technology of Teaching.* Englewood Cliffs, NJ: Prentice-Hall.

———. 1974. *About Behaviorism.* New York: Knopf.

Thorndike, Edward L. 1898. *Animal Intelligence: An Experimental Study of the Associative Processes in Animals. Psychological Review Monograph* 2 (supp. 8).

———. 1913. "The Psychology of Learning." *Educational Psychology,* vol. 2. New York: Teachers College Press.

Bloom, Benjamin S. (1913–1999)

Benjamin Bloom led an effort by educational psychologists in the mid-1950s to develop a comprehensive system of describing and assessing educational outcomes. The development of educational objectives—often according to Bloom's taxonomy—has become a critical part of almost any curriculum development effort or instructional design process.

Although he was made famous throughout the world by the taxonomy of educational objectives that bore his name, it was not his only noteworthy work: Bloom also championed the relatively progressive idea that most students had more potential than educators gave them credit for, and he pushed the establishment to question some long-held assumptions about the roles of educators and education in helping students learn. By any measure, Benjamin Bloom had an astonishing impact on the practices and philosophies of educators around the world during the latter half of the twentieth century. His work ranged from the purely practical to the philosophical; his ideas, dynamic though they were during his lifetime, can hardly be easily categorized.

Born in 1913 to an immigrant family in Pennsylvania, Bloom went to Pennsylvania State College to become a teacher. He received his bachelor's and master's degrees there by 1935. He wanted to continue his education, however, and determined to become a protégé of progressive educator Ralph Tyler by studying under him at the University of Chicago. Bloom also worked under Tyler's direction on the famous Eight-Year Study, a laboratory school arrangement whereby individual schools could assess students without using traditional grades and students' acceptance to participating universities would not be affected.

During his doctoral studies and after their completion in 1942, Bloom worked on staff at the University of Chicago Board of Examiners, overseeing the creation, administration, and scoring of the exams taken by undergraduate students in fulfillment of their degrees. He became an instructor in the university's department of education two years later and eventually became the university examiner, directing all of the school's assessment efforts. His early research, influenced by his testing work, focused on how educational outcomes could be categorized in terms of their cognitive complexity; this led to what became arguably his best-known contribution to the field.

In the late 1940s, along with colleagues from around the world, Bloom began an effort under the auspices of the American Psychological Association to create a systematic and organized model of educational objectives. *The Taxonomy of Educational Objectives: The Classification of Educational Goals (Handbook 1: Cognitive Domain)*, published in 1956 and known by many simply as the *Handbook*, was hardly the work of one person, but it nonetheless came to be known simply as "Bloom's taxonomy."

The work in part reflected the dichotomy of perspectives that was Benjamin Bloom. Although he is commonly considered to be a behaviorist thinker, concerned with developing in students particular behaviors that reflect whether learning has occurred, his work also reflects a cognitive perspective, concerned with how students engage in learning and construct knowledge. Although the structure of the taxonomy focuses on measurable student behaviors resulting from instruction, Bloom looked at the taxonomy as a way to encourage educators to think critically and systematically about how to plan educational experiences to achieve certain goals.

The taxonomy is hierarchical in nature. Behavior that demonstrates mastery at each level is a prerequisite for moving to the next level, and objectives at higher levels are more cognitively complex. Likewise, the behavior that demonstrates competency at lower levels is necessarily more concrete and observable than that of the more complex levels. The taxonomy consists of the following six levels:

1. *Knowledge:* This lowest level of the taxonomy represents basic interactions with information, or the basic recall of data. The learner's primary cognitive activity is remembering—both basic facts as well as more complex ideas and theories. In practice, most educational activities fall into this category.
2. *Comprehension:* Objectives at this level focus on the translation, interpretation, and extrapolation of information. Learners are required to demonstrate broader recognition and understanding of information in formats unfamiliar to them.
3. *Application:* At this level of the taxonomy, students must correctly demonstrate the use of information, preferably in realistic environments.
4. *Analysis:* This level represents learners' abilities to deconstruct information and describe both the organizational structure and the relationships between component parts.
5. *Synthesis:* These objectives reflect a student's ability to put individual parts of previous learning together to form a new whole; Bloom referred to it as the most "creative" category of the hierarchy for that reason.
6. *Evaluation:* In this top level of the taxonomy, students begin to assess work by themselves and others. These judgments are not only cognitive ones based on previous learning; evaluation is also the point where learners begin to connect cognitive processes with affective behaviors like interest, attitudes, and values.

The *Handbook,* besides outlining the taxonomy in great detail, also focuses on the design and implementation of testing for each level of the taxonomy. Lower levels are relatively easy to assess, because behaviors are easily observable and can be objectively measured. As cognitive complexity increases, assessment becomes more difficult. At higher levels, students need more time and greater creative freedom to complete the tasks, and assessment becomes more inherently subjective. Thus large-scale testing at higher levels becomes difficult to do economically.

Bloom credited the taxonomy with filling a dire need in education—never before had there been such a systematic way to design instruction for particular outcomes. Since Bloom's groundbreaking work, the taxonomy, as well as the process of developing educational objectives for learning experiences based on it, have become ubiquitous in all types of instructional endeavors.

After the publication of the *Handbook*, Bloom grew into new areas of interest and research. During the 1950s education as a field—particularly as reflected in Bloom's work as a university examiner—was largely focused on the "sorting" role: determining the relatively small number of students who were the most intellectually capable, then allowing them access to further education. The vast majority of students, it was assumed, were in the middle range of intellectual capability; a few were much less capable.

Bloom started to question these fundamental assumptions about students' potential in the 1960s. He began doing research about a fundamentally different pedagogical method, which he called "learning for mastery": He theorized that most students could master basic educational skills if given appropriately focused instruction and the time they needed to learn. Bloom's method included formative assessments early in the instructional process to identify students' weaknesses and allow teachers to craft appropriate instruction; students who mastered the skill early could then go on to enrichment exercises or serve as peer tutors for those still working toward mastery. The process would take much longer than traditional classroom teaching practices initially, although the need to revisit subject matter on a regular basis would diminish over the long term. Students' mastery of concepts would render such repetition unnecessary.

Later in life, Bloom extended his mastery learning ideas and began to study the development of talent in high-performing individuals from a variety of fields. His discoveries, which have broad implications for education in general, were astounding to many: Those whom we think of as outstanding performers in their fields were, in fact, usually not prodigies as children. They achieved their levels of mastery by constant practice over time, coupled with instruction from increasingly more capable mentors as their abilities increased. Bloom deconstructed and analyzed the processes by which these individuals learned and found that their performance had reached the point of automaticity—they no longer thought about their actions. Many activities taught in school, Bloom theorized, could be taught so that they became as familiar to students as walking; in this way, many more learners could then go on to higher levels of cognitive activity.

Gabriel Reedy

See also Bloom's Taxonomy; Instructional Objectives

References

Anderson, Lorin W., and Lauren A. Sosniak, eds. 1994. *Bloom's Taxonomy: A Forty-Year Retrospective.* Chicago: National Society for the Study of Education.

Bloom, Benjamin S., ed. 1956. *Taxonomy of Educational Objectives: The Classification of Educational Goals (Handbook 1: Cognitive Domain).* New York: Longmans, Green.

———. 1972. "Innocence in Education." *School Review* 80: 1–20.

———. 1978. "New Views of the Learner: Implications for Instruction and Curriculum." *Educational Leadership* 35(7): 563–576.

———. 1986. "Automaticity: The Hands and Feet of Genius." *Educational Leadership* 43(5): 70–77.

Bloom's Taxonomy

Bloom's taxonomy (originally known as the taxonomy of educational objectives) is a classification of educational goals. The idea for an educational taxonomy was proposed at the 1948 convention of the American Psychological Association by a group of researchers led by educational psychologist Benjamin S. Bloom. The purpose of the taxonomy was to classify student behaviors that represented the intended outcomes of the educational process. Insofar as it was successful in accomplishing this objective, it was believed that the taxonomy would also facilitate communication among teachers, especially regarding test materials and ideas about testing. The original plan for the taxonomy included three domains of behavior: the cognitive, the affective, and the psychomotor. The cognitive and affective domains have been structured and described in two published handbooks (see Bloom 1956 and Krathwohl, Bloom, and Masia 1964). Although they did not develop a framework for the psychomotor domain, one has been outlined in print (see Simpson 1966; Harrow 1972).

The cognitive domain consists of six major classes of behaviors, listed here from the simplest to the most complex. They include *knowledge, comprehension, application, analysis, synthesis,* and *evaluation.* Bloom and the other developers of the taxonomy asserted that the objectives in any given class are likely to make use of, and build upon, behaviors found in the preceding class.

Knowledge includes, first, the recall of specific information, such as terminology, dates, events, people, and places. It also includes recalling ways and means of dealing with specific information—in other words, how to organize, study, judge, and critique it. Third, knowledge involves recalling the larger structures, theories, and generalizations by which the specific information is organized.

Comprehension represents the lowest level of understanding. More specifically, it refers to a type of understanding in which the individual knows what is being communicated and can make use of it but cannot necessarily relate it to other material or appreciate its full implications. There are three major types of comprehension. First, translation is the paraphrasing of a communication, or its conversion to another language. Second, interpretation is simply the explanation or summary of a communication. Third, extrapolation is an extension beyond the given data to determine its implications and consequences.

Application refers to the use of abstractions in specific, concrete situations. These abstractions may take several forms: general ideas, rules of procedures, generalized methods, or technical principles.

Analysis is the fourth class of behaviors in the cognitive domain. This includes the systematic application of the elements of a communication, of the relationships between those elements, and of the organization, arrangement, and structure that hold the communication together.

Synthesis involves the joining of separate elements to create a whole. This can involve the creation of a unique communication, such as the recounting of a personal experience. It might also include the development of a plan of work or a proposed set of operations. Finally, synthesis may involve deriving a set of abstract relations, in order to classify or explain some data or phenomenon, or to make deductions about it.

Evaluation is the most complex class of behaviors in the cognitive domain of the taxonomy. It refers to judgments about the value of certain materials and methods for some purpose. These judgments are based, first, upon internal evidence, such as logical accuracy and consistency. They are also based upon external criteria or standards of evaluation.

Several years after the development and structuring of the cognitive domain of the taxonomy, three of the project's original members published a second handbook that outlined the *affective domain.* The affective domain is composed of five classes of behaviors. Listed from the simplest to the most complex, they include *receiving* (attending), *responding, valuing, organization,* and *characterization* by a value or value complex. Like the levels of the cognitive domain, each class of the affective domain should build upon the behaviors found in the preceding class.

The most basic outcome in the affective domain is for the individual to be open to receiving (or attending to) certain phenomena and stimuli. The three subcategories of receiving are the awareness or consciousness of something's existence, the willingness to receive or tolerate it rather than avoid it, and the ability to control one's attention or select those things to which one pays attention.

Responding to the phenomenon, rather than just passively receiving it, is the second objective. The concern here is with the individual's acquiescence in responding, willingness to respond, and emotional satisfaction with the response.

Valuing, the third class of objectives, refers to the internalization by an individual of a set of specified, ideal values. This involves an individual's acceptance of, preference for, and commitment to a particular value or values. Beyond simply internalizing values, the individual also finds the need for some kind of value organization. The organization of values into a system first requires conceptualizing each value, then organizing the value system according to an ordered relationship.

The most complex class of objectives within the affective domain is the characterization by a value or value complex. At this level, the individual has already received and responded to certain phenomena, created a value hierarchy, organized it into a consistent system, and allowed that value system to guide his behavior. The two concerns here are the degree to which the value system controls the individual's behavior (or how widely the individual has generalized the value system), as well as the integration of the value system—including the individual's beliefs, ideas, and attitudes—into a total philosophy of life or outlook on the world.

Teachers can use both the cognitive and affective domains of the taxonomy to make decisions about assignments, readings, and teaching strategies. The taxonomy can also prove useful for making choices among the many technological tools and media applications currently available to educators. Teachers may choose, for example, to utilize a video documentary to enable students to build and expand a knowledge base about a particular phenomenon, such as the civil rights movement in the United States. In addition to viewing the documentary, students may be asked to conduct an Internet search for more information about the movement. That undertaking could utilize a popular search engine such as Google or an academic database like JSTOR. One way in which the teacher may choose to assess students' comprehension of the material presented in the documentary is by asking them to summarize important dates, events, places, and people or to describe and explain the causes of the movement.

Students might be asked, further, to apply their conclusions about the causes of the civil rights movement to some other social movement (the women's movement, for instance) in an effort to formulate some general ideas or principles about the rise of social movements. The application of students' ideas could be accomplished through the presentation of another documentary on the women's movement, or through another Internet-based homework assignment that involves searching for information on the women's movement. In the process of applying their conclusions about the civil rights movement to the women's movement, students would be forced to thoroughly analyze the relationships between the causes of the respective movements as presented in the two documentaries.

Teachers could also utilize technology to assess students' progress from one level of the taxonomy to the next. Students might be required—individually or in small groups—to prepare a slide presentation to the class, utilizing an overhead projector or computer software to demonstrate their knowledge about and comprehension of the important aspects of the two social movements. Again, students' knowledge about the movements could come from the documentaries shown in class and/or from individual homework assignments in which they are required to search for relevant information on the Internet.

An in-class presentation might be utilized in which students must present an application of their thoughts about the rise of the civil rights movement to the rise of the women's movement. A thorough analysis of each movement's precipitating factors could also be accomplished using overhead slides or a computer-based presentation. A final course project might involve the synthesis of all of these ideas into one theory about the rise of social movements, as well as the critical evaluation of the ability of that theory to explain the rise of different movements. Again, one can see how a computer-based in-class presentation could be beneficial here.

As students move from one class of behaviors to the next during a project such as that outlined above, they must use technology and media applications (in-class documentaries, Internet searches, overhead projector and computer presentations) to demonstrate their mastery of each level. In this way, Bloom's taxonomy contributes to a classroom in which active learning is the means of education. Using the taxonomy as a guide to instructional design, teachers are better able to decide what media applications and technological tools are most useful to accomplish the objectives outlined at each level. The taxonomy also serves as a guide for teachers in evaluating students' progression from one class of intended behaviors to the next highest class.

Though the implications of Bloom's taxonomy for educational technology have only recently begun to be explored, the taxonomy itself has been widely influential in a variety of academic disciplines since its development. In an effort to incorporate new knowledge and ideas into the original framework, one of the original project's members recently published a revision of the taxonomy (see Anderson and Krathwohl 2001). Benjamin Bloom, the originator of the idea of the taxonomy and the editor of the original handbook on the cognitive domain, did not live to see this revision. He died in 1999, shortly before the book's publication.

Michael DeCesare

See also Behaviorism; Bloom, Benjamin S.; Cognitive Psychology; Instructional Design; Instructional Objectives

References

Anderson, Lorin W., and David R. Krathwohl, eds. 2001. *A Taxonomy for Learning, Teaching, and Assessing: A Revision of Bloom's Taxonomy of Educational Objectives.* New York: Longman.

Anderson, Lorin W., and Lauren A. Sosniak, eds. 1994. *Bloom's Taxonomy: A Forty-Year Retrospective.* Chicago: University of Chicago Press.

Bloom, Benjamin S., ed. 1956. *Taxonomy of Educational Objectives: The Classification of Educational Goals, Volume 1: Cognitive Domain.* New York: David McKay.

Harrow, Anita J. 1972. *A Taxonomy of the Psychomotor Domain: A Guide for Developing Behavioral Objectives.* New York: David McKay.

Krathwohl, David R., Benjamin S. Bloom, and Bertram B. Masia. 1964. *Taxonomy of Educational Objectives: The Classification of Educational Goals, Volume 2: Affective Domain.* New York: David McKay.

Simpson, B. J. 1966. "The Classification of Educational Objectives: Psychomotor Domain." *Illinois Journal of Home Economics* 10(4): 110–44.

Bruner, Jerome S. (b. 1915)

It would be a difficult, if not impossible, task to find another figure that has had more of an impact on educational technology and instructional methods than Jerome Bruner. During the latter part of the twentieth century, when research and educational practices were dominated by the behaviorist perspectives of psychologists and scholars, a new movement was born that offered educators and researchers a new way to approach learning. This movement, now known as the cognitive revolution, owes much of its inception and energy to Jerome Bruner.

Born in New York in 1915, Jerome Bruner was the youngest of four children. His father, a watchmaker, died when Jerome was a boy, and the youngster then traveled extensively with his mother. At the age of seventeen he entered Duke University. In 1937 he completed his baccalaureate degree from Duke and then moved to Harvard University, where he completed a Ph.D. in psychology in 1941. His work in the field of psychology, specifically in cognition, would later alter instructional practices in classrooms across the country.

In the 1940s, when Bruner began his work, the perspective that most educators held about learning and cognition was influenced greatly by behaviorist principles. These principles dictated that learning was the process of an observed behavioral change in a student and followed a basic pattern: The instructor introduced a stimulus, the student generated a response, and the teacher then reinforced the response. If there was no observed behavior, then learning didn't occur.

The behaviorist perspective came about as an effort to elevate psychology into the realm of a true science by relying only on empirical data for theory development and research. Internal cognitive processes were largely ignored. In the midst of this prevailing perspective, Bruner published *The Process of Education* (1961). This work would serve as the catalyst for new research aimed at exploring the internal processes involved with learning and cognition such as mental structures and schemas. Bruner's research was novel in that he used creative methodologies to study internal processes experimentally so that empirical data could be generated. This seemed to inspire other researchers who followed Bruner into new territories of cognitive research. Interested primarily in how learners make use of categories, representations, and the formation of concepts, Bruner continued his research throughout the 1960s and was one of the founders of the Cognitive Research Center at Harvard.

Bruner served on the President's Science Advisory Committee during the terms of John F. Kennedy and Lyndon Johnson. His contributions helped guide policy and practice in the U.S. public education system. Bruner was also one of the principal designers of Head Start, the program that was created to better prepare young children for entry into K-12 education. In 1966 he published *Toward a Theory of Instruction,* in which he outlined four critical aspects of instruction: (1) consider the student's predisposition toward learning; (2) consider ways of structuring knowledge so that it is easily grasped by the learner; (3) consider how to best sequence the content so as to maximize learning; and (4) consider the nature and pacing of rewards and punishments. From Bruner's ideas about learning came several educational practices that are still in use today, such as the spiraling curriculum, learner analysis, and the idea that learning is an active, social process.

In the 1970s Bruner began to study other aspects of learning such as language acquisition and scaffolding. His ability to look at learning with fresh ideas led him to define discovery learning, which holds that students learn best through discovery and that they are problem-solvers, interacting with the world and developing and testing hypotheses. He is also given credit for pioneering constructivism, which has become a predominant perspective in education with many conceptual definitions and has gained rapid momentum over the years, particularly in educational technology. Bruner's brand of constructivism holds that new ideas or concepts are constructed by the learner based upon prior knowledge. When information is acquired by the learner, it is placed into a cognitive structure that assists in organizing and assigning meaning to it (Bruner 1983, 1986).

Never one to become stagnant, Bruner next turned his attention to the social and cultural influences on learning and the interaction of narrative on modes of thought. According to Bruner, narratives provide us frameworks for assigning meaning to the information we are processing. Through our own stories and the stories of others we experience the world and form the bulk of our reality (Bruner 1991).

Bruner's recent work has focused on the interaction of mental development, culture, narrative understanding, and interpretation. He is concerned with the passing of culture, specifically through legal praxis, legal codes, and schooling (Bruner 1996).

Bruner has held several academic posts during his career at Harvard, Oxford, Princeton, Cambridge, and New York University. He holds honorary doctorates from Yale, Sorbonne, Berlin, Rome, Columbia, and elsewhere. His work has won him the Balzan Prize in 1987 and the CIBA Gold Medal in 1974. Bruner's current academic home is New York University, where he is a research professor of psychology and a senior research fellow at the school of law.

George O. Hack

See also Cognitive Apprenticeship; Cone of Experience; Constructivism; Elaboration Theory of Instruction

References

Bruner, Jerome. 1960. *The Process of Education.* Cambridge, MA: Harvard University Press.

———. 1966. *Toward a Theory of Instruction.* Cambridge, MA: Harvard University Press.

———. 1983. *Child's Talk: Learning to Use Language.* New York: Norton.

———. 1986. *Actual Minds, Possible Worlds.* Cambridge, MA: Harvard University Press.

———. 1991. *Acts of Meaning.* Cambridge, MA: Harvard University Press.

———. 1996. *The Culture of Education.* Cambridge, MA: Harvard University Press.

C

Campus Computing Project

Begun in 1990, the Campus Computing Project (www.campuscomputing.net) is the largest continuing study of the role of computing and information technology in U.S. higher education. The project is widely cited by campus officials and others across and beyond the campus community as the definitive source for information about and insight into campus information technology (IT) planning and policy issues that affect U.S. colleges and universities.

A key component of the project, the annual Campus Computing Survey, collects data on campus planning, programs, and policies linked to the deployment and utilization of computing and IT resources to support and enhance instruction and scholarship. The survey respondents, typically the chief academic computing or IT officials at their institutions, are individuals specifically responsible for and knowledgeable about the current direction of technology planning, policy, IT finances, and technology implementation efforts on their campuses.

The 2002 Campus Computing Survey was conducted during the summer of 2002. The questionnaire was mailed to the chief academic computing officers at 1,339 two- and four-year colleges and universities across the United States. Where it was not possible to identify a specific individual with a senior academic computing title, the questionnaire was sent to the senior academic officer. The survey results summarized below are based on data from 632 two- and four-year public and private colleges and universities across the United States, reflecting a response rate of 47.2 percent. (More than 30 private two-year colleges were included in the 2002

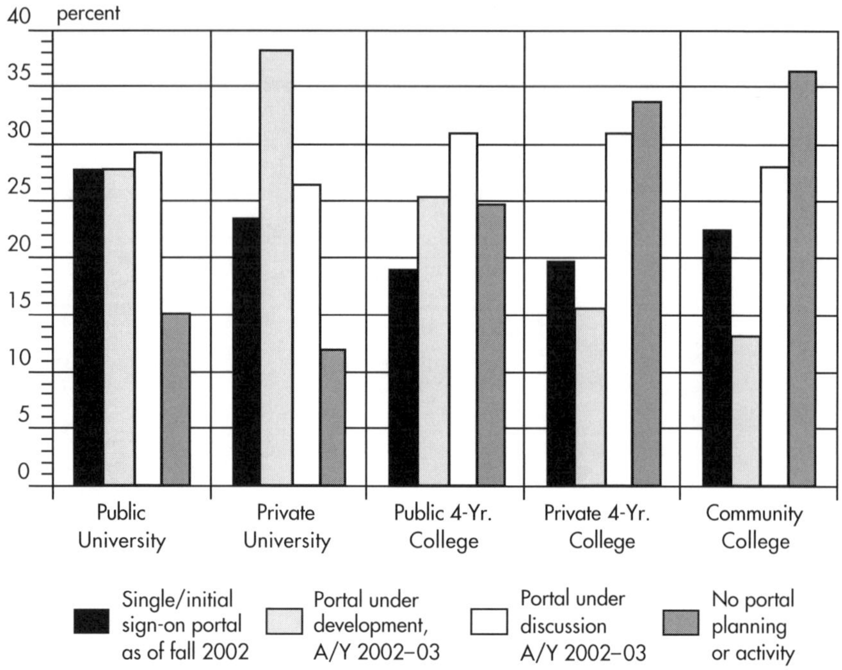

FIGURE 1: Campus Portals
(percentages by campus sector, 2002)

SOURCE: Used with the permission of the Campus Computing Project.

survey mailing. However, only eight institutions completed the questionnaire. The low response rate from this group should not be viewed as being representative of the larger population of the 615 private, nonprofit, degree-granting, two-year colleges in the United States.)

Campus Portals

The 2002 Campus Computing Survey reveals that U.S. colleges and universities are making steady progress in developing and deploying campus web portals (see Figure 1). Just over one-fifth (21.2 percent) of the campuses participating in the 2002 survey report that they have a "single/initial sign-on campus portal" up and functioning as of fall 2002. Another fifth (20.4 percent) report that their campus portal is "under development" or being installed in the current academic year. Just under one-third (29.5 percent) indicate that portal issues are now "under review/discussion" at their institutions, and a similar proportion (29 percent) indicates that there has been no portal planning or related portal activity at their institution.

For all practical purposes, the campus conversation about the role and value of web portals began in the late 1990s. Consequently, the 2002 survey data suggest that web portals are finally making the transition from an

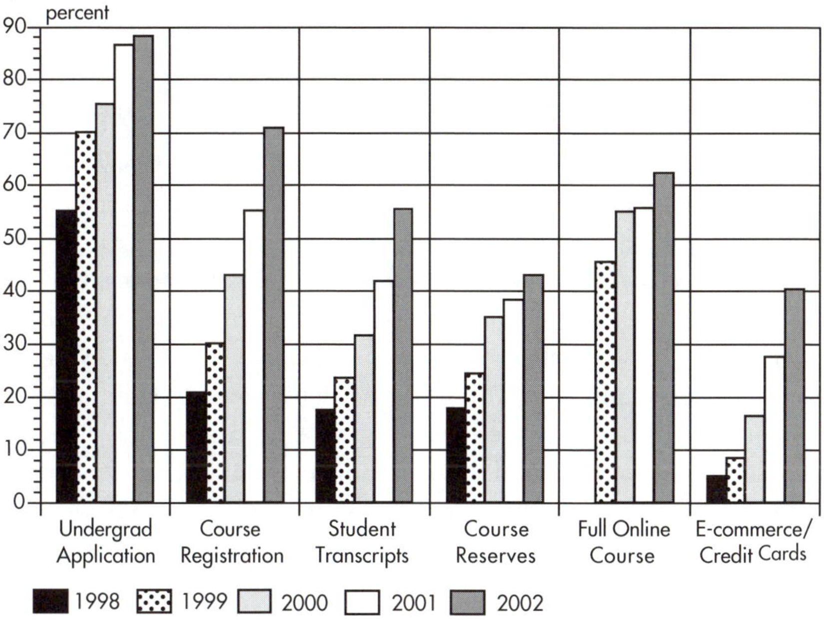

FIGURE 2: Trends in Website Services
(percentages by sector, 1998–2002)

SOURCE: Used with the permission of the Campus Computing Project.

abstract concept into a real institutional service. Moreover, the campus investment in portals can (and should) be seen in the broader context of a new institutional commitment to enhanced campus services across all sectors of higher education.

Other data from the 2002 survey reflect the campus commitment to web portals. For example, one-fourth (24.9 percent) of the campuses participating in the 2002 survey currently have a strategic plan for portal services; another third (32.7 percent) are now developing these plans. Survey respondents rate providing a campus portal for web-based student services at 5.5 on a 1–7 scale (1 = not important; 7 = very important), up from a scale score of 5.3 in 2001 and 5.2 in 2000.

E-commerce and E-service: Progress Marked by Continuing Challenges

The 2002 survey data also document some significant gains on a number of e-commerce and e-service measures across all sectors of higher education (Figure 2). For example, two-fifths (40.1 percent) of the campuses participating in the annual Campus Computing Survey can now process credit card payments from the campus website, up from 27.6 percent in 2001 and more than double the number in 2000 (18.6 percent). In contrast, just 5.1 percent of campuses participating in the 1998 survey could process online credit card transactions. More than two-thirds (70.9 percent) of the survey respondents report that their campus now offers online course registration, compared to just over half (55.4 percent) in 2001, 43.1 percent in 2000, and just a fifth (20.9 percent) in 1998.

Additionally, the proportion of campuses that now provide online access to student transcripts has tripled from 1998 to 2002, rising from 17.6 to 55.2 percent. Course reserves are now available online at more than two-fifths (43 percent) of the institutions participating in the annual Campus Computing Survey, up from just over one-sixth (17.9 percent) in 1998.

These numbers and other data shown in Figure 2 and provided in the annual Campus Computing Report (Green 2002a) reflect significant gains on a number of key e-commerce and e-service measures across all sectors of U.S. higher education. And as noted above in the discussion about campus portals, the gains reflect, in part, what should be described as a new institutional commitment to enhanced campus services across all sectors of higher education.

However, while all sectors of higher education have registered significant gains on the range and scope of online services offered to students, the 2002 survey data continue to provide ample evidence that some sectors are well ahead of others. Not surprisingly, as shown in Figure 3, public and private research universities typically offer more online services while community colleges continue to lag behind other sectors.

Additionally, while the gains in online campus services between 1998 and 2002 may seem striking, they are less impressive when seen in the context of the experience and expectations of U.S. college students—ages seventeen to sixty-seven—who come to campus to learn *about* and to learn *with* technology. For today's college students—only one-fifth of whom are "traditional" students (i.e., full-time undergraduates living on/adjacent to campus)—web-based services are best represented by their off-campus online experiences at retail websites, including banks and credit card companies. Retail and consumer sites provide increasingly customized services and support that are not available from college and university websites.

Indeed, consumer and commercial websites foster students' expectations about the kinds of services that should be on campus websites. Large numbers of college students—full-time undergraduates living in college dorms or campus-adjacent apartments, community college students who come to campus once or twice a week, and executive MBA students on campus two weekends a month, among others—can easily access information about their bank, credit card, and cell phone accounts on the web. Yet these same students often do not have access to parallel services from the colleges they attend (e.g., online transcripts, course registration, and financial account or financial aid information).

	Public University	Private University	Public 4-Yr. College	Private 4-Yr. College	Community College
Undergraduate application	98.6	91.2	90.2	89.4	79.6
Financial aid application	84.9	70.6	72.0	61.7	72.1
Course catalog	100.0	94.1	92.3	89.9	92.5
Program requirements	84.9	88.2	86.7	84.1	75.5
Course registration	91.8	85.3	79.0	52.4	80.3
Online courses	90.4	55.9	77.6	36.1	78.2
Library/card catalog	98.6	94.1	90.2	89.0	77.6
Course reserves	74.0	64.7	49.7	43.6	17.7
Student transcripts	75.3	73.5	61.5	40.5	61.2
IT support resources	98.6	88.2	83.2	76.7	53.1
Instructional software	74.0	64.7	49.7	41.4	24.5
College bookstore services	71.2	79.4	72.7	60.8	55.8
E-commerce (fee payment)	69.9	47.1	49.7	23.8	42.2

FIGURE 3: Website Services
(percentages by sector, 2002)

SOURCE: Used with the permission of the Campus Computing Project.

Other data from the 2002 survey also support the notion of a lagging e-commerce/e-service infrastructure. One key indicator: The 2002 survey respondents rated the campus capacity for e-commerce eleventh on a list of twelve technology infrastructure metrics that include network and telecommunications services, user support services, online reference resources, network security, and IT training for students and faculty (Figure 4).

Taken together, these data suggest that many campus websites and online campus services lag the consumer sector by as much as two years. Moreover, the budget cuts now affecting all sectors of U.S. higher education could impede institutional efforts to expand and enhance e-learning and online services.

Key IT Priorities

As in the past years, survey respondents identified assisting faculty to integrate technology into instruction as the single most important IT issue confronting their campuses over the next two or three years (Figure 5). One-fourth (24.4 percent) of the 2002 survey respondents tag instructional integration as the key IT issue for their institutions, down from one-

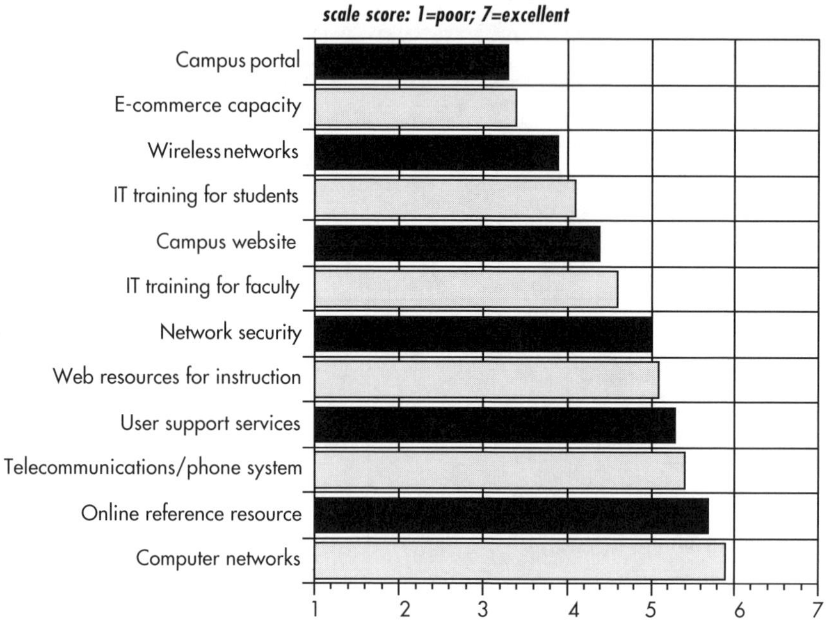

FIGURE 4: Rating the Campus IT Infrastructure
(all campuses, 2002)

SOURCE: Used with the permission of the Campus Computing Project.

third (31.5 percent) in 2001, and a dramatic drop from the two-fifths (40.5 percent) posted in 2000.

Ranked second in 2002 are enterprise resource planning (ERP) upgrade/replacement issues: Almost one-fifth (18.9 percent) of that year's respondents tag ERP issues as the key IT priority for their campus, up from 12.6 percent in 2001. IT financing rose to third place in 2002 at 15.1 percent of respondents, up from 11.7 percent the prior year. Providing adequate user support, which held the number-two position for many years, fell to fourth in 2002, dropping to 13 percent, down from 15.4 percent in 2001, 22.3 percent in 2000, and 25 percent in 1997.

Yet the aggregated data mask important differences in institutional priorities across the various sectors of U.S. higher education. For example, ERP issues ranked first in both public and private universities: Almost a third identify ERP replacement/upgrade issues as the single most important IT issue for their institutions (Figure 6). In contrast, assisting faculty with instructional integration of information technology remains the top issue for public and private four-year colleges and also for community colleges. ERP upgrade/replacement ranks second in both public and private four-year colleges; community colleges rank ERP upgrade/replacement third, just behind IT financing.

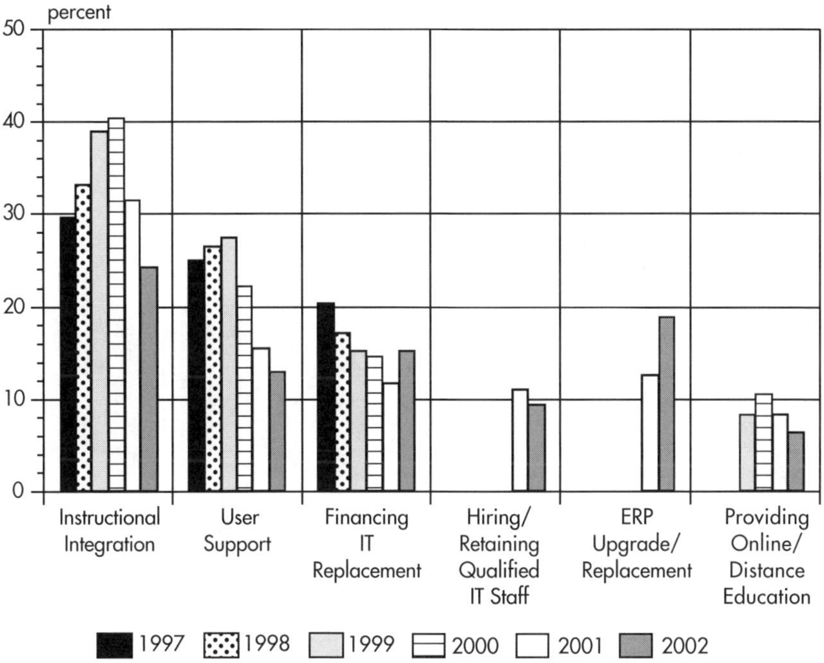

FIGURE 5: Single Most Important IT Issue
(percentages, 1997–2002)

SOURCE: Used with the permission of the Campus Computing Project.

The survey data also highlight the strategic issues of most concern to campus IT officials. The strategic issues shown in Figure 7—clarifying campus IT goals, IT training for IT personnel, web resources for instruction, course management systems—are the four issues that received the highest ratings from a larger list of twenty-nine items that cover instruction, personnel, budgets, and web services.

Not surprisingly, the data presented in Figure 7 highlight the continuing concern for instructional resources on the web. Finally, Figure 7 confirms the emergence of course management software (CMS) or learning management software (LMS) as a key component of the instructional technology infrastructure.

Yet somewhat surprising are the topics that are *not* high on the list of strategic campus IT issues. For example, notably absent among the top issues are: electronic commerce (scale score: 4.8), policies regarding intellectual property (5.2), and providing a campus portal for web-based student services (5.5). Perhaps one way to explain the lower ranking of these three items is that they are not exclusively IT planning and policy issues. Rather, portal planning, intellectual property issues, and e-commerce/e-service planning and implementation each involve multiple campus of-

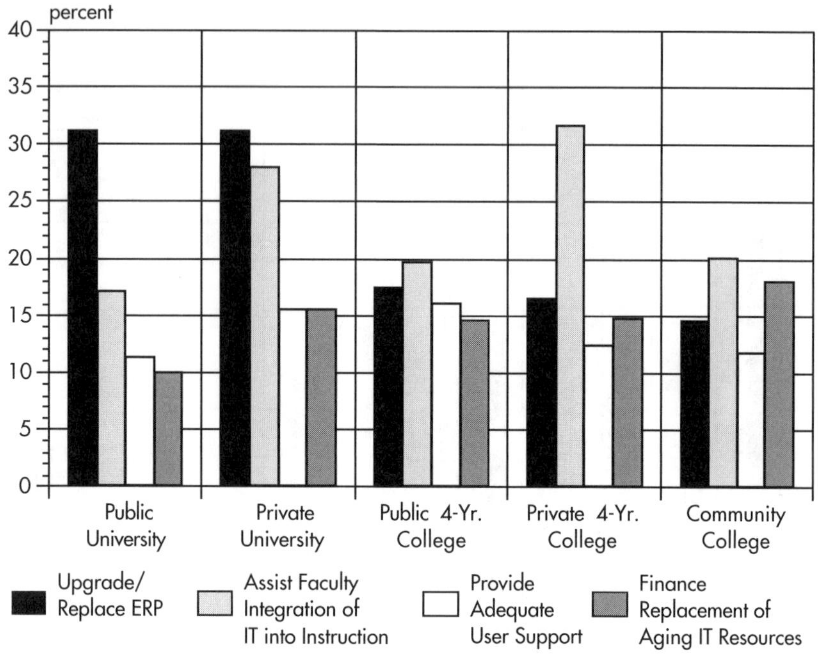

FIGURE 6: Single Most Important IT Issue
(percentages by campus sector, 2002)

SOURCE: Used with the permission of the Campus Computing Project.

fices: information technology, academic affairs, finance, and in the case of intellectual property, probably legal affairs. Consequently, because senior campus IT officials do not "own" these issues the way that their units own other issues—professional development for IT personnel or the IT infrastructure for e-learning (including the deployment of course management systems)—it may not be surprising that senior campus IT officials rank these "collaborative" issues lower than others.

Technology Planning

The 2002 survey data highlight the continuing challenge of IT planning in U.S. colleges and universities. More than two-thirds (70.6 percent) of the institutions participating in the 2002 Campus Computing Survey report a campus strategic plan for information technology, up from 63.3 percent in 2001 and just 48 percent in 1998.

At one level, these numbers suggest important and impressive gains in campus efforts to anticipate and to address a wide array of critical information technology challenges. Yet as in past years, additional data from the annual Campus Computing Survey suggest that many of these campus strategic plans may be incomplete.

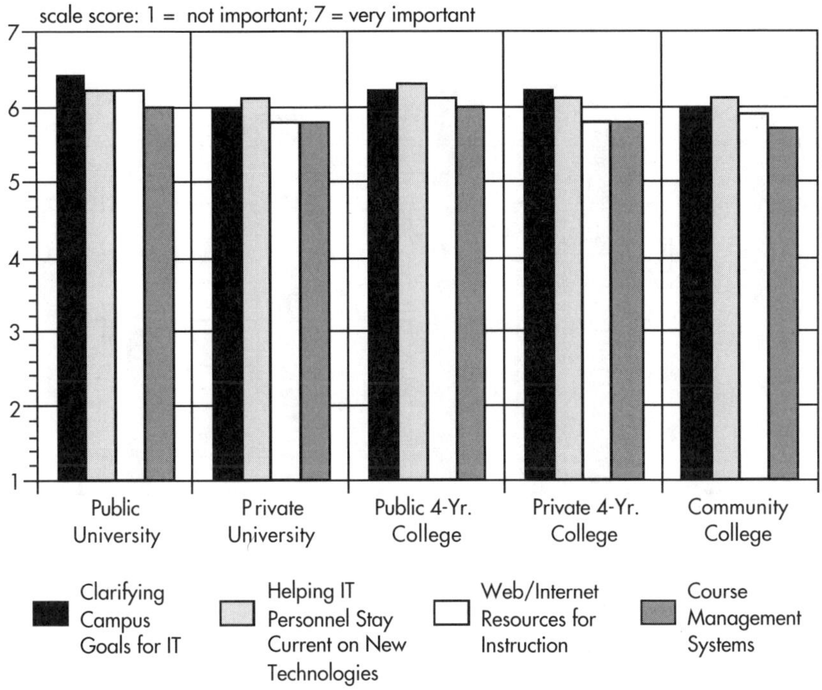

FIGURE 7: Key Strategic IT Issues
(percentages by sector, 2002)

SOURCE: Used with the permission of the Campus Computing Project.

Indeed, probe just a bit below the surface and it is clear that many of the institutions that claim to have a strategic plan for information technology frequently are missing some key components of an overall IT strategic plan. For example, just over half of the 2002 survey respondents (54.7 percent) report an IT financial plan that acknowledges the need to acquire and retire aging equipment. This compares to 52.2 percent reporting a financial plan in 2000 and just a fifth (21.9 percent) in 1994. Again, while the gains between 1994 and 2002 are impressive, the survey data also reveal that almost half (45.3 percent) of the institutions participating in the 2002 survey *do not* have IT financial plans.

Other metrics from the 2002 survey confirm that some campus IT planning efforts may be incomplete (Figure 8). For example, only one-seventh (13.4 percent) of the campuses participating in the 2002 survey report strategic plans for electronic commerce (up from 11.8 percent in 2001 and more than double the 6.8 percent posted in 2000). A fourth (24.5 percent) report a plan for student portal services, compared to one-fifth (21.4 percent) in 2001 and one-eighth (12.6 percent in 2000). Less than two-fifths (37.1 percent) report a strategic plan for distance learning (up from 34.3 percent in 2001 and 29.2 percent in 2000). Although these metrics show

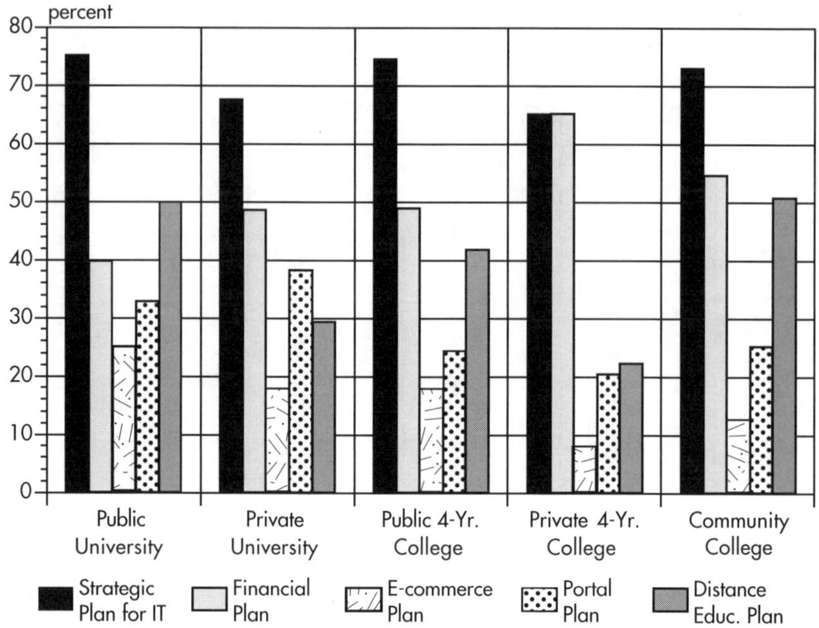

FIGURE 8: IT Strategic Planning
(percentages by sector, 2002)

SOURCE: Used with the permission of the Campus Computing Project.

impressive gains from 2000 to 2002, the numbers are still well below the percentage of institutions reporting that they have a strategic plan for information technology.

Rising Use of IT in Instruction

Yet even as campuses and campus officials continue to struggle with technology planning, the annual survey data continue to document the rising use of technology to support instruction (Figure 9). More than two-thirds (69.5 percent) of all college classes now utilize e-mail, up from (64.1 percent) in 2001, 59.5 percent in 2000, and 20.1 percent in 1995. Similarly, fully half (50.3 percent) of all college courses now use Internet-based resources (e.g., URLs in the course syllabus), compared to 10.9 percent in 1995, 33.1 percent in 1998, 42.6 percent in 2000, and 47.4 percent in 2001.

The 2002 survey data also highlight the growing role of webpages for individual courses: Survey respondents estimate that one-third (34.8 percent) of all college courses now have a webpage, compared to 30.7 percent in 2000, 22.5 percent in 1998, and 9.2 percent in 1996. Concurrently, the 2002 data indicate that more than one-fourth (26.9 percent) of all college faculty have a personal webpage not linked to a specific class or course, compared to 19 percent in 1999. (See Figure 10.)

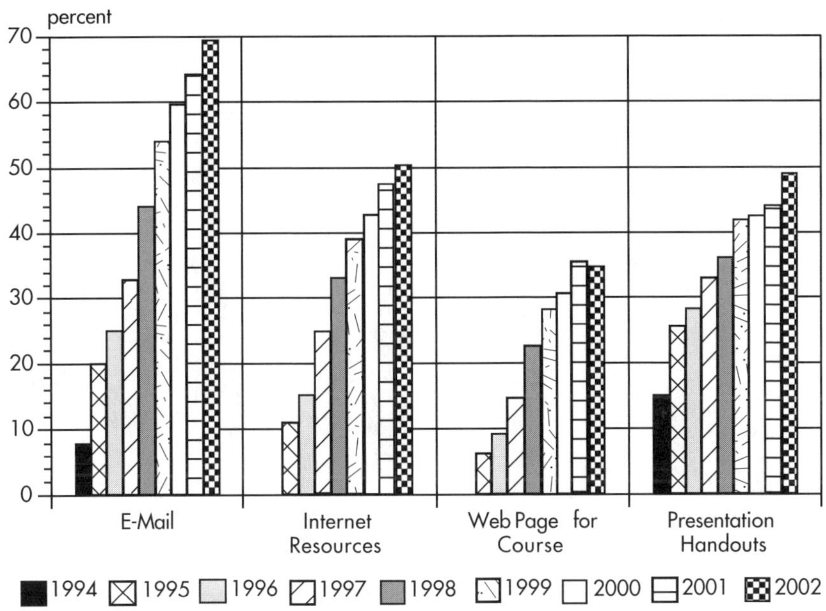

FIGURE 9: Rising Use of Technology in Instruction
(percentages for all institutions, 1994–2002)

SOURCE: Used with the permission of the Campus Computing Project.

As in past years, the 2002 survey reveals important variations in the deployment of IT resources across sectors. For example, almost half (49.0 percent) of all courses in public research universities have an individual webpage for class materials and resources, up from 33.7 percent in 1999 and more than double the number from 1997 (17.7 percent). In contrast, while the number has also doubled over five years in private four-year colleges (from 15.2 percent in 1997 to 31.1 percent in 2002), it is almost a two-fifths below the level in private universities (51.6 percent, up from 28.8 percent in 1997). The pattern for using e-mail as a course resource is similar: large gains within sectors but significant variation across sectors.

Course Management Software

The 2002 survey data again confirm the increasingly important role of course management software or learning management software as a core instructional resource. Overall, the percentage of college courses that use a CMS/LMS tool has risen from 14.7 percent in 2000 to 26.5 percent in 2002. While the numbers vary by sector (Figure 11), the growing deployment of (some might say campus dependency on) CMS is consistent across all sectors.

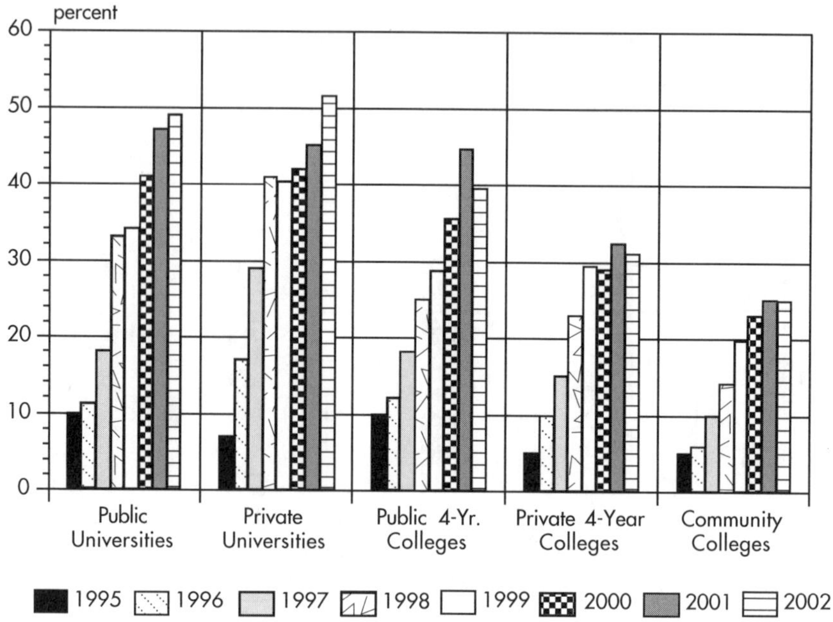

FIGURE 10: College Courses with Website
(percentages by sector, 1995–2002)

SOURCE: Used with the permission of the Campus Computing Project.

The survey data also indicate that almost half (47.5 percent) the institutions participating in the 2002 survey report a strategic plan for CMS/LMS deployment, up from 41.8 percent in 2001 (Figure 12).

Concurrent with the rising use of CMS/LMS has been the willingness of campuses to establish a single CMS/LMS standard for their institutions. Fully four-fifths (82.1 percent) of the survey respondents report that their institution has established a single product CMS/LMS standard as of fall 2002, up from 55.4 percent in 2000 (Figure 12).

Taken together, the survey data may actually underestimate the percentage of institutions that have established broad campus CMS/LMS standards. For example, some institutions have reached a kind of "CMS détente," allowing one or two academic units to use one CMS/LMS product while the rest of the campus uses another one. Alternatively, the distance education program may have opted for one particular product while the rest of the institution may use a CMS/LMS from a different provider.

CMS/LMS deployment standards represent a rare example of institutional policy moving ahead of campus practice: Rather than supporting several CMS products or waiting for market forces to determine the "winning" application, the vast majority of campus officials and instructional

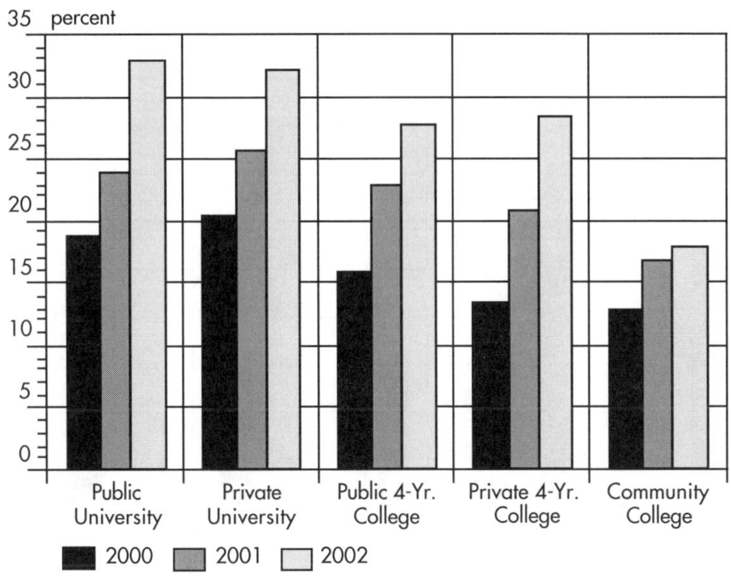

FIGURE 11: Classes Using Course Management Software
(percentages by sector, 2000–2002)

SOURCE: Used with the permission of the Campus Computing Project.

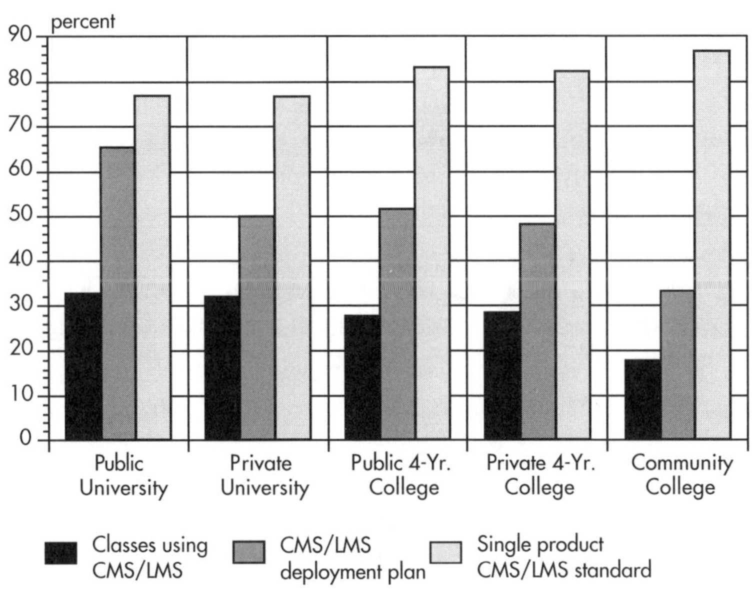

FIGURE 12: CMS/LMS Deployment, Planning, and Standards
(percentages by sector, 2002)

SOURCE: Used with the permission of the Campus Computing Project.

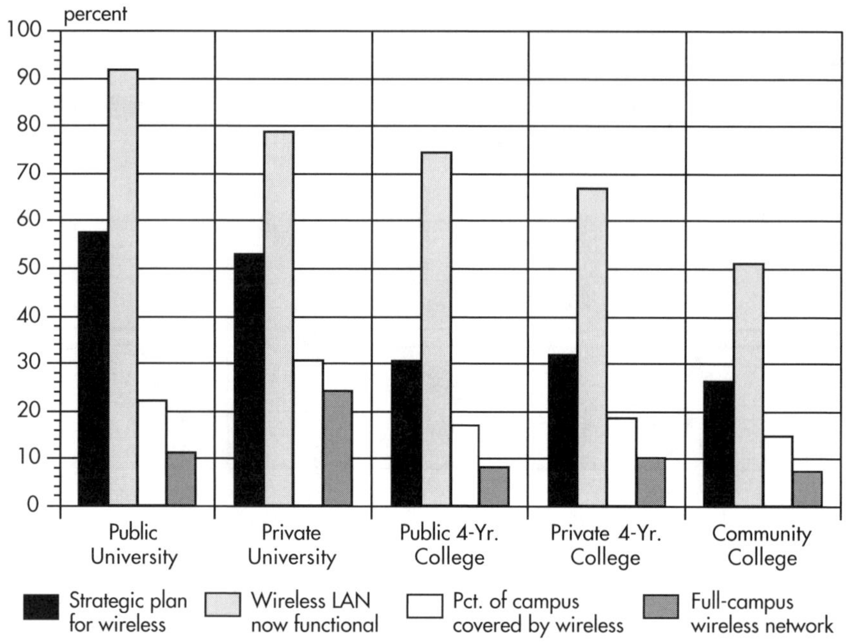

FIGURE 13: Wireless Planning and Deployment
(percentages by sector, 2002)

SOURCE: Used with the permission of the Campus Computing Project.

technology committees are selecting a single CMS product for their institutions. These decisions reflect an effort to make a clear statement about the role of CMS applications in instruction. Additionally, the movement to early campus standards should help promote broader and faster deployment and also simplify and facilitate user support services.

Moving toward Wireless

The 2002 survey data track the growing movement toward wireless networks across all sectors of U.S. higher education. Fully two-thirds (67.9 percent) of the surveyed campuses report wireless local area networks (LANs) as of fall 2002, up from half (50.6 percent) in 2001 and less than one-third (29.6 percent) in 2000.

One-tenth (10 percent) of the survey respondents indicate that *full-campus* wireless networks are up and running at their institutions as of fall 2002, compared to 6.4 percent in 2001 and 3.8 percent in 2000. Another one-tenth (9.5 percent) report that their institutions should have working, full-campus wireless LANs by fall 2003. One-third (34.7 percent) of the campuses have a strategic plan for wireless networks, up from one-fourth (24.3 percent) in 2001 (Figure 13). Across all sectors, for institutions with wireless LANs, the 2002 data indicate that wireless services cover about a

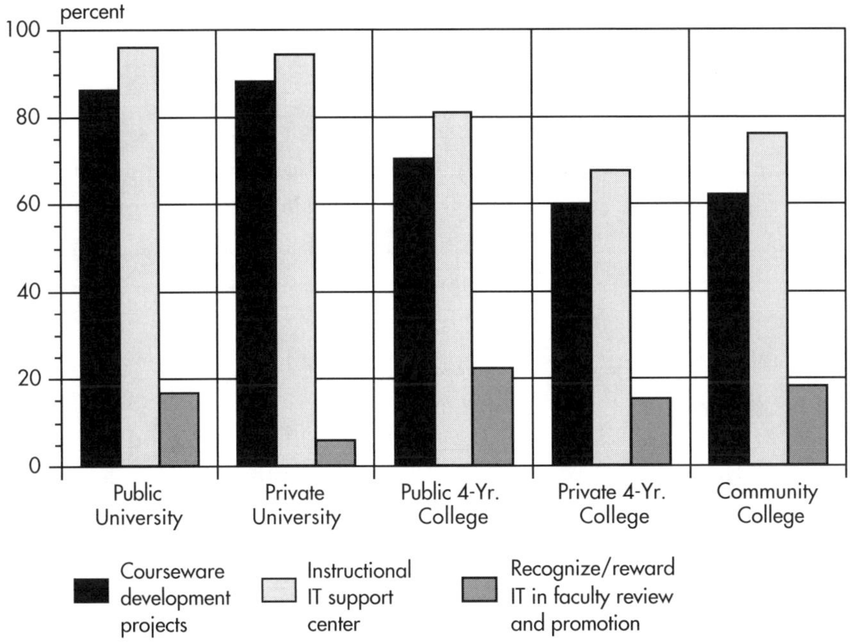

FIGURE 14: IT Support versus Faculty Recognition and Reward
(percentages by sector, 2002)

SOURCE: Used with the permission of the Campus Computing Project.

fifth of the physical campus (18.3 percent), almost double the level of wireless campus access/coverage reported in the 2001 Campus Computing Survey (10.9 percent).

Faculty Recognition and Reward

One of the continuing (and unfortunate) ironies of campus efforts to promote the instructional use of information technology is the fact that comparatively few campuses provide formal recognition and reward for faculty efforts at instructional integration (Green 2002b). Although the majority of institutions provide some support for faculty efforts to develop instructional applications and have campus technology centers to support instructional integration, few institutions recognize faculty IT efforts in the review and promotion process. (See Figure 14.)

Consistent with past years, the pattern for 2002 is similar across all sectors: Institutional support for courseware development projects is common, as is the campus presence of technology support centers to assist faculty with instructional integration. The percentage of campuses reporting that IT is now to be part of the review and promotion process has increased from 13.4 percent in 1999 to 17.4 percent in 2002. However, despite these gains, only a small minority of colleges and universities have a formal program that provides recognition and reward for individual faculty efforts at instructional integration.

Additional Information About the Campus Computing Project

Readers interested in additional information about the Campus Computing Project should consult the project's website (www.campuscomputing.net), which provides more information about the annual survey reports, plus reprints of articles that draw on the survey data.

Also of potential interest may be the emerging international network of affiliated scholars from universities in Brazil, China, Hong Kong, Japan, South Korea, and elsewhere who are developing their own campus computing projects based on the research model and methodology developed in the United States. The Asian Campus Computing Project, which includes China, Hong Kong, Japan, and other Asian nations, maintains a website that provides additional information about the individual and collective campus computing projects in these countries (www.accsonline.net).

Kenneth C. Green

See also Computer-Mediated Communication; Courseware; Diffusion of Innovations; Technology across the Curriculum; Technology Planning; Web Portals; Web-Based Course Management Systems; Wireless Networks

References

Green, Kenneth C. 2002a. *Campus Computing, 2002: The 13th National Survey of Computing and Information Technology in American Higher Education.* Encino, CA: Campus Computing, December 2002 (www.campuscomputing.net).

———. 2002b. "Coming of Age in Academe." *Converge Magazine* (October) (www.convergemag.com/magazine/story.phtml?id=29178).

CaseNEX

CaseNEX is the first web-based environment to use case-study methods to challenge students to examine theories they have learned or are learning and to apply these theories to situations they may encounter in the future via interactions with cases (Bronack and Kilbane 1998). In his 1985 presidential address to the American Educational Research Association, Lee Shulman advocated the use of case methods in teacher education (Shulman 1986). Cases, according to Shulman, are more than simply reports of events; they are stories placed in a larger theoretical context. He pointed out the obvious fit of case methods in teacher education as a natural consequence of the movement toward viewing teaching as situated cognition, decisionmaking, and reflection. Shulman also associated the use of cases with technology, looking ahead to such innovations as simulations and videodiscs for extending capacities of professional practice. Researchers and practitioners have begun to instantiate these ideas in teacher education. The CaseNEX website provides course packages and a variety of tools for designing and delivering teaching cases within and among groups of preservice and in-service teachers.

Case methods have been used successfully in business schools, law schools, and medical training for years as an alternative to direct instruction, seminar, and didactic lecture strategies (Merseth 1996). Most case methodologies, whether in teacher education or other professions, share a common element: Student discussions form the foundation of case-based learning. In small groups or instructor-guided Socratic discussions, students must consider various issues in the case. Typically, expert analyses assuming various perspectives accompany cases and suggest that there are no "right" answers but instead a variety of acceptable perspectives that teachers can assume as they learn to make reasoned, ethical decisions about teaching and learning. Although it is impossible to address completely the complexity of teachers' jobs, case methods can provide realistic environments through which various problem-solving strategies and tactics can be applied and investigated.

Cases may be presented using a variety of media formats, including text, graphics, video, audio, or some combination of these. Most teaching cases are text-based. Text-based cases can be an efficient way to communicate complex classroom scenarios, but they do not reflect the richness of real life, as do cases presented in visual forms. Newer technologies allow case studies to communicate the intricacies of classroom practice with various media. Cases communicated through video formats, for example, can encourage viewers in ways that text-based cases cannot. Videocassette, laserdisc, CD-ROM, and more recently the World Wide Web are popular choices to communicate these vivid presentations of real people in real situations.

The emergence of hypermedia and the World Wide Web offer the latest vehicle for delivering cases. These tools enable the delivery of multiple technologies that, when combined, can create and enhance the verisimilitude of cases and the human interaction deemed essential in case-based teaching and learning. Whereas video and text formats are generally linear, hypermedia enables students to choose their own pathways through the case materials. Collections of digital documents cross-referenced by links allow viewers of a case to move easily from one text page to another, or to other media such as audio, video, or graphics. Students can manipulate the case materials, allowing them to determine their own questions and frames of inquiry. Such student-constructed analyses mirror more closely the actions of classroom teachers (Merseth and Lacey 1993). Novices can also build their own knowledge about teaching. The innovative instructional design of such multimedia cases reflect important dimensions of the classroom, including multidimensionality, simultaneity, and immediacy (Merseth and Lacey 1993). The WWW also allows for a highly interactive environment that enables richer transactions among people than any of its technological predecessors. Computer networks enable synchronous and asynchronous conversations that contribute to the construction of knowledge by bringing various perspectives to bear on case analyses and discussions. Such interactions would otherwise be hindered by space and time boundaries. (See Figure 1.)

CaseNEX offers approximately thirty multimedia case studies combining text with video and audio files, pictures, and other graphics. CaseNEX currently provides cases about life in K-12 classrooms in the United States, Cuba, India, South Africa, and Norway. Participants from approximately twenty universities in four countries and fifteen school districts in the United States use CaseNEX. Participants use the Internet and communications technologies to learn to analyze multimedia cases and to connect with participants at different geographical sites. They meet physically at local sites with an instructor or team of instructors who use case methodology to guide and assist learning.

A set of integrated tools supports the CaseNEX site. These enable participants from several countries and regions in the United States to analyze the cases online and to interact with each other in a variety of ways. Participants use an internal messaging system (CaseMail) to set up synchronous conferences, group messaging (AOL Instant Messenger), and videoconferencing (iVisit) programs to discuss cases and share educational approaches. A feature called Knock Knock allows users who are logged on at the same time to identify one another. Students' pictures and bios make the connections even more personal. The time-independence of asynchronous, online discussion groups offers students opportunities to reflect, evaluate, summarize, and communicate perspectives about their shared case experiences. Electronic journals make it possible for users to record reactions to cases and receive individual feedback from instructors.

Studies of CaseNEX highlight several important factors that characterize students' involvement with cases. T. W. Kent (1997) collected data over a two-year period on the use of the first web-based multimedia teaching case to be used in CaseNEX. This case, entitled Project Cape Town, combines text, graphics, audio, and video to describe four events taken from four South African schools experimenting with racial integration. Kent identified several navigational patterns through Project Cape Town. Asking participants to do what comes naturally when navigating Project Cape Town, he characterized three user styles—impulsive, response-focused, and balanced—based on the degree of structure imposed by participants moving through a case. The impulsive style of use is characterized by

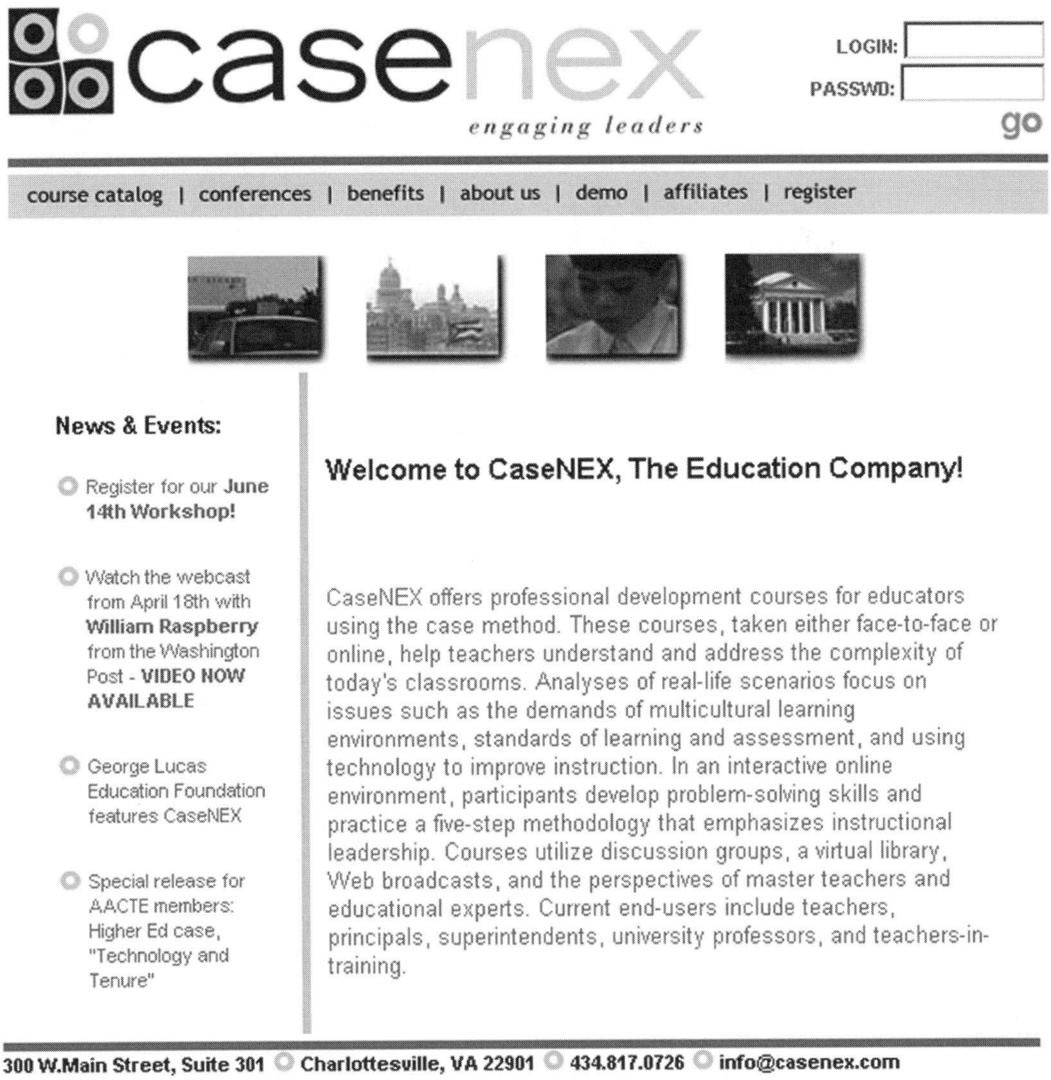

FIGURE 1: Sample Web page from CaseNEX Project.

SOURCE: Used with the permission of the CaseNEX Project.

breaking routines and by following lines of interest related to the content presented in the resource. In contrast, a response-focused user considered answering all of the questions a priority, postponing access to much of the material until after completing the four case events and their questions. The balanced style reflected a more thorough, systematic exploration of all areas and resources of Project Cape Town. Kent also found that the media increased the realism of the case and that most users considered the nonverbal information valuable in helping shape their analyses. Kent urged other researchers to investigate cultural factors influencing student thinking and students' responses to cases.

S. C. Bronack (1998) took Kent's challenge to investigate the possible effects of multimedia cases on teachers' thinking. More specifically, he compared the decisionmaking skills and level of concern demonstrated by teachers enrolled in CaseNET (the name prior to summer 2000) with those of teachers not participating in the online community. He also examined the effects of course participation and nonparticipation on teachers' concerns about web-based instruction. About 95 percent of Bronack's participants believed that the program helped them bridge the gap between theory and practice. Approximately the same percentage agreed or strongly agreed that CaseNET helped prepare them to face classroom situations similar to those described in cases. The results suggested that participation in CaseNET positively influenced participants' concern about all seven items of innovation, yet no statistically significant difference was found between the treatment groups' performances. Although the CaseNET participants' analysis scores showed a slight improvement over those of nonparticipants, the gain was not great enough to differentiate one group from the other. Bronack suggested that the small number of participants and limitations of the instrument used to measure participants' analyses might be explanations for the failure to find significant results. Understandably, the Bronack study raised questions about instructor effects on the success and execution of the CaseNET experience.

Like Bronack, C. Kilbane (2000) studied the efficacy of the CaseNET problem-solving method in an experimental study. She examined effects of learner control on participants' experiences with a multimedia case and investigated the problem-solving proficiency of participants enrolled in CaseNET and two additional groups of students not participating in the online community. Kilbane's results suggest that preservice teachers in CaseNET were able to learn the problem-solving strategy and to apply it to a multimedia case study. The experimental group was better at identifying issues and applying knowledge from various sources than were participants in the other groups. Participants in CaseNET used professional knowledge from the case resources (lesson plans, student progress reports, etc.) to identify problems and propose actions but were limited to using knowledge from case resources they knew how to locate and interpret. (See Figure 2.)

To date, no studies of multimedia cases have compared variations in the CaseNEX treatment or the match of treatment to participants. Can teacher educators use cases to encourage teachers to engage intellectually in solving real-life educational and moral problems? Can the CaseNEX method, or any other web-assisted approach, be delivered to students virtually without diminishing significantly the quality of the experience? Moreover, do case-based strategies influence deeper structures of moral reasoning?

cases | journal | syllabus | discuss | case mail | log out

getting started | class materials | talking to each other | help

Why Can't Ricky Read?

A CaseNEX Case from Kevin Flanigan

This case is the property of the The University of Virginia and should not be copied or distributed without permission.

Contents

Scene 1

Ricky is embarrassed as he tries to read aloud in social studies class.

Scene 2

Ricky's mom, after seeing the "F" on Ricky's social studies test, e-mails the special education teacher, Mrs. Hardy.

Scene 3

Mrs. Hardy and Mr. Thompson, the social studies teacher, discuss Ricky's problems.

Scene 4

A Child Study Meeting is held. Everyone has a different solution.

Discuss This Case

Check Out Readings

Write In My Journal

Ask My Instructor

How Do I Use This Case?

Scene One

Ricky looked at the word - it was one of those long ones that seemed to constantly stalk him. He made a game attempt to sound it out, "Dem, dem, dem, oak, dem-oak . . ."

"It's democracy, you idiot," whispered Joey. A group of boys in the back erupted in laughter.

"Joey, get out of my room and go directly to the office!" commanded Mr. Thompson. "I'll expect to see you sitting outside of Dr. Barnett's office after class. I will not tolerate that type of behavior in here. Susan, why don't you start at the next paragraph." Joey slowly got up and sauntered toward the door, obviously relishing the "bad boy" image he had been so carefully cultivating since summer.

REALPLAYER

Watch "Why Can't Ricky Read?" Scene 2"

FIGURE 2: Sample Web page from CaseNEX Project.

SOURCE: Used with the permission of the CaseNEX Project.

Case-method teaching and learning provide opportunities for the discussion of complex educational dilemmas that challenge one's current level of moral understanding. With the development of ever-newer and faster technologies, as well as the growing technological adeptness of teachers and teacher educators, it may be reasonable to consider the conditions of web-assisted and distance case-study methods of teaching and learning that are critical for facilitating teacher development. Currently, research is under way to examine the complex and dynamic relationship of live and virtual case method contexts, case discussions, and the moral development of teachers.

Marsha A. Gartland
Robert F. McNergney

See also Active Learning; Situated Cognition; Teachers: Preparation and Training

References

Bronack, Stephen C. 1998. *Analyzing Multimedia Cases: Teacher Development in a Web-Based Environment.* Unpublished doctoral diss., University of Virginia, Charlottesville.

Bronack, Stephen C., and Clare R. Kilbane. March 1998. *CaseNET: Teaching Decisions via a Web-Based Environment.* Paper presented at the Society for Information Technology (SITE), Washington, D.C.

Kent, Todd W. 1997. *Project Cape Town: A Description of the Use of a Multimedia Web Case.* Unpublished doctoral diss., University of Virginia, Charlottesville.

Kilbane, Clare. 2000. *Preservice Teachers' Use of Multimedia Cases.* Unpublished doctoral diss., University of Virginia, Charlottesville.

Merseth, Katherine K. 1996. "Case and Case Methods in Teacher Education." In J. Sikula, T. Buttery, and E. Guyton (eds.), *Handbook of Research on Teacher Education: A Project of the Association for Teacher Educators.* New York: Macmillan Library Reference.

Merseth, Katherine K., and C. A. Lacey. 1993. "Weaving Stronger Fabric: The Pedagogical Promise of Hypermedia and Case Methods in Teacher Education." *Teaching and Teacher Education* 9(3): 283–299.

Saunders, S. L. 1991. *Evaluating a Videotaped Case of a Beginning Elementary School Teacher and Describing Its Use.* Unpublished doctoral diss., University of Virginia, Charlottesville.

Shulman, Lee. 1986. "Those Who Understand: Knowledge Growth in Teaching." *Educational Researcher* 15(2): 4–14.

Clark, Richard E. (b. 1940)

Richard E. Clark, Ed.D., is a professor of educational psychology at the University of Southern California, where he has been a member of the faculty and has held various administrative posts since 1978, including division head for educational psychology and technology and director of professional studies and community programs. He has been elected a fellow of the American Psychological Association, the American Psychological Society, and the Association for Applied Psychology, exceptional recognition for someone who does not have a degree in psychology.

He is best known among educational technology professionals as a critic of the theory that media have direct causative influence on specific types of learning. His 1983 publication "Reconsidering Research on Learning from Media" touched off a controversy sometimes summarized as the media-methods debate. Throughout the 1980s and early 1990s he continued to defend his position through point-counterpoint articles in professional journals and panel discussions at academic conferences.

Clark has devoted his efforts since then to exploring what *does* work—the positive side of his media critique. This quest has led him into study, in turn, of human cognition, instructional methods, instructional design processes, motivation theory, and performance interventions. In 2001 he was focusing on writing several books to synthesize these studies into a cognitively based theory of designing learning environments incorporating motivational features to maximize the transfer of learning to real-world accomplishments.

Clark's professional quests have been shaped by his life experiences, like most other people. Born September 15, 1940, in Howell, Michigan, near Detroit, he was the eldest of nine children, so he took on parental-type responsibilities early. From his mother he learned the indispensability of making lists and planning ahead in order to cope with the demands of taking care of others as well as oneself. He attended a small rural elementary school that featured multigrade organization, not as a progressive experiment but as a necessity. The inquisitive child was able to move ahead without hindrance, reading above his grade level right from the start. As he says, "I didn't get bored with school until high school." High school was more conventional, leading to the more conventional sort of indifference to academics. However, during his teen years Clark suffered a back injury that led to a lifelong disability that altered the course of his life. A doctor warned him that he had better prepare to make his way in the world with his mind since his physical capabilities would be limited. From that point on young Dick began to plan for postsecondary education, the first in his family to do so.

After a false start at the University of Michigan he found a place and a mentor—Bob Dye—at Western Michigan University; he majored in history and political science, graduating in 1963. Looking for a practical career and needing financial aid, he took advantage of an offer of a full scholarship to enter the new Annenberg School for Communication at the University of Pennsylvania to study journalism. By the time he completed a master's degree he was married and had a son, so finding a job became a top priority. Coincidentally, the founder of the Annenberg School, Walter H. Annenberg, was also the owner of the Philadelphia TV station WFIL, which was undergoing a strike at the time. Annenberg hired virtually all the J-school's graduates to fill vacant jobs at the station, including Clark, who became an associate producer. Such an entry-level job did not provide enough income to support a family, so Clark took another position, working mornings at WFIL and evenings at WHYY, the public TV station in Philadelphia.

After about a year, an opportunity arose out of connections made while producing the public TV series *The Compleat Gardener.* It led to the agricultural extension program at Rutgers University, where Clark had responsibility for a radio network and other communications operations. Shortly, though, Bob Dye, his mentor at his alma mater, Western Michigan University, beckoned. There he became director of broadcasting, including oversight of a campuswide dial-access audiovisual delivery system. The position also included an appointment as assistant professor in general studies. As he now seemed to be on a career path in higher education rather than commercial television, Clark realized that he would need to undertake doctoral studies at some point.

After exploring options for doctoral study, Clark in 1967 decided on the mass communications program at Indiana University, attracted by the chance to work with Keith Mielke, then a leading researcher on the effects of television on children, later senior researcher for the Children's Television Workshop. After a year, Clark qualified for a National Defense Education Act fellowship and transferred to the Ed.D. program in the educational media (soon to be instructional systems technology [IST]) department, entering what was to become his ultimate professional home, the school of education.

In IST he was influenced by researchers such as Malcolm Fleming, who encouraged young researchers to have confidence and pursue their own interests, and Gavriel Salomon, who pursued hard and important questions. His classmates included a number of future leaders of the field, including Sivasailam Thiagarajan, Thomas Schwen, W. Howard Levie, Diane Dormant, and Harold Stolovitch.

On completing doctoral studies, Clark was recruited to Stanford University, in 1971 joining the staff in the Research and Development Center on Teaching, later becoming one of the founding leaders of the ERIC Clearinghouse on Educational Media and Technology (renamed *Information* and Technology in 2001). Although he had an adjunct appointment in the school of education, this was not a tenure-track position, so when Syracuse University asked him to become an associate professor, later full professor, and chair of the instructional technology program in 1974, he left Stanford for upstate New York.

The Syracuse program, then known as the area of instructional technology, had recently gone through a major curriculum revision, but Clark felt that inquiry was not sufficiently embedded in the graduate program. In a detailed memo to the faculty he argued,

> My personal concern is to communicate my strong desire that we grow away from the training of people who are primarily concerned with technical skills in developing instruction, evaluating programs, managing resource centers, producing films and television programs, etc. That we change our focus to grow towards the training of people who are more skilled in inquiring about problems and their solutions. . . . All student activities therefore that involve the actual development of instruction or production of films, etc. should be conducted in an atmosphere of *constant critical discussion of the usefulness of the concepts being acquired and the process being employed.* (Ely 1998, 59; emphasis added)

The theme of "constant critical discussion" could be said to be one of the hallmarks of Clark's career. He consistently exemplifies a questioning attitude, looking at fundamental problems and seeking evidence to reach a conclusion. This stance is neatly summarized in his article (Clark 1984, 230) responding to propositions about the central concerns of instructional technology: "As responsible professionals, our only ethical choice is to ensure that we accept the evidence for the products we advocate."

That search for evidence led Clark into the center of the debate that prompted a hundred websites and became virtually synonymous with his name: the media-methods debate. The debate revolved around the issue of whether media in themselves affect learning. Throughout his graduate studies Clark had struggled with this question, unable to fully accept the intuitively obvious supposition that media, such as films and TV programs, have a dramatic impact on audiences in ways that books and lectures cannot. It's probably fair to say that his insistent quest for proof of this supposition was found to be annoying to his colleagues, although tolerated and sometimes encouraged by his teachers.

A decade later Clark was still unsatisfied with the answers being proposed, so he took a sabbatical in 1981 and devoted himself full-time to combing through every bit of the research literature related to media and learning. It was an intense and exhaustive effort. The result was his famous critique, "Reconsidering Research on Learning from Media," published in the prestigious *Review of Educational Research,* which has become the most frequently cited source in its field. In it he presented evidence for the hypothesis that instructional methods have been confounded with media and that methods are what influence learning. He offered the analogy that different media were similar to the different delivery systems developed by pharmacists to introduce medicine into the body. Tablets, liquid suspensions, suppositories, and injections are all different "media," but their effect is dependent on whatever the "active ingredient" is. Pepto-Bismol soothes traveler's dysentery whether it is taken in tablet or liquid suspension form. What matters is the active ingredient, bismuth. The "medium," of course, has an effect on the speed of the effect, the cost of the effect, and the convenience of the effect, but it does not *cause* the effect. That is, media selection ought to be based on logistical considerations—availability, expense, production requirements, and the like—because there is always more than one medium that has the capacity to provide the cognitive experience needed for effective learning.

This hypothesis was not new or original, as Clark went to pains to point out, as it had been offered earlier (e.g., Mielke 1968; Schramm 1977). The greater attention attracted by Clark's version may be attributed to the times, the venue, or, more likely, the strength of the claim made by Clark. He made the explicit and clear claim that there were *no* learning benefits possible and urged that researchers not continue to waste effort on the question until a "new theory" was developed.

This unpopular position was not one that Clark arrived at on purpose. He reports that he started out expecting to find evidence that media *did* make a difference:

> In 1980, when I began the two years of focused reading of media research which resulted in the original publication, I was taking a challenge from Bob Heinich (the former editor of *AV Communication Review*) to develop a specific taxonomy of media and learning outcomes. I was also working with Gabi Salomon on what became our *Handbook of Research on Teaching* review of media studies. . . . In those days before electronic mail, we were sending each other ten or more single-spaced letters of detail and argument as our manuscript developed. I began with the expectation that media were a significant element in any educational reform which sought achievement gains. The problem was that as I reviewed the evidence it seemed clear that it did not support my expectations or intuition. (Clark 1991, 35)

A flurry of responses to Clark's 1983 article were published between the mid-1980s and 1990. Most of those attempted to suggest qualifications to the hypothesis or to define terms more precisely in order to explicate the constructs of media and method. It was not until 1991 that someone accepted the challenge to propose a new theory. Robert Kozma's (1991) retort focused on the distinction between "learning *from* media" (Clark's phrase) and "learning *with* media" (Kozma's phrase). Kozma, principal scientist at the Center for Technology in Learning of SRI International (at the time of the publication he was an associate professor in the school of education at the University of Michigan), presented a new theoretical framework—basically the constructivist one—that envisions the learner actively collaborating *with* the mediated message to construct meaning. His argument essentially was that capabilities of a particular medium, in conjunction with methods that take advantage of these capabilities, *interact with and influence the ways learners represent and process information* and may result in more or different learning when one medium is compared to another for certain learners and tasks (Kozma 1991, 179; emphasis added).

This reframing of the issue stimulated even more debate, which raged in the pages of instructional technology journals and in panel discussions at professional conferences for several years, culminating in two special issues of *Educational Technology Research and Development* (vol. 42, nos. 2 and 3 [1994]) devoted exclusively to this debate. Clark's conclusion to this climactic debate is captured in the title of his contribution to the special issues, "Media Will Never Influence Learning" (Clark 1994). He contends that Kozma did not directly address the distinction between medium and method or how the effects of those interacting variables might be separated or provide convincing evidence to support his theory. On the other side, Kozma has continued to espouse his viewpoint:

> But it seems to me that the interaction between medium and method (that is, the extent to which they *share* the variance) is the crux of the whole argument. The most powerful attribute of any medium is its ability to enable and constrain methods. The methods you can use with computers are very different than the methods you can use with video and this is because of the unique capabilities of the computer vis-à-vis video. (Kozma 2000)

After 1994 Clark decided to move on from defending an essentially negative point back to his original quest to discover what *does* make a difference in instruction. His platform after 1978 has been as a tenured full professor in the school of education at the University of Southern California. From 1992 to 1996 he was head of the division of educational psychology and technology. In response to invitations from institutions around the world he has traveled widely, with extensive immersion as a teacher and consultant, particularly in Ireland, Germany, the Netherlands, and Indonesia. In the early 1990s he took leave from Southern Cal to spend two years residing in Dublin, Ireland, creating and managing his own training consulting firm, Atlantic Training, Inc. He managed to outcompete some of the largest and most prestigious international consulting firms to win training contracts with the Irish Electricity Supply Board and the European Patent Office. These provided the opportunity to test his ideas about "what works" in training and education.

His focus in this latter period has been on the cognitive processes underlying teaching and learning, seeking instructional methods that facilitate the sorts of cognitive processes that lead to long-term retention and transfer to real life. Along the way he discovered the polar opposite of robust instructional methods, coining the term "mathemathantics" to refer to instructional methods that "kill" learning, that is, experimental treatments that turn out to be significantly *inferior* to the control treatment (Clark 1989).

This pursuit of powerful interventions has led to an increasing appreciation of the importance of motivation and the development of Clark's own model of motivation (Clark 1999). This, in turn, opens the door to the world of interventions beyond instruction, leading to the domain of human performance technology. So his most recent research has been on the larger issue of what sorts of interventions in the whole working environment make a difference for human performance. His commitment to this larger issue led to the founding of a new doctoral program (human performance in the workplace) at Southern Cal in 1996. This program has attracted national renown as the most prominent doctoral program devoted to human performance technology. Some fifty scholars have completed doctoral degrees in this program.

These vast and highly influential accomplishments have been spurred by a passionate conviction that one can accomplish great things by focusing attention on basic, urgent problems and then investing one's total effort on the pursuit of a solution. His advice to young researchers: "Believe in your vision, but follow the evidence!"

Michael Molenda

See also Instructional Technology; Performance Support; Research on Media and Learning; Schramm, Wilbur; Television and Learning

References

Clark, Richard E. 1983. "Reconsidering Research on Learning from Media." *Review of Educational Research* 53(4): 445–459.

———. 1984. "'Where's the Beef?': A Reply to Heinich." *Educational Communication and Technology Journal* 32(4): 229–232.

———. 1989. "When Teaching Kills Learning: Research on Mathemathantics." In H. N. Mandl et al. (eds.), *Learning and Instruction: European Research in an International Context,* vol. 2. London: Pergamon.

———. 1991. "When Researchers Swim Upstream: Reflections on an Unpopular Argument about Learning from Media." *Educational Technology* (February): 34–40.

———. 1994. "Media Will Never Influence Learning." *Educational Technology Research and Development* 42(2): 21–29.

———. 1999. "Motivation Systems." In D. Langdon, K. Whitesides, and M. McKenna (eds.), *Intervention Resource Guide: 50 Performance Technology Tools.* Somerset, NJ: Jossey-Bass.

Ely, Donald. 1998. *An Evolution of Educational Technology: Celebrating the Syracuse Program at Fifty.* Syracuse NY: Instructional Design, Development, and Evaluation, School of Education, Syracuse University.

Kozma, Robert B. 1991. "Learning with Media." *Review of Educational Research* 61(2): 179–211.

———. 2000. "Bob Kozma's Response on the Great Media Debate." Available online at hagar.up.ac.za/rbo/construct/kozma.html (January 24, 2001).

Mielke, K. W. 1968. "Questioning the Questions of ETV Research." *Educational Broadcasting* 2: 6–15.

Schramm, Wilbur. 1977. *Big Media, Little Media: Tools and Technologies for Instruction.* Beverly Hills, CA: Sage.

Cognitive Apprenticeship

Cognitive apprenticeships are modeled after traditional apprenticeships. Whereas traditional apprenticeships involve learning a visible activity or skill, cognitive apprenticeships involve using mentors to model processes that are often invisible, such as problem-solving, comprehension, and computation. Apprenticeships can be distinguished from traditional instructional methods because they provide opportunities for practice rather than specifications for practices (Berryman 1991).

Traditional Apprenticeship Learning

In traditional apprenticeships, apprentices learn to become part of a community of expert practitioners by modeling the performance of the teacher. Some characteristics of traditional apprenticeships include: learning related to bodily-kinesthetic endeavors; learning for the immediate value of getting a job done rather than for a symbolic goal such as a diploma; learning that moves from simple to more complex; learning with standards that develop naturally from and are embedded within the work environment; and learning in which teaching is often undetectable because instruction is guided by what the apprentice does rather than by what the teacher says (Berryman 1991). In some ways the traditional apprenticeship model is not transferable to many of the skills needed to succeed in modern society because such skills are frequently cognitive in nature and thus cannot be learned solely through observation and modeling.

Cognitive Apprenticeship Learning Models

Cognitive apprenticeship can be described as follows:

> Cognitive apprenticeship supports learning in a domain by enabling students to acquire, develop and use cognitive tools in authentic domain activity. Similarly, craft apprenticeship enables apprentices to acquire and develop the tools and skills of their craft through authentic work at and membership in their trade. Through this process, apprentices enter the culture of practice. So the term apprenticeship helps to emphasize the centrality of activity in learning and knowledge and highlights the inherently context-dependent, situated, and enculturating nature of learning. (Brown, Collins, and Duguid 1989a, 39)

The cognitive apprenticeship model (Collins, Brown, and Newman 1989) strives to maintain as many of the key characteristics of traditional apprenticeships and aims to explain a process by which a novice becomes a member of a community of expert practice. The cognitive apprenticeship model is intended to cross the traditional boundaries between academic and vocational education because each requires knowledge and expertise in the other's domain. For example, an automobile mechanic must now be able to understand and operate the many computer-based systems associated with newer models. In the past, automobile mechanics has been classified strictly as a vocational discipline, computer systems strictly as an academic discipline.

The cognitive apprenticeship model has four building blocks that help to define and implement an effective learning environment for students of all ages: *content, methods, sequence,* and *sociology* (Collins, Brown, and Newman 1989)

Content

Four types of content are typically required for a successful cognitive apprenticeship model. First, students must be exposed to the concepts, facts, and procedural knowledge of a subject. This is often the only type of content considered in traditional instructional methods and is typically taught out of context. This type of content must be integrated with problem-solving strategies used by experts, metacognitive strategies such as planning, setting goals, evaluating, and monitoring one's own learning, and learning strategies that enable students to recognize relationships among fields, relate prior experiences and knowledge to new knowledge, and learn how to learn versus learning how to memorize for a test.

Methods

In a cognitive apprenticeship model, instructional methods must move beyond the traditional lecture model in which the teacher acts as a dispenser of knowledge and the students as sponges charged with soaking up this knowledge. Students should be given opportunities to observe, invent, discover, and collaborate, with the teacher serving as a facilitator who offers hints or guiding questions and monitors individual as well as class progress toward the desired goals.

Sequencing

Learning should be structured in such a way that students participate in increasingly complex tasks. Many skills and knowledge are required to operate in any domain, and skills should be learned in a logical sequence in which students gradually build upon what has previously been learned.

Sociology

The learning environment should reflect an environment as close to the real world as possible. Students retain and transfer knowledge best when it is learned within an authentic context. For all domains this includes the need to learn to collaborate with others to solve problems and carry out solutions.

Educational Technology and Cognitive Apprenticeships

Educational technology has been used to implement cognitive apprenticeship learning models. The use of computer-mediated communication to facilitate such strategies has assumed many names in recent years (e.g., distance mentors, cybermentoring, telementoring, and teleapprenticeships). The use of multimedia to facilitate cognitive apprenticeships has also received attention (Casey 1996).

The Electronic Emissary (Harris 1998) is one of the most in-depth examples of this strategy in K-12 schools. The Emissary maintains a database of volunteer subject-matter experts and matches these experts with K-12 students and classrooms from around the world who are studying in their areas of expertise. For example, a student interested in how weather forecasters use Doppler radars could communicate with a professional forecaster to learn the nuances of this community of experts.

The Teaching Teleapprenticeship Model (www.ed.uiuc.edu/TTA/) is another cognitive apprenticeship model that uses electronic networks to connect key individuals in the education of preservice and in-service teachers. The project is funded by the National Science Foundation and aims to improve mathematics and science teaching by connecting teacher education students, practicing teachers, and university faculty via electronic networks that enhance and expand traditional face-to-face mentoring.

Kara Dawson

See also Active Learning; Electronic Emissary; Telementoring

References

Berryman, Sue E. 1991. *Designing Effective Learning Environments: Cognitive Apprenticeship Models.* ERIC Document 337 689, pp. 1–5.

Brown, John S., Allan Collins, and Paul Duguid. 1989. "Situated Cognition and the Culture of Learning." *Educational Researcher* 18(1): 32–42.

Casey, C. 1996. "Incorporating Cognitive Apprenticeship in Multi-Media." *Educational Technology Research and Development* 44(1): 71–84.

Collins, Allan, John S. Brown, and Susan E. Newman. 1989. "Cognitive Apprenticeship: Teaching the Crafts of Reading, Writing, and Mathematics." In L. B. Resnick (ed.), *Knowing, Learning, and Instruction: Essays in Honor of Robert Glaser* (pp. 453–494). Hillsdale, NJ: Lawrence Erlbaum Associates.

Harris, Judy. 1998. *Virtual Architecture: Designing and Directing Curriculum-Based Telecomputing.* Eugene, OR: International Society for Technology in Education.

Cognitive Flexibility Theory (CFT)

Cognitive flexibility theory is a theory of learning and instruction that was developed to address four main goals:

Goal 1: Helping people to learn important but difficult subject matter. Often the most central topics are the ones we find hardest to teach and learn. Too often a superficial understanding of concepts and memorization of facts substitute for more meaningful and deeper understanding.

Goal 2: Fostering adaptively flexible use of knowledge in real-world settings. A key goal of CFT is to prepare people to apply their knowledge flexibly, adapting prior understandings and experiences to fit the needs of new situations that often differ radically from the initial conditions of learning ("transfer" in cognitive and educational psychology). As such, CFT is a theory that puts a premium on *assembling* knowledge and experience as required by some new problem, text, or case (rather than relying on finding a stored "prescription" for how to think and act). We call these ensembles of knowledge and experience, drawn and then assembled from different learning occasions of the past, *schemas of the moment.* Again, CFT is a theory that emphasizes an adaptively creative response to new situations rather than the more mechanical following of routinized schemes that already exist in memory.

Goal 3: Changing underlying ways of thinking. CFT seeks to change not only the specific knowledge that an individual acquires but also—and more important—the underlying worldviews (epistemological beliefs, habits of mind) that an individual employs when approaching the acquisition and use of knowledge. The kind of knowledge one constructs, and the way that knowledge is deployed, depends on the "lenses" that filter one's view of the world. More often than not, we find that individuals' worldviews are premised on assumptions of simplicity (single right answers, compartmentalized knowledge components, knowledge in the abstract and not tied to contexts of application, etc.) that interfere with the development of complex understanding and the ability to apply knowledge to a wide variety of real-world contexts. In order to deal with important complexities, one must have a mind-set to open perception, to look for multiplicity, for interconnectedness, for subtlety and nuance. Systems based on CFT are designed to achieve these aims.

Goal 4: Developing hypermedia learning environments to promote complex learning and flexible knowledge application. The tenets of CFT are the basis for the design of computer learning environments that provide the kind of nonlinear and multiperspectival organization of material necessary to achieve the first three goals. To paraphrase one researcher, a flexible medium can make possible a flexible cognitive "message."

As our work and everyday lives become increasingly complex and more rapidly changing, the kinds of skills described in the first three goals have become most essential for us to instill in students and professionals. And yet they are the most difficult for us to teach (Feltovich, Coulson, and Spiro 2001; Feltovich, Spiro, and Coulson 1997; Spiro, Coulson, Feltovich, and Anderson 1988). In the twenty-first century, as all organizations become more horizontal and less vertical, with responsibility and individual initiative required at even low levels of job hierarchies (where workers are closer to the realities on the ground and thus have a better sense of the situations that must be responded to), getting a good job will require these skills and abilities. Fortunately, we now have technology, driven by learning theories that target difficult learning, that can enable the attainment of essential skills. It is a time of fortuitous confluence in needs and resources to address those needs.

Background and Principles

CFT had its origin in the mid-1980s (Spiro, Vispoel, Schmitz, Samarapungavan, and Boerger 1987; Spiro et al. 1988). That was a time when a dominant model of learning was *schema theory* (e.g., Anderson 1978; Ausubel 1968; Bartlett 1932; Bransford et al. 1977; Minsky 1975; Schank and Abelson 1977; Spiro 1980), an approach that placed a premium on using organized packets of knowledge in memory ("schemas," or "schemata," after Immanuel Kant; also sometimes "frames" and "scripts") as a basis for understanding and applying knowledge. The problem we noted was that one could not have a prestored schema for everything that one might encounter (Spiro and Myers 1984). Thus the notorious difficulty in producing "transfer," the reconfigured use of old knowledge in new situations that differ from the initial conditions and contexts of learning.

CFT was developed as a successor to schema theories, one that would replace rigidly prepackaged knowledge structures with more open and adaptable ones—knowledge that would be applicable across the wide range of situations in which it might be required. This is a challenging goal because most domains of knowledge and all domains of professional practice (medicine, engineering, business, teaching) are made up of events or cases (real-world occasions of using knowledge) that are at best irregularly related to each other (we call these *ill-structured domains*). In most real-world domains, wide-scope abstractions and general principles do not account for enough of the variability in the way knowledge has to be used. Instead one must have experience with a large number of cases to see "how things go," to see the different ways that conceptual knowledge is combined and applied in real contexts.

In order to produce such more open and widely applicable knowledge structures, it became clear that learning could not proceed in a single direction, organized into neat categories. Instead, CFT employs the metaphor of criss-crossing a landscape (Spiro et al. 1987, 1988; Spiro, Feltovich, Jacobson, and Coulson 1992a, 1992b; the metaphor was first used by Wittgenstein 1953, in the preface to his *Philosophical Investigations,* in a more limited context). In teaching and learning, one proceeds from case to case (example to example) following different routes of organization on successive traversals of the knowledge landscape. Sometimes one returns to the same site (case), but coming from a different direction, bringing a different set of perspectives. Thus different facets of each case are highlighted when juxtaposed to varying other cases (and seeing those multiple facets is essential in producing transferable knowledge). Thus, in CFT, *revisiting is not repeating.*

We soon came to realize that the best way to deal with this kind of nonlinear and multiperspectival learning and instruction is by capitalizing on the random access capabilities of computers (Spiro et al. 1988; we have at times referred to this kind of nonlinear teaching as "random access instruction"; see Spiro and Jehng 1990). Thus in 1987 we began an effort, continuing to the present day, to design cognitive flexibility hypermedia systems (CFHs), based on the tenets of CFT, for domains as diverse as medicine, high school biology, literary interpretation, military strategy, and teacher preparation. By using the criss-crossed landscape approach in the design of case-based learning environments we have been able to have some success in producing the often elusive finding of learning transfer (see, e.g., Jacobson and Spiro 1995). (To see hundreds of discussions, applications, and tests of CFT, some of which accurately render CFT and some of which do not, use a search engine such as Google with the following keywords: Spiro cognitive flexibility.)

Since its earliest days some of the principle tenets of CFT and its applications to CFH learning environments have been the following (Spiro et al., 1987, 1988, 1992a, 1992b, 1997; Spiro and Jehng 1990; Spiro, forthcoming):

Multiple knowledge representations. Knowledge that has to be used in many ways has to be represented in many ways. Whenever one sees a complex situation with a different conceptual "lens" or from a different perspective, new and important features of the situation are revealed. Thus CFT proposes a "principled pluralism" (Spiro 2001). This is not an anything-goes mentality but rather one in which new ways of assembling and applying knowledge from different perspectives that students provide must be accompanied by associated justifications with evidence drawn from the facts of the case. CFHs aid in this assembly and justification process. Furthermore, by constructing open knowledge structures and instilling a cognitive ethos of internal dialogue among alternative perspectives, applications of CFT permit individuals to become better prepared to participate in group and collaborative learning and prepare individuals to simulate the benefits of groups when they are alone and must think for themselves (Feltovich, Spiro, Coulson, and Feltovich 1996).

Interconnectedness. Conceptual and case knowledge cannot be "boxed" into separate mental compartments or presented in separate "chapters" always organized with the same "table of contents." In CFHs, it is as if the material was organized with a near infinitude of different organizing bases, different tables of contents. And each organization juxtaposes material in ways suitable to application to different kinds of cases, thereby supporting the goal of widely applicable or transferable knowledge. This is an important aspect of CFHs' *nonlinearity* of teaching and learning.

Context-dependency and conceptual variability. Conceptual knowledge is essential, but in real-world situations concepts are used in somewhat different ways, at different times, and in different combinations with each other. CFHs put a premium on illustrating the variable uses of concepts across contexts in order to prepare people to better apply conceptual knowledge on their own.

Cases and minicases. In order to prepare people to apply knowledge across a wide variety of real-world cases—especially when general principles do not provide an adequate basis for action—one must have experience with *many* cases. In order to accelerate the acquisition of that experience, CFHs make extensive use of minicases and employ the trope of *synecdoche* at the core of instruction. A synecdoche is a part-whole relationship; unlike metonymy (another part-whole trope), in which the parts of the whole do not resemble each other, in a synecdoche there is a microcosm-macrocosm relationship. By looking closely at a dense, well-chosen strip of a larger case, one gets a view of how things go over larger segments of the domain (much as is the case with fractals). To paraphrase another researcher, when we do not have the time or ability to show people an entire world of knowledge, the *only* way to see a whole world may be in a "grain of sand." Of course, the world that is seen in any grain of sand will be incomplete and somewhat imperfectly rendered; in ill-structured domains, any microcosmic view will only partially capture the features of the macrocosm. Thus, as in all CFT applications (whether involving minicases or cases), criss-crossing in different directions is required to gradually build a more complete and accurate synoptic view of the complex whole. However, that building process moves forward with greater rapidity when appropriately chosen minicases, ones that are both rich and at the "crossroads" with many other parts of the domain's landscape, are chosen for instruction.

Another benefit of organizing instruction around minicases (and then building to larger case structures) is that it solves a difficult problem in instructional sequencing: The traditional incrementalist approach, building from simple to complex, does not work because the early simplifications interfere with the later attainment of complex understandings (Fetovich et al. 1989; Spiro et al. 1988; Spiro, Feltovich, Coulson, and Anderson 1989; Feltovich et al. 2001). But to start with unbridled complexity would be confusing and discouraging to the learner. CFHs begin with "bite-sized chunks" of complexity, minicases that retain the basic features of complexity (e.g., multiplicity, interconnectedness, context-dependency, etc.). So the worldview of complexity is instilled from the beginning (instead of having an inappropriately oversimplified mind-set initially established and then having to be undone) but beginning with cognitively manageable amounts.

Thus there are at least three important benefits of a focus on minicases: (1) *experience acceleration* (more information is quickly conveyed about "how the world goes"); (2) a *new logic of instructional sequencing* that avoids initial oversimplifications that prove harmful in building more advanced understandings; and (3) *early establishment of complex underlying habits of mind* or ways of thinking. (There are many other benefits of the use of minicases, a list too long to go into here. See the extensive discussion in Spiro and Jehng 1990 and Spiro, forthcoming.)

New Directions

Currently, CFT has been developing most rapidly in the area of digital video cases for professional training, especially teaching. There is a large volume of work being done with video cases, and it is very valuable. However, that work has mainly involved showing, talking about, and archiving teaching cases. As important as these activities are, it seems clear that the bandwidth, both of the technology and the user, is not being sufficiently exploited. CFT is being applied in new ways to permit more *learning* to occur with video cases. Some of the special emphases of this work are the following (for a more detailed presentation of these new directions in CFT, see Spiro, forthcoming).

One of the most important aspects of the current work is the concentrated effort to *change underlying ways of thinking* (habits of mind, worldviews) toward those more compatible with complexity and flexibility (Spiro et al. 1996; we referred to these as "*prefigurative* schemas" in Feltovich, Spiro, and Coulson 1989 and Mishra, Spiro, and Feltovich 1996). Habits of mind are hard to change. We are employing various kinds of video special effects to alter perception, to get people to see more in a video case, to anticipate more complexity and then be able to detect it and work with it, learning to harmonize and organically balance the multiple agendas of complex activity. Our special effects are intended to *shake up* established ways of thinking, to call attention in the most salient ways possible to the various kinds of complacency of perception, thought, and action that so frequently are observed. When a learner habitually looks at one part of a scene, we put a spotlight (figuratively, and sometimes literally) on the periphery of habitual vision to redirect the gaze and make it more encompassing. When some pat way of thinking is induced during the course of a case, we use video techniques to create *dissonance,* to get people to say "But wait! It's not that simple!"

Another important aspect of current CFT media applications is their return to an earlier emphasis of CFT on making use of *experiential* modes of representation to convey *conceptual* information (Spiro 1980; Spiro, 1982a, 1982b; Spiro, Crismore, and Turner 1982). We make greater use of nonverbal processing channels to convey additional complexity, including information about the trajectory of the ebb and flow of conceptual themes as they intertwine in the course of real-world cases. We call these representations *metaphorical experiential symbol spaces* (MESSes)—overlearned correspondences between conceptual information and perceptual symbols. Experiential modes of representation can be overlaid upon each other and upon conceptual/verbal understandings to capture more complexity without exceeding limits of cognitive capacity (think of how much complexity is encoded in a single image of a face; or how we can be saying one thing while seeing and feeling something else simultaneously).

Finally, and perhaps most important of all, is the current CFT focus on *experience acceleration.* It is commonly believed that it takes ten years of experience for a professional to attain a reasonable degree of expertise (Feltovich et al. 1997). We need to shorten that amount of time. By the efficacious arrangement of minicases and the use of new video techniques, that can be accomplished. In CFT, cases are revisited in different contexts to bring out other aspects of their multifacetedness. Since these revisited video scenes rapidly become *overlearned,* it becomes possible to present them in greatly attenuated form in subsequent criss-crossings, greatly accelerating the rate of case-based learning. (We call the learning that is related to previously overlearned parts of video cases that are *not* re-presented in subsequent presentations—that only employ increasingly smaller "stand-ins" for the rest of the case—the "along-for-the-ride" effect.)

This new media paradigm, accelerated criss-crossing of densely packed, distinctive extracts from overlearned minicases (case *signatures*), utilizes the kind of *quick-cutting* (sometimes interspersed with commentary) that is so pervasive in contemporary film, television, and advertising, especially that targeted for more youthful audiences (see also Stephens 1998). Although frequently bemoaned as an attention-attenuator, when appropriately applied for educational purposes it simply involves a *different* kind of attention, one better suited to complex nonlinear learning—and one that people thirty and younger are already highly accustomed to and often prefer. We expect this mode of presentation will become codominant with traditional linear and verbal instruction and training modes in the near future. We are utilizing this mode in a species of CFHs that we call *experience acceleration systems* (see Spiro, forthcoming).

The goal of all of this new work is to permit professionals in training to enter the messy world of practice better prepared to deal with the complexities they will face.

A Theory That Builds Bridges

Cognitive flexibility theory is not a competitor with other theories. It is, in a sense, a metatheory, a theory that builds bridges across theories and across curricular chasms that are too often left unreconciled.

CFT's principled pluralism bridges different paradigms and perspectives, allowing each to complement the others. CFT offers an organic *unity of diversity.* This is not an abstractive unity, which loses the particulars of the individual; rather it is a nonreductive unity in which a whole is formed that retains the diversity that was the basis for forming that unity. An apt analogy is the memory of a particular person's face, which we experience as a physiognomic whole (that is *that* person's face), but which still retains the cognitively recoverable diversity of features (and relationships among features) that comprise the incredible complexity of any face. CFT brings wholes and their complex conceptual aspects together, each contributing to our understanding of the other.

CFT bridges knowledge and practice by embedding each within the other, so the common problem of seeing the connection between the "basic science" and "clinical practice" parts of curricula (to use the analogy of medical education, equally applicable in any domain of professional training and most school subject areas) is mitigated.

CFT bridges cognitive and individualistic approaches, on the one hand, and collaborative, sociocultural, and situated views, on the other. CFT is inherently contextual in its method. And the kind of *open* knowledge structures it fosters are just the kind of noninsular understandings that best support and interact with collaborative and situated learning. A group is made up of individuals, and those individuals must possess the right habits of mind for the group to function at its best, habits of mind that CFT instills. And a goal of CFT is to make it possible for individuals to function alone, as sometimes we all must, with an internalization of the benefits of multiperspectival group dialogue—the group simulated within the individual mind.

Finally, CFT builds bridges between *standards-based* models and *constructivist* approaches. Too often teaching toward standards leaves students with memorized knowledge that is neither deep nor usable. Constructivist approaches too often leave the acquisition of some important aspects of knowledge to chance. With CFT, one can satisfy the goals of standards-based and constructivist approaches simultaneously while allowing the strengths of each to compensate for the weaknesses of the other. CFT marries standards with deep understanding and adaptive flexibility of knowledge application. Because CFT always ties constructive processes to a landscape of actual occurrences in the world (as best that objectivist ideal can be achieved or at least approximated—whether it is scientific data or the "data" of a literary text), it is a *realist* constructivism and thus in tune with the goals of standards advocates.

All of these things are necessary if we are to become a society of individuals who can master the complexity of the world around us, deal with the constant change we face, and perform the kinds of twenty-first-century jobs that will require independent, adaptive, and creative thought.

Rand J. Spiro
Brian P. Collins
Jose Jagadish Thota
Paul J. Feltovich

See also Cognitive Psychology; Constructivism; Hypertext; Learner-Centered Environment

References

Anderson, R. C., Rand J. Spiro, and M. C. Anderson. 1978. "Schemata as Scaffolding for the Representation of Information in Discourse." *American Educational Research Journal* 15: 433–440.

Ausubel, David P. 1968. *Educational Psychology: A Cognitive View.* New York: Holt, Rinehart, and Winston.

Bartlett, Frederic C. 1932. *Remembering.* London: Cambridge University Press.

Bransford, John D., K. W. Nitsch, and J. J. Franks. 1977. "Schooling and the Facilitation of Knowing." In R. C. Anderson, R. J. Spiro, and W. E. Montague (eds.), *Schooling and the Acquisition of Knowledge* (pp. 31–56). Hillsdale, NJ: Lawrence Erlbaum.

Feltovich, Paul J., Richard L. Coulson, and Rand J. Spiro. 2001. "Learners' Understanding of Important and Difficult Concepts: A Challenge to Smart Machines in Education." In P. J. Feltovich and K. Forbus (eds.), *Smart Machines in Education* (pp. 349–376). Cambridge, MA: MIT Press.

Feltovich, Paul J., Rand J. Spiro, and Richard L. Coulson. 1989. "The Nature of Conceptual Understanding in Biomedicine: The Deep Structure of Complex Ideas and the Development of Misconceptions." In D. Evans and V. Patel (eds.), *The Cognitive Sciences in Medicine* (pp. 113–172). Cambridge, MA: MIT Press.

———. 1997. "Issues of Expert Flexibility in Contexts Characterized by Complexity and Change." In P. J. Feltovich, K. M. Ford, and R. R. Hoffman (eds.), *Expertise in Context: Human and Machine* (pp. 125–146). Cambridge, MA: MIT Press.

Feltovich, Paul J., Rand J. Spiro, Richard L. Coulson, and J. Feltovich. 1996. "Collaboration within and among Minds: Mastering Complexity, Individually, and in Groups." In T. Koschmann (ed.), *Computer Systems for Collaborative Learning* (pp. 25–44). Hillsdale, NJ: Lawrence Erlbaum.

Jacobson, Michael J., and Rand J. Spiro. 1995. "Hypertext Learning Environments, Cognitive Flexibility, and the Transfer of Complex Knowledge: An Empirical Investigation." *Journal of Educational Computing Research* 12: 301–333.

Minsky, Marvin. 1975. "A Framework for Representing Knowledge." In P. H. Winston (ed.), *The Psychology of Computer Vision.* New York: McGraw-Hill.

Mishra, Punyashloke, Rand J. Spiro, and Paul J. Feltovich. 1996. "Technology, Representation, and Cognition: The Prefiguring of Knowledge in Cognitive Flexibility Hypertexts." In H. van Oostendorp and A. de Mul (eds.), *Cognitive Aspects of Electronic Text Processing* (pp. 287–305). Norwood, NJ: Ablex.

Schank, Roger C., and Robert P. Abelson. 1977. *Scripts, Plans, Goals, and Understanding.* Hillsdale, NJ: Lawrence Erlbaum.

Spiro, Rand J. 1980. "Constructive Processes in Prose Comprehension and Recall." In R. J. Spiro, B. C. Bruce, and W. F. Brewer (eds.), *Theoretical Issues in Reading Comprehension* (pp. 245–278). Hillsdale, NJ: Lawrence Erlbaum Associates.

———. 1982. "Subjectivite et memoire" [Subjectivity and memory]. *Bulletin de psychologie* (special issue "Language Comprehension") 35: 553–556. Reprinted in W. Kintsch and J. Le Ny (eds.) in 1983 as *Language Comprehension.* The Hague: North-Holland.

———. 1982. "Long-Term Comprehension: Schema-Based versus Experiential and Evaluative Understanding." *Poetics: The International Review for the Theory of Literature* 11: 77–86.

———. 2001. "Principled Pluralism for Adaptive Flexibility in Teaching and Learning to Read." In R. F. Flippo (ed.), *Reading Researchers in Search of Common Ground.* Newark, DE: International Reading Association.

———. Forthcoming. "Accelerating Experience, Changing Ways of Thinking, and Mastering Complexity for Adaptive Practice: New Directions in Cognitive Flexibility Theory." *Educational Technology.*

Spiro, Rand J., and Jihn-cheng Jehng. 1990. "Cognitive Flexibility and Hypertext: Theory and Technology for the Nonlinear and Multidimensional Traversal of Complex Subject Matter." In D. Nix and R. J. Spiro (eds.), *Cognition, Education, and Multimedia: Explorations in High Technology* (pp. 163–205). Hillsdale, NJ: Lawrence Erlbaum.

Spiro, Rand J., and A. Myers. 1984. "Individual Differences and Underlying Cognitive Processes in Reading." In P. D. Pearson (ed.), *Handbook of Research in Reading.* New York: Longman.

Spiro, Rand J., Avon Crismore, and Terence J. Turner. 1982. "On the Role of Pervasive Experiential Coloration in Memory." *Text* 2: 253–262.

Spiro, Rand J., Paul J. Feltovich, and Richard L. Coulson. 1996. "Two Epistemic World-Views: Prefigurative Schemas and Learning in Complex Domains." *Applied Cognitive Psychology* (special issue on Reasoning Processes) 10: S 51–61.

Spiro, Rand J., et al. 1987. "Knowledge Acquisition for Application: Cognitive Flexibility and Transfer in Complex Content Domains." In B. C. Britton and S. Glynn (eds.), *Executive Control Processes.* Hillsdale, NJ: Lawrence Erlbaum.

Spiro, Rand J., et al. 1988. "Cognitive Flexibility Theory: Advanced Knowledge Acquisition in Ill-Structured Domains." *Tenth Annual Conference of the Cognitive Science Society* (pp. 375–383). Hillsdale, NJ: Lawrence Erlbaum. Reprinted in R. B. Ruddell and M. R. Ruddell (eds.), *Theoretical Models and Processes of Reading,* 4th ed. Newark, DE: International Reading Association, pp. 602–616.

Spiro, Rand J., et al. 1989. "Multiple Analogies for Complex Concepts: Antidotes for Analogy-Induced Misconception in Advanced Knowledge Acquisition." In S. Vosniadou and A. Ortony (eds.), *Similarity and Analogical Reasoning* (pp. 498–531). Cambridge, MA: Cambridge University Press.

Spiro, Rand J., et al. 1992a. "Cognitive Flexibility, Constructivism, and Hypertext: Random Access Instruction for Advanced Knowledge Acquisition in Ill-structured Domains." In T. Duffy and D. Jonassen (eds.), *Constructivism and the Technology of Instruction* (pp. 57–75). Hillsdale, NJ: Erlbaum. Reprinted from a special issue of the journal *Educational Technology on Constructivism.* Reprinted in L. Steffe and J. Gale (eds.), *Constructivism in Education* (pp. 85–107). Hillsdale, NJ: Lawrence Erlbaum.

Spiro, Rand J., et al. 1992b. "Knowledge Representation, Content Specification, and the Development of Skill in Situation-Specific Knowledge Assembly: Some Constructivist Issues as They Relate to Cognitive Flexibility Theory and Hypertext." In T. Duffy and D. Jonassen (eds.), *Constructivism and the Technology of Instruction* (pp. 121–128). Hillsdale, NJ: Erlbaum. Reprinted from a special issue of the journal *Educational Technology on Constructivism.*

Stephens, Mitchell. 1998. *The Rise of the Image, the Fall of the Word.* Oxford: Oxford University Press.

Wittgenstein, L. 1953. *Philosophical Investigations.* New York: Macmillan.

Cognitive Psychology

Cognitive psychology is the branch of psychology that attempts to understand how the mind works. More specifically, cognitive psychologists are interested in how the mind processes information, with a focus on underlying processes of attention, perception, learning, and memory. Cognitive psychology found strong roots as a reaction against behaviorism in the middle of the twentieth century, when behavioral scientists could not account for certain aspects of human behavior. Instead of seeing the mind as a collection of stimuli and corresponding responses, cognitive psychologists describe the mind as a processor with modules designed for various purposes. Cognitive psychology is closely related to educational technology because the computer has become the most recent metaphor for the way in which the mind processes information. Cognitive psychologists also use technology to try to simulate cognitive events. Finally, educational technologists draw on findings from cognitive psychology in the design, development, and implementation of technologies for teaching and learning.

In the early 1900s, John Watson began the behaviorism school of psychology. Arguing that the study of mental processes was unscientific, Watson and others (drawing on the experiments of Ivan Pavlov and Edward Thorndike) studied physiological responses to environmental stimuli. B. F. Skinner continued this line of research with a tremendous amount of experimental data, arguing that almost every aspect of human behavior is nothing more than a set of conditioned responses. Rewarding or punishing certain behaviors could program—or reprogram—these responses.

However, in the middle of the twentieth century, opponents of behaviorism provided evidence of various behaviors that could not be explained by simple conditioned responses. Among the most prominent of these opponents was the linguist Noam Chomsky. He wrote a review of Skinner's book *Verbal Behavior* in 1959 that argued behaviorism could not account for certain aspects of language. He specifically cited accounts where subjects would utter *original* statements that were completely grammatically correct. This innate language component could not be explained by external stimuli or conditioned responses; nor could behaviorism explain anomalies such as improvisation and creativity in music and art. Along with being unable to account for these phenomena, researchers became frustrated with the methodology of behaviorism that limited their studies to observable events and behaviors outside the body and mind. These circumstances paved the way for the so-called first cognitive revolution.

To suggest that cognitive psychology started in the 1950s would be incorrect. The first psychologists, such as Wilhelm Wundt, William James, and Hermann Ebbinghaus, as well as early philosophers such as Aristotle and Plato, were interested in cognitive processes long before behaviorism. However, it is generally argued that the cognitive revolution gave birth to cognitive psychology or the interest in cognition—an exploration of the processes by which knowledge is gained.

There are numerous lines of converging and diverging thought that originated from this revolution—all of which explore the ways in which knowledge is perceived, stored, retrieved, and communicated. Therefore, attempting to summarize, synthesize, and select major theorists and theoretical underpinnings in cognitive psychology is a difficult task. Cognitive psychology is often thought of as fitting hand-in-glove with information-processing theory, but the cognitive movement also gave birth to or aided in the development of such theories and areas of study as cognitive development, constructivism, schema theory, situated cognition, and cognitive science. George Miller and Ulric Neisser are credited for aiding the revolution with their important work, but theorists such as Jerome Bruner, John Dewey, and Jean Piaget were also pioneers in helping alter the zeitgeist. Some of these theories and their philosophers focused on the internal processes and rules we are born with. Others valued how knowledge is acquired. Most current psychologists agree that we need to draw on both arenas.

Needless to say, cognitive psychology has given rise to a number of novel approaches to learning and teaching, as well as a number of educators, psychologists, theorists, and philosophers who have pushed how we view cognition. The most recent event in the history of cognitive psychology is the advent of the second cognitive revolution. An interest in the individual and how the individual processes information is one tie that binds all of the earlier cognitive approaches. Philosophers and theorists, drawing on Russian psychology and psychologists such as Lev Vygotsky and Alexander Luria, are now beginning to value context and the cultural and social-cultural adaptations, semiotics, and tools that play a role in our cognition. This second cognitive revolution has given way to social-cultural, cultural-historical, social constructivist, distributed cognition, and situated cognition theories, as well as new psychologies such as cultural psychology, narrative psychology, and discursive psychology.

Due to the reciprocal and substantial relationship with neuroscience, philosophy, anthropology, education, psychology, sociology, and so on, understanding the role of cognitive psychology in educational technology is also a daunting task. However, there are three significant relationships we can identify. The first, and perhaps the most recognized, comes from viewing the computer as metaphor for the mind. In an information-processing approach to cognition, the mind is viewed as an information-processing machine—much like that of the computer. The body takes in information from the senses, processes it (decoding), and then produces output. Acceptance of this metaphor has prompted an interest in artificial intelligence, where humans try to get machines to perform human mental tasks.

A second relationship, drawing on the first, is the use of computers to try to explore and understand human mental processes. Computers, and other technologies, have been used to measure such things as attention, memory, and sensory perceptions. They have also been used in areas such as reading to study behaviors like eye movement.

Perhaps the most salient for our discussion of educational technology is the third significant relationship, which adopts a very broad definition of cognitive psychology. With the cognitive revolution, new theories of learning were devised and applied to new perspectives on teaching. For instance, theorists now recognize the importance of scaffolding and the need for feedback. Technology became a way to support the kind of pedagogical approaches that may not have been possible in traditional educational settings. For instance, certain cognitive approaches suggest that students need personalized and individualized feedback. Most teachers do not have enough time or enough assistance to complete such a task in class. However, certain technologies have been built to support immediate (just-in-time) and yet individualized feedback. Other pedagogical perspectives have suggested that discursive practices with students from outside the student's culture would enrich learning. Online technologies allow this type of learning to take place—learning that would have been *possible* but not *probable* otherwise.

This third relationship is double-sided. Educational technologies make it easier to apply findings from cognitive psychology and its related fields. However, educational technologists also draw on cognitive findings to design and develop new educational technologies. For instance, we know that information must be divided into chunks to be most easily appropriated. We also know that multimodal sensory inputs (text with graphics and video) can be more reinforcing than just text alone. Thus, the relationship between educational technology and cognitive psychology is beneficial, due in part to its recursive relationship.

Richard E. Ferdig

See also Behaviorism; Bloom, Benjamin S.; Bloom's Taxonomy; Bruner, Jerome S.; Cognitive Apprenticeship; Constructivism; Gagné, Robert Mills; Instructional Design

References

Bransford, John, Ann Brown, and Rodney Cocking. 1999. "How People Learn: Brain, Mind, Experience, and School." Available online at stills.nap.edu/html/howpeople1.

Pinker, Steven. 1997. *How the Mind Works.* New York: Norton.

Collaboration and Negotiation Tool for Case-Based Instruction (CoNet-C)

The Collaboration and Negotiation Tool for Case-Based Instruction is an online discussion tool specifically designed to support case-oriented small-group discussion. The tremendous interest in group work and learning with computers has led many scholars to explore how networked technology can be used to create learning environments that promote discussion and collaboration. Many instructors and designers are familiar with generic Internet technologies for collaboration, such as bulletin-board systems, chat technologies, and e-mail. Yet few guidelines are available to help software designers and instructional designers develop online collaborative environments that support specific teaching and learning activities, such as problem-based learning or case-based instruction. The major goal of the CoNet-C project is to provide a feasible online discussion environment to support case-based instruction through the achievement of the following objectives: to facilitate students in participating in meaningful small-group peer interactions, thereby enhancing their online learning; to assist teachers in managing and guiding small-group online discussions efficiently, thereby enhancing their online teaching; and to allow researchers to conduct experimental study of online discussions efficiently, thereby enhancing their online research.

The Internet has become one of the most popular delivery methods for distance teaching and learning in contemporary society. Current models of instruction emphasize opportunities for students to articulate and reflect on their knowledge, interact with peers and teachers, and approach material from multiple perspectives. Educators and researchers have been trying to incorporate existing instructional theories and knowledge into online instructional environments in order to improve online learning and teaching. Among various instructional strategies, online discussion has been one of the most popular instructional methods. In non–face-to-face teaching environments like online instruction, online discussion provides an essential and unique way that teachers can interact with their students or have students interact with one another. Because of the importance of online discussion, most online course management systems widely used in colleges (e.g., WebCT®, Blackboard®) provide online discussion tools that allow teachers to create discussion forums for certain topics and assignments. In the discussion forums, students can exchange opinions with their peers.

Recently, with the emergence of blended or hybrid courses at the postsecondary level (i.e., courses that blend in-class and online learning experiences), even onsite instructors who teach their classes in regular classrooms often use online discussion as a supplementary instructional method. Many instructors believe that online discussion can be an effective instructional method to promote small-group interaction when they have larger-sized classes with limited time for in-class discussion. Therefore, online discussion can become an important instructional method in regular as well as online classrooms.

Despite the popularity and importance of online discussion, existing tools have limitations because of their inherent general-purpose functionality. Thus it may be difficult for teachers to create a certain discussion setting in which they can guide meaningful discussion in a way that corresponds with their specific instructional methods and goals. For example, an online discussion setting in which teachers want students to communicate with each other to complete group projects should be different from a discussion setting in which teachers want students to review various small cases or problems and discuss them as a small group. Therefore, it is important to develop discussion tools that provide customized online discussion capabilities that are compatible with a variety of instructional methods.

Goals of the Project

The overall goal of the CoNet-C project is to develop an online discussion tool aimed at supporting case-oriented discussion. In the design and development of the online discussion tool, at least three major groups of users were considered: learners, instructors, and researchers. CoNet-C is designed to be adaptive and tailored to these three different user roles. In terms of learners, CoNet-C helps learners focus on the given problems or cases for discussion and to participate in small-group discussions more effectively. The discussion tool is also designed for teachers to manage online discussions easily. For example, teachers can assign a certain problem or case to individual students in small groups for a particular period of time. Finally, CoNet-C includes features designed for researchers who are conducting experimental study in online discussion. Researchers can manipulate online treatment schedules and collect data through a special interface. CoNet-C is an extension of the Collaboration and Negotiation Tool (CoNet) project, which was originated by Susan M. Land at Pennsylvania State University in 1997. Since then, various versions of discussion tools have been developed for different orientations of online instruction, such as project-based instruction and case-based instruction at the college level.

A Prototype of the Discussion Method

CoNet-C was implemented in the context of an undergraduate course in turfgrass management taught by Alfred J. Turgeon at Penn State. In this online class, students studied a series of online lesson modules at the introductory level of turfgrass management. While completing the modules, they worked on five open-ended case problems related to turfgrass management. Turgeon created and used a modified jigsaw discussion method for his turfgrass management class, which became a prototype for the interface design of CoNet-C. With this approach, each student is assigned to a small group of four to six members before the discussion begins. The instructor assigns different cases or problems to different members in a group. Each member generates her own solution to the given problem or case. Then, within a small group, each member shares the assigned problem and the solution with her peers. Each student should lead the discussion by replying to all questions from peers or by defending her position from group members' challenges. After finishing all discussion, all members within a group revise their initial solutions based on what they have learned from others in that group. After sharing final solutions with peers, they close the session of discussion.

The interface of CoNet-C has three main interfaces: learner, teacher, and researcher. These are designed to support discussion activities that are reflected in the jigsaw method described above.

Learner Interface

The main emphases in the design of the learner interface were to minimize navigation problems during discussion and to generate a customized interface. Thus, we tried to design an interface that could seamlessly integrate students' dual roles of both problem-solvers and peer-reviewers in discussion.

Once students log on to CoNet-C, they are led to a customized main page, where they can review the assigned problems and interact with their small group. Students can participate in a certain forum of discussion by navigating through the group members' names and discussion sessions. This navigation structure, based on members' names and discussion sessions, can reduce students' confusion in locating a particular forum and prevent interference among topics or forums.

Through the customized main page, students can find their own rectangular boxes at the top of the screen; the remaining boxes are their peers' areas. Within her own area, the student can review the assigned problems, create and revise her own solution to them, and lead discussions about her solutions with peers. By clicking on others' solution areas, she can review peers' solutions to the given problems and discuss them. For all interactions, CoNet-C identifies the learner's authority to either read or write in a certain area. For example, a student can revise her own solution; however, the same person cannot change her peers' solutions, only review them. Therefore, this interface helps students focus on discussion topics and to interact with peers efficiently.

To assist students in discussing their solutions and to provide constructive feedback to peers, a series of guiding questions was designed to serve as prompts. These guidelines were embedded into the discussion tool to facilitate learners' generation of critical questions and comments that might help improve peers' solutions. Some question prompts were generic (e.g., "Could you please give specific examples of that?") and others were more specific ("How would your answer change if all chemical application had to be made by licensed professionals?") in order to support a range of responses.

Instructor Interface

The main focus in the design of the instructor interface is to minimize the instructor's load in managing discussion activities, such as forming small groups, assigning problems to individual students, scheduling discussion sessions, monitoring discussions, and so on.

CoNet-C provides a comprehensive interface for supporting instructors to manage discussion; through this interface, instructors can easily form small groups, assign particular questions to individual students, open discussion sessions in a certain period of time according to their class schedules, and monitor all of the small groups' discussions. Most important, CoNet-C provides a database so instructors can create and revise discussion cases or problems and assign them to individual students. Once instructors create a problem or case database, the problems/cases can be utilized for discussion in classes.

Researcher Interface

Researchers share the identical interface with instructors; however, there are some advanced functions for research purposes. The main focus in the design of the researcher interface is to assist research activities such as treatment control and data collection. Besides forming small groups, researchers can assign individual students or groups into either experimental or control groups, and they can manipulate different interventions for different groups. Thus during certain discussion sessions, students can receive different treatments online. Also, all student interactions and their evolution processes in forming solutions are kept in a computer database. Thus researchers can retrieve all necessary information for their particular research purposes. For example, while students can see only the latest version of their solutions to given problems, researchers (and instructors) can retrieve all previous versions of their solutions and analyze how students' solutions have evolved throughout their peer discussions.

CoNet-C System Structure

There are three basic structures of the CoNet-C system. Three major databases hold students' information, discussion problems/cases, and peer interactions. By incorporating these databases, instructors create particular discussion settings in which students participate in small-group peer discussions. While students participate in assigned discussions, CoNet-C refers to these databases continuously to provide customized discussion interfaces for particular users.

Design Experiment and Implementation

CoNet-C has been successively tested in college online classrooms and redesigned based on feedback from students, instructors, technical supporters, and other stakeholders. This broadened approach is based on the notion of a "design experiment" whereby technology innovations are studied within real contexts and where the innovations, students, and teachers can adapt themselves through iterative implementations to enhance learning and teaching. Through this approach, an online tool was developed to assist instructors in organizing effective discussion environments; practical ways were also found to help instructors incorporate new technology innovations into their daily classroom settings. For example, criteria for grading need to be tied directly to participation in discussion. Criteria should include points for responding in a timely manner and for providing the expected quantity and quality of postings. Cases or problems to be discussed must be open-ended enough to encourage diverse points of view. Providing case problems that were either too simple or focused (i.e., only one correct answer) led to limited interaction and discussion. Studies suggest that students infrequently referred to the supplemental guiding questions; therefore future implementations should make this feature more prominent, as some students reported difficulties in the process of generating constructive feedback. In addition, students who used the guidance perceived that it made asking questions easier, which in turn improved the quality of their feedback. Most students reported that they preferred the CoNet-C interface to a generic discussion tool because it was easier to write and revise solutions for each separate case discussion.

The instructor reported that the case/problem database saved time in assigning problems to students because it reduced the need to retype or repost cases/problems. He also perceived CoNet-C to be a tool that supports teaching and learning in addition to online discussions. Future research will be designed to extend the contexts in which CoNet-C is studied to support generalization and to collect additional data regarding how students learn from asynchronous discussions and the problems and processes they experience in this endeavor.

Ikseon Choi
Susan M. Land

See also Collaborative Technologies; Computer-Mediated Communication (CMC); Distance Education; Online Learning Communities; Web-Based Instruction

References

Aronson, Elliot, and Shelley Patnoe. 1997. *The Jigsaw Classroom: Building Cooperation in the Classroom.* New York: Longman.

Choi, Ikseon. 2002. "Effects of Peer-Challenge Support on Learning During On-Line Small Group Discussion." Doctoral diss. Pennsylvania State University. Available online at etda.libraries.psu.edu/theses/available/etd–0712102–211229.

Hawkins, J., and A. Collins. 1992. "Design-Experiments for Infusing Technology into Learning." *Educational Technology* 19(1): 63–67.

Collaborative Technologies

Collaborative technologies defines an area of research and development that examines the overlap between collaboration, a strategy for working together, and technology—the practical application of knowledge to extend human capabilities. Effective collaboration requires four key elements: (1) verbal and visual communication; (2) written communication; (3) coordination/management of the collaboration process; and (4) management of resources used or created throughout the collaboration. Collaborative technologies are designed to facilitate communication, coordination, and management in a variety of situations and to reduce constraints or compensate through technology for aspects of the collaborative process that are more difficult when participants are at a distance. Modern collaborative efforts in business, industry, education, medicine, or the military are used to cope with external competition that in many organizations is both local and global. Those organizations that utilize the capabilities provided by new collaborative technologies will increase their odds of survival.

Imagine a collaborative team working on a project in which all team members are in the same cluster of offices. There is a shared file cabinet where documents are stored, signed out, updated, and returned. Project questions can be answered by a quick walk down the hall to meet with a fellow team member. Team meetings are brief, focused, and effective. Project milestones are posted on the wall where everyone can see them daily.

Now imagine a collaborative team working on a project where team members never meet in the same physical space (e.g., a small company employing contract professionals, or a large company with team members from different parts of the organization—marketing, design, manufacturing, human resources, etc.—situated in different parts of the country, or a multinational company with team members scattered around the world in different time zones and speaking different languages). These teams will need to utilize robust collaborative technologies to engage in the project activities listed above.

There are many collaborative technology products and systems currently available with combinations of features and capabilities. In general, there are systems that use intranets (a private, secure network) and applications that use the capabilities of the Internet.

Team members using the same corporate network, even those in different buildings and different cities, could use a network-specific collaborative product. Several major companies offer products in the intranet area, and new products are constantly being introduced. Applications and systems that support collaboration over the Internet are particularly useful when the team does not have a shared corporate network. New systems and applications to facilitate Internet-based collaboration are also being introduced frequently. For intranet or Internet approaches to work it is critical that the communication and computer systems be compatible with each other and that all members of the collaborative team have the skills and knowledge necessary to utilize the technology as a full participant. The next section identifies selected technologies that contribute to an effective collaborative environment using the four key elements as a framework.

Verbal/Visual Communication

The most mature technology in this category is voice communication. This includes telephone conference calls, Internet telephony, and voice-messaging systems that provide integration with mobile devices such as cell phones and pagers. Although the technology for video telephones has been available for decades, problems with image quality and the need for high bandwidth have slowed adoption. The most common technology for visual communication is videoconferencing. This technology ranges from one-to-one conference calls using desktop video systems, one-to-many using a broadcast system, or many-to-many where all participants are visible to all other participants. In the latter situation, the system should provide the participants with the ability to electronically "raise a hand" to get the attention of the current speaker or individual in control of the collaborative session.

Document cameras are used to display a digital video image of a three-dimensional object for all participants as well as support for scanning and storing of two-dimensional images. The verbal/visual communication technology should also support the sharing of multimedia presentations (video clips, slide shows) using applications that are available to all participants. Visual communication between participants is enhanced through the use of a virtual whiteboard that allows participants to share information in real time.

Written Communication

Electronic correspondence (e-mail) is essential, although some correspondence may still be shared using postal or messenger services. Thus the system should be able to send and receive Internet e-mail as well as e-mail from a corporate network, manage messages (including the ability to recall or unsend a message), provide a history of messages received, display message status (received, opened, replied, forwarded), set up folders, search messages, and sort messages by various criteria. The system should support specific workgroup conferencing (small groups) with the same features as the general e-mail application, including the ability to attach specialized documents (spreadsheets, scanned documents, blueprints, etc.) and executable files (computer code) to an e-mail message. Members of the collaborative group should be able to use distribution lists to send the same message to all team members. Online (threaded) discussion forums can be used to share ideas and concerns on all aspects of the project, and when the written correspondence is a web document, the system should support hyperlinks within files and messages. Clicking hyperlinks should launch a web browser and take the reader to the linked content.

Coordination and Collaboration of the Cognitive Process

Documents developed using project-management applications need to be available on the network so that all participants can see the tasks to be performed, responsible individuals, overall timelines, and specific milestones or deliverables. The project-management system should be flexible enough to adjust project elements based on actual reports from participants, the shared version on the network should be the most current, and changes by one member to a specific task should be automatically communicated to all participants or subprojects linked to that task. The system should maintain a log of all activities related to the projects and provide a mechanism whereby team members can vote electronically on specific topics and/or questions. A shared scheduling system should be used by project members to arrange meetings and calendar events, (holidays, vacations, etc.). Information in the collaborative system should be open to authenticated members using as a minimum user ID and password. Biometric devices (thumbprints, retinal scans) and smart cards with embedded computer chips may replace this form of security. When data is moved between members of the collaborative system (i.e., over the Internet) some form of data encryption should be available.

Management of Resources Used or Created throughout the Collaboration

Participants should be able to view, create, modify, review, mark up, and revise documents. The system provides a document storage capability (a database with project documents) and automatic data backup and archiving that would allow multiple users to work on different versions of the same document at the same time, offer an "oops" capability (i.e., undo changes), as well as full restoration to an earlier version. There should be an audit trail of changes to identify who did what and when to the document.

Team members should be able to search the files, notes, and/or discussions stored in the collaborative environment using a variety of keyword identifiers. Computer applications used to produce documents or components of the project should be stored on the shared server and be available to team members. The system should support the use of electronic forms with drop-down selections to make data entry easier and less prone to errors. Some users may need to manipulate an application or system from the command line. Advances in graphical user interfaces have reduced the need for this functionality except in situations in which the participants have a high level of programming expertise.

A collaborative system with all of these features would require abundant bandwidth or a dedicated communications network, as well as high-end computer processing power. These systems are usually the most expensive.

John R. Savery

See also Communities of Practice; Computer-Mediated Communication; Knowledge Management; Performance Support; Telecollaboration and Teleresearch

References

Association for Information Systems (aisel.isworld.org).

Collaboration and Virtual Office Resource Center, ProSci Research (www.collaboration-tools.com/distance-learning.htm).

Collaborative Technologies Research Center, Computer Science and Mathematics Division, Oak Ridge National Laboratory (www.csm.ornl.gov/ctrc).

Communication Theory

As a field of research, communication is relatively young. In the 1940s researchers departed from sociology, anthropology, and psychology to develop the first theory that focused on the *messages* exchanged between people. This theory (communication theory) was the formal beginning of the study of communication processes. Communication theory describes communication as a process in which individuals exchange messages and interpret the meaning of those messages. Today, communication theory is still used as the basis of how communication is viewed and researched.

In 1949 researchers Claude Shannon and Warren Weaver forwarded the first model of communication, making the field of communication distinct from other social sciences. The Shannon-Weaver model of communication has been slightly modified over the years; however, the basic components and appearance of the model remain the same. It is important to note that although the Shannon-Weaver model was initially intended to describe the media communication process, such as the use of the radio to transmit messages, the model is now used to describe virtually all forms of communication, including face-to-face communication.

Based on the Shannon-Weaver model, communication theorists today generally identify eight components of the communication process: source, receiver, encoder/decoder, message, channel, field of experience, noise, and environment. The *source* is the person who sends the message, and the *receiver* is the person who receives the message. Many researchers note that we rarely take turns sending and receiving messages; instead we are in the constant state of sending and receiving. The *messages,* often referred to more broadly as "symbols," need to be sent through some sort of channel. Thus if a teacher (source) has a message for his students, the teacher must determine a channel, or method, to send the message. The message might be sent through talk, or a note, or an e-mail. Some messages are sent nonverbally, that is, without words. Examples of nonverbal messages include body language, pictures, facial expressions, and dress.

These symbols must be converted into meaning through the processes of *encoding* and *decoding.* As Shannon and Weaver first intended in their theorizing, encoders and decoders can be centered in the technology of the medium. For example, an operating system on a personal computer encodes and decodes the symbols of computer language into pictures and words that the typical user can understand. Today, we look at individuals as also having an encoding and decoding process in communication. Each message an individual sends and receives is interpreted, or decoded and encoded, for better understanding.

Every person has a *field of experience* that shapes the interpretation, or encoding and decoding, of messages. A person with technology experience might choose e-mail as a primary channel of communication. However, a person with little computer experience might feel frustration when trying to retrieve an e-mail. The frustration felt by the source of the message can influence how that person interprets the message. Culture is an important component when considering a person's field of experience. For instance, in some cultures eye contact is an appropriate communication behavior, whereas in other cultures eye contact is a display of disrespect. Therefore, the norms and customs of differing cultures can influence how people perceive messages.

The *environment* in which the communication takes place can also influence the sending and receiving of messages. Although communication sometimes takes place face to face, with the source and the receiver sharing the same environment, sometimes communication takes place over time or space, and the participants have different environments. During a phone conversation, for example, one person can be in a noisy house, unable to attend to the messages, while the other person can be in a quiet environment ideal for holding a conversation. Environment influences the communication process.

Communication is not a precise science, and messages are not always received exactly as they are intended. Anytime there is interference in the communication process, *noise* occurs. Interference, or noise, can take place through any of the elements in the communication process. Noise can be within the source or the receiver. There is noise if a person's field of experience leads to a misinterpretation of a message. Noise can also occur in the channel. For example, a person's cell phone might cause noise if there is too much static to hear the messages clearly. Noise is the result of what we commonly know as miscommunication.

When considering the parts of the communication process, it is important to note the synergy between the elements. Each element has an impact on the other elements. When one chooses the channel to send the message, one must also consider the environment, the message, the receivers, and the receiver's field of experience. Advertisers are experts at considering the impact each component has on the other components. When advertisers need to market a product, they consider the audience they are targeting and find the appropriate channel in light of the audience's experiences and environment. Not one of these communication elements stands independent of the other elements.

In addition to the Shannon-Weaver model of communication, the middle of the twentieth century saw the rise of the *magic bullet* (or *hypodermic needle*) theory. The magic bullet theory describes the process of sending messages through a medium as similar to sending out a bullet to an audience, or injecting an audience with a message. Although the theory was first developed in the nineteenth century, the increased attention to propaganda during World War I and World War II brought the magic bullet to the forefront of communication theory. During this time, it was believed that sending out a message through a war poster, for example, would hit an audience and bring about a desired change. Thus in this early view communication was essentially thought of as being one-way: sender to receiver. Today, however, with Nielsen ratings, opinion polls, and interactive media technology, we view communication as a two-way process. The two-way process enables us to determine if a message brings about a desired response. Therefore, if a politician's campaign is not translating into support, the campaign committee can analyze the audience feedback, change the campaign, and resend the new message.

The *diffusion of innovation theory* further describes the process of changing a message that does not receive the desired response. The founder of diffusion of innovation theory, Everett Rogers, explained that diffusion is the process by which an innovation is communicated through certain channels over time among members of a social system. An "innovation" can be any product, concept, or idea that is perceived as new. A college course taught at a distance via computer, for example, is an innovation that is being adopted by many institutions. Rogers (1995) developed a diffusion of innovation model composed of the following six main concepts:

1. Communication channels are the means by which a message is transmitted from one person to another.
2. When messages concerning a specific innovation are transmitted from a source to a target member, *the innovation-decision process is initiated.* This process is a sequence of decision points through which a target member passes. The sequence starts with *knowledge* or awareness of a potential innovation. The target member is then *persuaded* by information from a source and forms a favorable or unfavorable attitude toward the innovation. The next step is the *decision* to adopt or reject the innovation. If an adoption decision is made, the target member then *implements* the innovation. Finally, *confirmation* occurs when the individual seeks reinforcement for the innovation-decision that he or she has made.
3. *Homophily* refers to the extent to which people are similar.
4. Perceptual *innovation attributes* are relevant to the innovation-decision process: (a) relative advantage (i.e., the advantages of adopting the innovation versus the costs of adoption); (b) compatibility (is the innovation consistent with the values, past experiences, and needs of the target member?); (c) complexity; (d) trialability; and (e) observability (the degree to which the results of an innovation can be observed by target members).
5. *Adopter categories* describe the degree to which an adopter is relatively early or late in adopting relative to other members in a social system.
6. Opinion leaders are people within a given social system who are able to influence other individuals' attitudes or behaviors in a desired way with relative frequency.

Given the diffusion of innovations framework, we can follow how an idea or product is initiated, adopted, or rejected.

The role of current communication theory, in general terms, helps us to answer the *how* and *why* questions about the communication experience.

More specifically, the goals of communication theory can be to *explain, understand, predict,* and *cause social change* (West and Turner 2000). For example, communication theory can *explain* our patterns of TV viewing or explain how people interpret visual images. Communication theory might also help one *understand* why we are persuaded by a politician's speech, or why people sometimes do not believe bad news. At times, we are able to *predict* something based on communication theory. Based on certain theories, we can somewhat predict if a couple will stay together, or if a particular student will have a favorable experience with a particular teaching style. Finally, communication theory can promote *social change,* as when it is used to guide principles of education entertainment. Some theories uncover problems and recommend methods to remedy those problems. For example, the government of a country struggling with an AIDS epidemic might include safe-sex story lines in a popular soap opera to educate the audience. In this case, communication theory provides a method to change the behavior of TV viewers and, in turn, promote social change.

Technology and communication theory are explicitly linked in numerous ways. As we use technology to replace face-to-face communication, we need to modify prior conceptions of the communication process and develop new theories to reflect the increasing adoption of technology in daily life. The prevalence of technology will also demand communication researchers to analyze which innovations are adopted by an audience and how those innovations will work as methods of diffusion. The adoption of communication technology is both an innovation and a method of diffusion that impacts all parts of the communication process, from sender to receiver.

Dawn L. Carusi

See also Diffusion of Innovations; Entertainment-Education; Instructional Communications

References

Berger, Charles R. 1977. "Interpersonal Communication Theory and Research: An Overview." In B. D. Ruben (ed.), *Communication Yearbook 1* (pp. 217–228). New Brunswick, NJ: Transaction.

Rogers, Everett M. 1995. *Diffusion of Innovations*, 4th ed. New York: Free Press.

Shapiro, M., et al. 2001. "Information Systems Division: Intrapersonal, Meaning, Attitude, and Social Systems." In W. B. Gudykunst (ed.), *Communication Yearbook 24* (pp. 17–49). Thousand Oaks, CA: Sage.

Watzlawick, Paul, Janet H. Beavin, and Donald D. Jackson. 1967. *Pragmatics of Human Communication.* New York: Norton.

West, Richard, and Lynn H. Turner. 2000. *Introducing Communication Theory.* Mountain View, CA: Mayfield.

Communities of Practice

The term "community of practice," although recently coined, refers to an age-old social structure that occupies a mediating position between the individual experience of learning and the reproduction of organizations and societies at an aggregate level. The term is used to refer to a perspective on learning, and to specific communities that are observable in many settings, including the online world. The perspective itself, and careful observation of specific communities, suggest important design considerations for online learning.

The term was first proposed by Jean Lave and Etienne Wenger in their 1990 book *Situated Learning: Legitimate Peripheral Participation.* It arose from an examination of apprenticeship systems and included five examples from different cultural and social settings: midwives in the Yucatan, tailors in West Africa, naval quartermasters, meatcutters, and sober alcoholics. The book has been influential as a theory that situates learning in everyday social processes and considers institutions of learning (e.g., schools) as a special case. Although it does not define the term "communities of practice," the book makes the argument that a community of practice is the primary unit for the propagation of knowledge and the reproduction of culture.

In a subsequent book, Wenger (1998) extends the conceptual framework to argue that communities of practice are also the primary unit for the development of new knowledge. He argues that in the case of apprenticeship the competence of the community pulls the experience of the novice until the novice has the full experience of competence; and conversely, the new experience of community members (e.g., in other communities or other domains or with new technologies) can cause the community to reexamine its assumptions and grow its practice: to learn. The book also contains an ethnographic study of claims processors in a large insurance company that serves as a reference point for the development of the theory.

During the late 1980s and 1990s, work in many different settings developed the practice of cultivating and supporting intentional communities of practice as part of the efforts to improve quality of life at the job, to improve product quality (and worker performance), and, finally, as part of an emerging discipline of knowledge management. John Seely Brown and Paul Duguid (1991) give a conceptual perspective and describe some practical outcomes in the context of a copier-repair community. Others (Wenger, McDermott, and Snyder 2002) gather considerable experience of community development and describe how a community-of-practice perspective can be a unifying view that provides a design trajectory toward human-centered knowledge management.

The Well and other online communities—where the principal practice was to converse and belong—made the idea of online communities quite fashionable in the 1990s. As the Internet grew and as online communities proliferated, it is useful to look at some of them from a community-of-practice perspective. Large-scale development projects (collectively, the open source movement) brought hundreds of programmers together to develop large systems such as Linux and Perl. The social hierarchy, the distributed responsibility, the steady innovation, and the visibility of individual performance produced many successful design and development projects and large-scale systems. They also produced a larger community that served the apprenticeship needs of the less experienced and moved the competence of the communities as a whole to new levels. Eric S. Raymond ("The Cathedral and the Bazaar," available online) describes the process.

Extrapolating from early adopters to online communities of practice that would serve other populations is not simple. Among other things, the practice of code development can be visualized and observed with existing Internet technologies, whereas most other practices, such as claims processing or midwifery, cannot. Grafting Internet technologies onto the ongoing practices of a community is not a trivial task (Wenger, McDermott, and Snyder 2002).

Theorizing about and carefully observing face-to-face communities of practice suggest ideas that may be useful to the design, production, and use of information and communication technologies for learning. Consider the following issues:

The association between learning and the classroom is questioned: "The class is not the primary learning event. It is life itself that is the main learning event. Schools, classrooms, and training sessions still have a role to play in this vision, but they have to be in the service of the learning that happens in the world" Wenger (2001). The physical boundaries of the classroom do not exist on the Internet. How can the design of online resources for learning support focused inquiry without re-creating the sequestration of a physical classroom?

If we consider practice to be a lens that organizes knowledge and learning, being clear about what practice a learning activity is to support becomes very important. The fact that the Internet makes the communication of information cheap and fast does not necessarily make the visualization of practice any easier. How can the engagement between the practice and the practitioner be made permeable so that peripheral community members have real access to a community's real practice without disrupting the work of practitioners?

From the perspective of legitimate peripheral participation, a community's thick boundaries are themselves meaningful social spaces for a student to traverse. These boundaries are not the province of admissions officers and the registration process but rather of learners who are making sense of the practice as they interact with other learners. The design of information and communication technologies for teaching must therefore consider the importance of social contact between all of the participants in the process. Social contact around and outside the classroom must be explicitly supported and provided for.

Wenger (1998) argues that learning, meaning, and identity are inextricably linked. This implies that the development of competence (i.e., learning) is necessarily reflected in a person's sense of identity. Thus Wenger argues that communities of practice are the means of producing our sense of personhood. The design of online learning environments must therefore consider whether the expression of competence through the completion of a test at the end of an instructional episode is an adequate expression of a new identity.

Wenger's (1998) is a middle-out theory that sees communities of practice between the internal experience of a person and the objective institutions of society. A consequence of this argument is that "no community of practice is fully capable of designing its own learning; and no community of practice is fully capable of designing the learning of another." This conclusion suggests two things: the importance of transparency and historical context in the design of learning resources, and the importance of negotiability in the exchange between learner and teacher.

John D. Smith

See also Computer-Mediated Communication; Knowledge Management; Learner-Centered Environment; Online Learning Communities; Performance Support; Situated Cognition

References

Brown, John Seely, and Paul Duguid. 1991. "Organizational Learning and Communities-of-Practice: Toward a Unified View of Working, Learning, and Innovation." Available online at www.parc.xerox.com/ops/members/brown/papers/orglearning.html.

Lave, Jean, and Etienne Wenger. 1991. *Situated Learning: Legitimate Peripheral Participation.* New York: Cambridge University Press.

Raymond, Eric S. 2000. "The Cathedral and the Bazaar." Available online at catb.org/~esr/writings/cathedral-bazaar/cathedral-bazaar/.

Wenger, Etienne. 1998. *Communities of Practice: Learning, Meaning and Identity.* New York: Cambridge University Press.

———. 2001 "Supporting Communities of Practice: A Survey of Community-Oriented Technologies." Available online at www.ewenger.com/tech.

Wenger, Etienne, Richard McDermott, William Snyder. 2002. *Cultivating Communities of Practice.* Cambridge, MA: Harvard Business School Press.

Community Technology Centers (CTCs)

Community technology centers are organizations whose primary focus is providing technology access to underserved populations in rural and urban environments. CTCs across the country service a wide range of individuals, from young children to senior citizens. Some have organized curriculums, whereas others are open to the public for walk-in use.

As the CTC movement grows, so do the funding opportunities. One major source of funding is the government, which offers grants through multiple departments. The U.S. Department of Education's CTC program (www.ed.gov/offices/OVAE/AdultEd/CTC/index.html), which appropriated $64.95 million for the 2001 fiscal year, is one such funding opportunity. Grants through this program are awarded to individuals or organizations interested in establishing or expanding CTCs in urban and rural communities. Another source of funding comes from private foundations. The Microsoft Giving Program (www.microsoft.com/giving/) provided $104 million in cash and software to nonprofit organizations that aim to increase access to technology in economically distressed communities. Similar funding opportunities are available from other sources.

As with any type of training, there is a wide range of philosophies concerning the best way to instruct users on the effective utilization of technology. Some focus on teaching specific technology skills that can easily translate into employment opportunities. Others are more concerned with developing general technology understanding that can be applied to any specific technology skill. Both of these philosophies are being implemented in two main categories: CTCs that exist within an established organization, and those that stand alone.

CTCs located within existing organizations offer community development services in addition to technology access and education. One place they can be found is within housing facilities such as the Neighborhood Networks program from the Department of Housing and Urban Development (www.hud.gov/nnw/nnwindex.html). Another common location for technology centers is within public libraries, such as the National City Public Library in California (its Community Computer Center offers open access as well as introductory computer courses to the public). A third location for community technology centers is within faith-based organizations. The Association of Christian Community Computer Centers (www.ac4.org) consists of more than 250 CTCs that are housed in Christian organizations.

CTCs located within an established organization offer services that may include, but are not limited to, preschool and family programs, afterschool and summer activities, adult education, workshops, job preparation, and career development, as well as technology access. CTC staff may direct the services and programs independent of other departments within a single organization. However, a major goal is to integrate technology into other community services. For example, the East Side House Settlement (www.eastsidehouse.org/technology.htm) in the South Bronx of New York City serves as an example of an organization offering technology access and education as well as other services. East Side House provides computers, telecommunications, and technology training through early-childhood, after-school, vocational, senior, and teacher education programs. The technology center includes five computer labs that are well equipped with up-to-date software. Students are instructed in word processing, spreadsheets, and databases for academic and professional purposes. Project READ is a literacy training program offered by the Redwood City Public Library in California. It provides computer-aided literacy instruction to participants through a variety of software packages. Students can access the software in a modern computer lab.

CTCs may also function as stand-alone technology centers. Such centers may carry out other services offered by traditional community centers through the learning of technology skills. Many stand-alone CTCs use technology to enhance reading skills, self-esteem, and team-building through curriculum-based computer courses. For example, Playing2Win (www.playing2win.org), located in Harlem, New York, contains two computer labs. It offers a variety of courses to clients, from basic keyboarding skills to multimedia design. Playing2Win also provides walk-in access for a small fee. A small group of high school students within this CTC comprise what is called HarlemLive (www.harlemlive.org). This program teaches web design, video production, and business skills through the creation of an online magazine of the same name. Another stand-alone CTC is Street Level Youth Media (streetlevel.iit.edu) in Chicago. This CTC has multiple locations throughout the city that allow teenagers to create multimedia video projects. The main goal of this program is to provide disadvantaged youth safe and fun opportunities to learn with the hope of keeping them away from gang involvement.

Tamara R. Pearson
Aisha I. Wood

See also Digital Divide

References

U.S. Department of Education (www.ed.gov/offices/OVAE/CTC) Community Technology Centers Network (www.ctcnet.org).

Component Display Theory

Component display theory is an instructional design theory in that it is a method for designing and developing instruction (Reigeluth 1999). The key contribution of component display theory was to extend the model of computer algorithms and data into the instructional world by metaphor, where instructional strategies or teaching methods are treated as algorithms and domain content is modeled as data. This separation of content and strategy is frequently regarded as the theory's key contribution.

Although its theoretical foundations are earlier research in concept learning (Merrill 1994), component display theory grew to maturity and was first implemented in the context of the Time-Shared Interactive Computer Controlled Information Television (TICCIT) project. In order to fully understand component display theory, one must turn first to TICCIT.

In 1972–1973 Victor Bunderson at the University of Texas was awarded a large grant to study learner control in the context of interactive television. Together with MITRE Corporation, Bunderson began work on the TICCIT project. Shortly after the grant award, David Merrill convinced Bunderson to bring TICCIT to Brigham Young University, where they would work together.

Building a principled mechanism that effectively allowed learners to control their path through instructional content proved more difficult than originally imagined. Eventually Merrill and a small committee of graduate students were tasked with solving the learner control subsystem problem. Because the team felt that the problem was one of systematically generating alternative strategy options, they worked from an algebra metaphor in which different types of instruction could be represented as variables to be added together, or concatenated, to effect specific learning outcomes. This meant devising a taxonomy of choices: Between what alternatives, or categories of strategy building blocks, would learners choose when exercising learner control? Merrill privately arrived at a two-by-two matrix that provided a separation of content and strategy and called a public meeting of all department faculty and students to announce his discovery. (This meeting was the first time any other members of the team heard the component display theory terminology.) TICCIT message authoring templates were created for each of the primary presentation forms, which made authoring content very quick. Later, several screen templates were also designed so that the screen layout and screen sequencing (in cases of multiple screens of messages) could be determined by formula through associations between authoring templates and screen templates. In this way content authored correctly would simply "run" in the screen templates.

In addition to the primary presentation forms, which are the core of component display theory, the TICCIT system also offered varying levels of difficulty for the learner to choose: easy, medium, and hard versions of the generality statement, and easy, medium, and hard (to understand) instances. Learner control of the difficulty of instruction and other system features was enabled by a specialized keyboard, with keys labeled "Rule," "Example," "Easy," "Hard," and so on.

Merrill's team produced a number of sequences, including what they called the all-American instructional strategy: a rule statement followed by an example statement followed by practice, with help available at any point along the way. TICCIT help was implemented as a huge state-based decision tree: If a learner missed a hard practice item, had he seen the hard rule and hard example? Had he successfully completed the medium-difficulty practice? Remediation was carried out by sending the learner back to the immediately previous uncompleted screen.

Initially, the project called for several such predetermined sequences of strategies (such as rules, examples, and practice opportunities) to be designed and implemented. Merrill's breakthrough on the learner control problem came when he realized that the block sequence of strategies was too coarse a grain and that learners needed control over the individual strategies (e.g., choose an example now and a rule then) instead of the larger sequences of strategies. It was like deciding to deliver bricks on demand rather than walls on demand. This resulted in the system opening up and learners gaining control at the level of individual strategies.

Component display theory was an extremely significant contribution to the field of educational technology, as it represented one of the first successful attempts to separate instructional strategy from instructional content. CDT is also an intellectual parent of many other important instructional theories, including C. M. Reigeluth's *Elaboration Theory* and Merrill's later *Instructional Transaction Theory.* Instructional Transaction Theory further specifies the format in which the instructional content is to be expressed (knowledge objects) and the ways in which prespecified strategies (instructional transaction shells) should operate on knowledge objects. In this regard, Merrill, Bunderson, and other members of the TICCIT team foresaw the late 1990s learning objects movement.

David Wiley
Andy Gibbons

See also Elaboration Theory of Instruction; Instructional Design; Learning Objects

References

Merrill, M. David. 1994. *Instructional Design Theory.* M. David Merrill, David G. Twitchell (ed.). Englewood Cliffs, NJ: Educational Technology Publications.

Reigeluth, Charles M. 1999. "What Is Instructional Design Theory and How Is It Changing?" In Reigeluth (ed.), *Instructional Design Theories and Models: A New Paradigm of Instructional Theory* (pp. 5–29). Hillsdale, NJ: Lawrence Erlbaum Associates.

Computer-Assisted Instruction (CAI)

Computer-assisted instruction occurs when an instructional program is delivered to a learner using a computer. CAI is sometimes considered a type of computer-based instruction (CBI), which refers to any form of computer use in an educational setting, including instructional programs, tutorials, simulations, instructional management, supplementary exercises, programming, and productivity software applications such as word processing and spreadsheets. CAI and CBI are often synonymous, the former sometimes in a more restrictive sense to refer to drill-and-practice, tutorial, or simulation software used for stand-alone learning activities or as supplements to teacher-directed instruction. CAI may also describe the instructional program itself or the delivery of the instructional program by a computer. CAI is sometimes known as educational software and courseware when packaged as a comprehensive curriculum with management and assessment features. In a typical CAI session the student sits in front of a computer, which presents information on the screen. The student reacts to the information presented by working with the mouse and/or keyboard. The pace of instruction may be controlled by the student, who may have control over the sequence of instruction. At certain points in the program, the student responds to questions posed, and the program notifies the student whether the response was correct or incorrect. In more complex CAI settings, the program may also keep track of the number of correct and incorrect responses and adapt the sequence of instruction according to performance throughout the program.

The primary advantages of CAI are that it allows learners to work at their own pace, controls the flow or sequence of instruction, and provides immediate feedback. More sophisticated forms of CAI adapt instruction to individual learner needs by varying lesson content, instructional sequence, and level of difficulty for each lesson as well as revising the types of feedback. These revisions may be accomplished while the learner is completing a lesson. For example, the computer may select and present math problems at varying levels of difficulty in response to an initial diagnosis of each learner's mathematics ability. Once an assessment is made, the control processes of the instructional program utilize feedback from the learner to continuously refine the estimate of the learner's progress.

CAI had its beginnings in the 1950s when educational researchers attempted to solve learning problems by applying the techniques of behavioral analysis as theorized by B. F. Skinner through programmed instruction (PI). The concepts of PI were then applied to crude teaching machines that first appeared in the late 1950s and early 1960s. PI and teaching machines were used throughout the 1960s by colleges, public schools, and the military. PI never achieved a high degree of popularity in schools, possibly because it was monotonous and did not fit well with group-oriented, fixed-schedule school settings. Early efforts to use computers in instruction, however, emerged from the guiding principles of PI. In its early forms CAI was an integration of computer technology and the PI movement.

Among the original CAI models to emerge was a project under the direction of Patrick Suppes at Stanford University. The Stanford Project was begun in 1963 to develop a tutorial system for instruction in elementary mathematics, language arts, and reading. By the end of the second year of operation approximately 400 students received daily CAI in either reading or mathematics. As a direct consequence of the Stanford Project and the need for curriculum-relevant CAI, Suppes formed a company (Computer Curriculum Corporation), which marketed courseware for minicomputer systems. Suppes advocated the creation of many articulated programs instead of isolated topical lessons to effectively deliver instruction.

Another early CAI project originated at the University of Illinois in 1960. That project, Programmed Logic for Automatic Teaching Operations (PLATO), designed computer hardware and software specifically to deliver instruction in a variety of subjects to a large base of learners simultaneously. PLATO researchers pioneered the use of color graphics, touch-sensitive screens, delivery modes, a high level of interaction between computer and learner, and learner control. PLATO became a registered trademark of Control Data Corporation and served as a model for many courseware delivery systems. In 1967 the University of Illinois established a research laboratory for the PLATO project, and it expanded into a large-scale computer-based educational system (PLATO IV).

As the costs of computing technology declined, the use of CAI became more feasible for classroom applications; the use of microcomputers, as well as educational software, in schools began to expand in the early 1980s. With the emergence of microcomputers and local area networks that connected many microcomputers to one server computer, educational software companies (including CCC and PLATO) migrated their courseware to microcomputer-based platforms. These systems became known as integrated learning systems (ILSs) and consisted of computer hardware, software, and courseware configured as a local area network. ILSs generally included a comprehensive package of software (courseware) that provided CAI on a network. The courseware included a management system that tracked individual learner progress and adjusted instruction accordingly. ILSs experienced enormous popularity and sales when several book publishing companies purchased ILS companies and invested heavily in reshaping them for the future. During much of the 1980s and 1990s ILSs accounted for a large portion of the CAI systems used in public schools.

A great deal of research has been conducted on the effectiveness and efficiency of computer use and on the effects of CAI on academic achievement as well as specific academic areas, high-level thinking skills, learning rates, learning retention, locus of control, and motivation. The research indicated that the use of CAI as a supplement to teacher-directed in-class instruction produced superior achievement and retention effects for students of different ages and abilities in different curricular areas. These comparisons indicated that CAI was more effective with lower-achieving students than with higher-achieving ones. The research comparing the effectiveness of CAI to standard methods of instruction, however, provided inconclusive results. Meta-analytic methods were used to compare the results from a number of CAI studies and concluded that the use of CAI produced higher achievement (see Bangert-Drowns, Kulik, and Kulik 1985; Becker 1992; Kulik 1994; Kulik and Kulik 1987).

Research findings that examined the effectiveness of CAI in terms of learner achievement and time required to learn material to a mastery level indicated that student learning rates were faster with CAI. Much of the research also examined attitudes about CAI, and CAI often produced positive student and teacher attitudes. In many studies teachers and learners often perceived CAI as appealing and having a positive impact on learning. While cost considerations were not a main focus of CAI research, some studies indicated that CAI was more cost-effective than some instructional methods such as tutoring and at least as cost-effective as classroom instruction. A body of largely qualitative research on CAI indicated that computer technology was of more educational benefit when its use was incorporated into the classroom practices of teachers and integrated with, and essential to, the curriculum. The primary value of CAI may be found in its potential to motivate students, increase access, and reduce the time needed to accomplish a given set of objectives.

The instructional use of computer technology is now better distinguished as learning *from* computers and learning *with* computers (see Salomon, Perkins, and Globerson 1991; Jonassen and Reeves 2001). Learning *from* computers occurs when the computer is the medium for the delivery of content through an instructional program. The complexity of the instructional strategies and the volume of curriculum content embedded within the instructional program may vary from single-focus and sequential to comprehensive and adaptive. Learning *with* computers occurs when computer technologies are used as tools to support teaching and learning. Learning from computers is associated with CAI, whereas learning with computers is associated with the use of computer and software resources that are not necessarily instructional (i.e., e-mail, web browsers, word-processing software, presentation software, etc.) as cognitive tools to support learning activities in active learning environments. In contrast to learning with and learning from computers, learning *about* computers refers to the development of skills required to operate and utilize computer hardware and software and is not considered to be CAI.

Steven Mills

See also Courseware; Intelligent Tutoring Systems; PLATO

References

Bangert-Drowns, Robert L., Joanna A. Kulik, and C.-L. C. Kulik. 1985. "Effectiveness of Computer-Based Education in Secondary Schools." *Journal of Computer-Based Instruction* 12: 59–68.

Becker, Henry J. 1992. "Computer-Based Integrated Learning Systems in the Elementary and Middle Grades: A Critical Review and Synthesis of Evaluation Reports." *Journal of Educational Computing Research* 8: 1–41.

Jonassen, David H. 1999. *Computers as Mindtools for Schools: Engaging Critical Thinking,* 2nd ed. Columbus, OH: Prentice-Hall.

Jonassen, David H., and Thomas C. Reeves. 2001. "Learning with Technology: Using Computers as Cognitive Tools." In D. H. Jonassen (ed.), *Handbook of Research for Educational Communications and Technology* (pp. 693–719). Mahwah, NJ: Lawrence Erlbaum Associates.

Kulik, Joanna A. 1994. "Meta-analytic Studies of Findings on Computer-Based Instruction." In E. L. Baker and J. H. F. O'Neil (eds.), *Technology Assessment in Education and Training* (pp. 9–33). Hillsdale, NJ: Lawrence Erlbaum Associates.

Kulik, Joanna A., and C.-L. C. Kulik. 1987. "Review of Recent Research Literature on Computer-Based Instruction." *Contemporary Educational Psychology* 12: 220–230.

Salomon, Gavriel, David N. Perkins, and Tamar Globerson 1991. "Partners in Cognition: Extending Human Intelligence with Intelligent Technologies." *Educational Researcher* 20(3): 2–9.

Computer-Assisted Language Learning (CALL)

Computer-assisted language learning is a rapidly evolving academic field that explores the role of computer technologies in language acquisition. The evolution of CALL mirrored developments in computer technology alongside the evolution of linguistic and instructional theories of language acquisition (Delcloque 2002; Warschauer 2002). Chronologically, CALL has evolved from a behavioristic model, to communicative and integrative models, to include finally a more collaborative approach. Implemented in the 1960s and 1970s, behavioristic CALL was informed by the behaviorist learning theories and used computers as tutors to engage learners in drill-and-practice activities with the intent of improving grammar and vocabulary. Emergent in the early 1980s, communicative CALL techniques were based on cognitive theories, which considered learning a process of exploration and discovery. These CALL programs encouraged learner interactions in

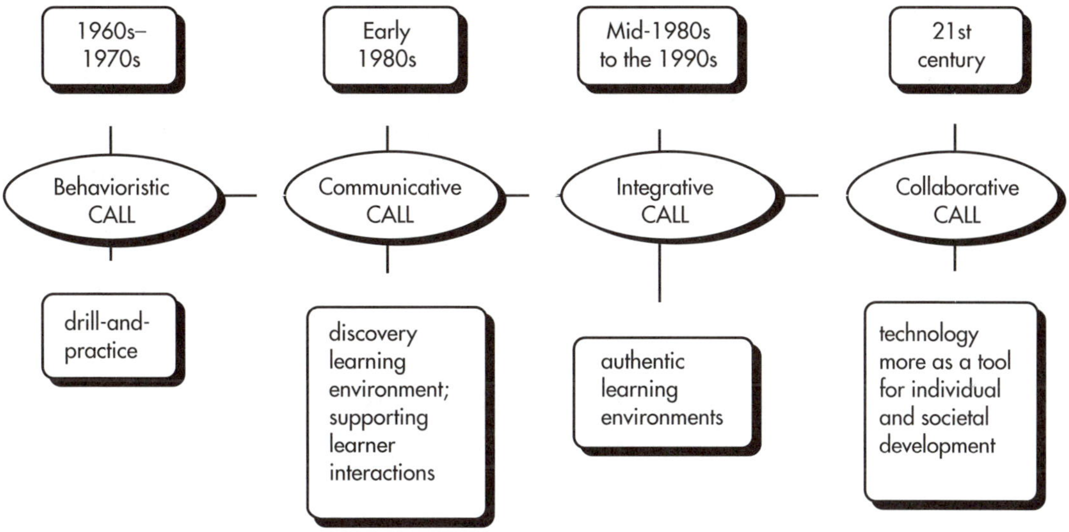

FIGURE 1: The Evolution of CALL

SOURCE: Created by the author.

processes that were to engage a broader schema of literacy, beyond mere grammatical forms and vocabulary. Supported by the socio-cognitive view of learning, integrative CALL referred to technology to create authentic learning environments, which *integrated* reading, listening, speaking, and writing skills in a target language. To this end, the current collaborative CALL identifies technology more as a tool for individual and societal development rather than just isolated language and literacy skills (Warschauer 2002). This approach is under the influence of sociocultural theories of education, where human learning and development are tied to collaborative purposeful activities mediated by tools and the social environment (Vygotsky 1978). Both integrative and developmental CALL support highly interactive and communicative language learning and include extensive use of the Internet. Figure 1 illustrates the evolution of CALL.

The development in CALL was influenced by technological development and the pedagogical shifts in language learning (Warschauer 2000). For instance, the evolution of CALL from being behavioristic to collaborative paralleled developments in computers from mainframe to personal computer, to the networked computer, to the World Wide Web. In particular, CALL has progressed in the context of the multifarious changes in language education such as the fostering of communicative ability, encouragement of creative self-expression, meaning negotiation, and the situation of language in authentic cultural contexts (Kern and Warschauer 2000). Although all forms of CALL are utilized, they now encompass more varied processes (from behavioristic to communicative models).

The Internet and the web had the greatest impact on language teaching and learning. Computer networks provide learners with opportunities to interact with speakers of the target language and share knowledge through multimedia authoring. The web also offers easy and immediate access to authentic language materials, including texts, audio, and video. It also supports the creation of virtual learning classrooms with various individuals and communication tools such as webcasts (i.e., live online streaming audio), audio and video conferencing, live chat, e-mail, listservs, discussion boards, and so on. A good example is the California State System's Virtual Language Learning (VLL) project, which aims at building virtual learning communities among instructors and learners in less commonly taught languages. The VLL engages instructors from four different campuses in collaborative course material development and refers to the "virtual language lab" software and other web communication tools for course delivery.

In addition, the wide collection of useful resources on the web can be easily used to support the short- and long-term production of the most current, creative, and engaging instructional activities (e.g., providing grammar and communicative exercises, developing students' online communication skills, and involving students in inquiry learning on current issues by using activities such as WebQuests). In particular, the web provides the possibility for teachers to engage students in creative language learning activities such as online journaling, collective editing, and creating animated grammar by using multimedia authoring tools such as Authorware, DHTML, and Flash. For assessment and evaluation, the web allows instructors to track learners' work and to measure their learning outcomes through interactive scenarios and tests, as well as authentic writing in the areas of formal writing (essays) and informal writing (discussion boards, e-mail). These activities, though powerful, need to be contextualized within pedagogical frameworks that function to both promote language acquisition and technology use and development in classrooms.

Pedagogical Advantages and Disadvantages

Research has evidenced the significant increase in student motivation when being engaged in real and meaningful learning activities. Motivation was found to be one of the best predictors of student achievement in language learning. The use of computer technology has the potential to increase language learners' self-esteem, develop vocational skills, and increase language proficiency and overall academic skills (Dunkel 1990). From a sociocultural perspective, CALL provides the tools and environment where students can engage in authentic language learning processes that can be communicative and collaborative. Stephen Krashen (1982) noted that individuals develop language more fully in comfortable environments where input of the target language is comprehensible (i.e., at the learners' proficiency level) and one level above the learners' proficiency level (i.e., the I+1 hypothesis). According to this model, in order to acquire language more holistically and naturally, there should be as many opportunities as possible for more natural and authentic learner interactions that are balanced with more direct instruction or language learning (i.e., overt teaching of grammar and rules).

CALL addresses theories of second language acquisition (SLA) and expands on them by taking into consideration comprehensible input of the target language as well as apperception, the students' ability to notice aspects of the input. For example, CALL activities and materials can provide input not only aurally but also textually and visually. As such, cues need to be included to ensure learners are focusing on the aspects important to comprehending the message. Carol Chapelle (1998) proposed a framework for developing relevant multimedia CALL, which has the following seven criteria that are drawn from the interactionist perspective of SLA:

- prominent significant linguistic characteristics
- modifications of linguistic input provided
- opportunities for both written and spoken comprehensible output
- opportunities for learners to notice errors in their own output
- opportunities for learners to correct their linguistic output
- modified interaction between the learner and the computer supported
- maximum opportunities for students to fully participate in second language learning tasks

The elements identified in this framework contribute to positive language acquisition. In this respect, the potential for CALL is positive when it is contextualized within SLA theory and a social learning environment that includes appropriate tools and support that facilitate language use and production.

In addition, CALL has the potential to create a space for dialogue where all participants have an opportunity to "speak" and the potential to provide input and be engaged in creative activities. These potentials are mostly achieved through computer-mediated communication (CMC), one element of CALL that refers to the use of e-mail and software programs such as the Daedalus Integrated Writing environment. CMC extends the use of technology beyond the classroom to virtual communities, which often cultivates new social relationships and supports "collaborative, meaningful, and cross-cultural human interactions" (Liu et al. 2003, 252). Through interactivity, collaboration, work-sharing, and real-time conferencing, online CALL can also support community-building among diverse groups of learners, which leads to new forms of socialization and acculturation (Warschauer 1997; Zhao 1996) and may foster the creation of "communities of inquiry capable of stimulating intellectual, moral, and educational growth among rich and poor alike" (Cummins and Sayers 1995, ix).

Yet CALL also brings negative effects that are common to many online learning environments, such as lack of human touch, isolated work, lack of equity in accessibility to technology (the so-called digital divide), and the possible growth of antisocial behavior from working in isolation (Pennington 1996). In CMC environments students may wander around and get lost in cyberspace. And as technology is not a sure and consistent tool, technical problems such as unreliable access to the web, support of oral production skills, and student resistance to working on computers are all areas that must be taken into consideration when weighing the positive and negative impacts of CALL (see www.history-of-call.org).

In addition, students' use of computers and related classroom technologies can be constrained or inhibited by sociocultural variables of teachers, such as their views of schools as institutions of social control, one that is managed through the direct instruction of skills. In such instances students feel inhibited to express themselves since they fear they are not meeting requirements, or are overly concerned with learning the technology skill (i.e., keyboarding) and are not fully engaged in the CMC process (Warschauer 1998). The approach of the classroom teacher in introducing the technology, as well as the values and beliefs the teacher holds about how to convey information and skills, the expectations teachers have of students, and the experience of the students in computer and technology skills are all factors that may hinder the potential of the community (Bailey 1996). So even when teachers intend to implement CALL and CMC activities, they are doing so within societal institutions that hold varying expectations for students dependent on their language, ethnicity, and social class. As a result, technologies rarely achieve the transformational effects intended, especially when used with language and ethnic minority students (Warschauer 1997). This is a crucial area in CALL in that its beginnings focused on the accuracy of language use and moved toward fluency and further to communicative activities (Warschauer 2000).

Scholarly Activities in CALL

The Computer Assisted Language Instruction Consortium (CALICO) and the European Association for Computer Assisted Language Learning (EUROCALL) are two of the leading professional organizations in CALL. According to EUROCALL's research policy statement (www.eurocall.org/research/research_policy.htm), scholarly work in CALL includes various activities and initiatives in research, development of materials, tools, and applications, and pedagogical practice.

It is noteworthy that under the influence of sociocultural theory (Vygotsky 1978, 1986), research in CALL has broadened from its earlier focus on the uses of hypermedia and multimedia in teaching a specific language to the uses of the web and virtual learning environments in supporting learner interaction. Online collaboration, communities of practice, and virtual learning communities are part of this, known as collaborative CALL.

Being inherently multidisciplinary, CALL research draws theories from SLA, cognitive science, linguistics, sociology, psychology, cultural studies, communication studies, and computer technology (CALICO 2002). Since the 1990s, research in CALL has begun to form its own theoretical and methodological paradigms. This is marked by CALL's standardized terminology, identified points of reference, and the inclusion of a significant number of scholarly activities (EUROCALL 1999). A well-recognized research method in CALL is to collect data from natural learning settings to test hypotheses generated by language acquisition theory (EUROCALL 1999). Another example is the newest socio-cognitive approaches in CALL research, which underscores the learners' meaningful interaction, negotiation, and knowledge construction in authentic learning environments. The overarching goal of CALL research is to address how technology supports the process of language teaching and learning, which will eventually lead to the development of CALL theory itself.

CALL research shares common methods as used in other educational and humanity research, including qualitative methods (e.g., case study, ethnography, grounded theory) and quantitative methods (e.g., descriptive, correlational, causal-comparative, and experimental). Qualitative methods have been used in studies that describe new learning environments and learner navigation patterns within these environments. Qualitative methods have also been used in studies that examine learner interaction with CALL programs and evaluate the program's instructional and technical design. By contrast, quantitative studies in CALL systematically investigate psycholinguistic and sociolinguistic variables on learning with technology and statistically analyze the effectiveness of instructional strategies (CALICO 1999).

At this juncture we need to consider users (specifically agency and identity) and broader conceptions of literacy beyond skills of SLA toward the ability of learners to be able to read, interpret information, and create knowledge from a variety of sources and to teach English to guide individuals in learning to write e-mail and use the Internet. In this environment "reading cannot be done without making critical decisions at every step of the way" (Warschauer 2000, 63). Thus, the future of CALL and

CMC requires educators to be able to teach from a critical literacy perspective that includes agency and identity as part of the language learning processes.

The Future of CALL

Some up-to-date topics in CALL are intriguing, such as language, identity, and the Internet, language in cyberspace, culturally appropriate interaction, and the Internet and minority language revitalization. The increasing reach of the Internet, which is considered one of the most democratic media recently developed, further complicates these issues. It not only enhances the powers of broadcasting and research but also promotes immediate dialogue between and among individuals and groups (Warschauer, forthcoming). The language used on the Internet, which is noteworthy since the number of non-English websites is growing rapidly, has exemplified this.

In *The Death of Cyberspace and the Rebirth of CALL,* Warschauer (2000, 61) states that "the significance of online communication lies not in its separation from the real world, but rather in how it is impacting nearly every single aspect of the real world." Therefore, as research and experiences with online communications grow, so too will the impact and our perceptions of CALL and CMC. Such technologies further the skills of reading, writing, and communicating, since online technologies and their use will become more important in the twenty-first century. In an examination of the future of information and communication technology, Warschauer (2000) identifies ten areas where information and communication technology will progress in the twenty-first century. Of these ten, the following will have the greatest impact on CALL:

- The emergence of new computer and online devices such as RCFoC Radio! (a "radio" show in three web-based audio-on-demand flavors: RealAudio technology from RealNetworks, ToolVOX from VOXware, and MP3).
- The Internet, which will change from being exclusive to being inclusive as a mass form of communication.
- The movement from text-based to audiovisual communication, such as the growing popularity of home video production.
- Language of the Internet will expand from English to being multilingual.
- Technology use will move from "nonnative" to "native"—with individuals growing up with technology as a native skill.

These expanded uses will contribute to an increase in the use of English as the language for communication in media and commerce and further the expansion in the number of nonnative English speakers' use of technology around the world. It is likely that future developments in the area of technology will have a direct impact on English teaching pedagogy, research, and use. This will result in the creation of new contexts, new literacies, and new identities in the formulation of texts, ideas, and interactions within and outside classroom and school environments. To prepare for this future, we need to go beyond seeing computers and languages as ends in and of themselves. Instead, they can become tools for empowering individuals to look through the lens of critical literacy where reading and writing go beyond the written word, creating a conduit for rewriting the word and the world in order to transform it (Freire and Macedo 1997).

Minjuan Wang
Karen Cadiero-Kaplan

See also Computer-Mediated Communication; WebQuests

References

Bailey, John. 1996. "Teaching about Technology in the Foreign Language Class." *Foreign Language Annals* 29(5): 445–464.

Chapelle, Carol. 1998. "Multimedia CALL: Lessons to Be Learned from Research on Instructed SLA." *Language, Learning, and Technology* 2(1): 22–44.

Cummins, Jim, and Dennis Sayers. 1997. *Brave New Schools: Challenging Cultural Illiteracy through Global Learning Networks.* New York: St. Martin's Press.

Delcloque, Philippe. 2002. "An Illustrated History of Computer-Assisted Language Learning: The History of CALL Web Exhibition." Available online at www.history-of-call.org (November 19, 2002).

Dunkel, Patricia. 1990. "Implications of the CAI Effectiveness Research for Limited English Proficient Learners." *Computers in the Schools* 7(1–2): 31–52.

EUROCALL Research Policy Statement. Joint Policy Statements of EUROCALL, CALICO, and IALLT. Arising from a research seminar at the University of Essen, Germany, April 30–May 1, 1999. Available online at www.eurocall.org/research/research_policy.htm (December 7, 2002).

Freire, Paulo, and Donaldo Macedo. 1987. *Literacy: Reading the Word and the World.* Westport, CT: Bergin and Garvey.

Kern, Richard, and Mark Warschauer. 2000. "Theory and Practice of Network-Based Language Teaching." In Mark Warschauer and Rick Kern (eds.), *Network-Based Language Teaching: Concepts and Practice* (pp. 1–19). New York: Cambridge University Press.

Krashen, Stephen. 1982. *Principles and Practice in Second Language Acquisition.* Oxford: Pergamon.

Liu, Min, et al. 2003. "A Look at the Research on Computer-Based Technology Use in Second Language Learning: A Review of the Literature from 1990–2000." *Journal of Research on Technology in Education* 34(3): 250–273.

Pennington, Martha, ed. 1996. *The Power of CALL.* Houston, TX: Athelstan Publications.

Scholarly Activities in Computer-Assisted Language Learning: Development, Pedagogical Innovations, and Research. Joint Policy Statements of CALICO, EUROCALL, and IALLT. Arising from a research seminar at the University of Essen, Germany, April 30–May 1, 1999. Available online at calico.org/CALL_document.html (October 1, 2002).

Sproull, Lee, and Sara Kiesler. 1991. *Connections: New Ways of Working in the Networked Organization.* Cambridge, MA: MIT Press.

Vygotsky, Lev. 1978. *Mind in Society: The Development of Higher Psychological Processes.* Ed. and trans. M. Cole, V. John-Steiner, S. Scribner, and E. Souberman. Cambridge, MA: Harvard University Press

———. 1986 [1934]. *Thought and Language.* Ed. Alex Kozulin. Cambridge: MIT Press.

Warschauer, Mark. 1997. "Computer-Mediated Collaborative Learning: Theory and Practice." *Modern Language Journal* 81: 470–481.

———. 1998. "Online Learning in Sociocultural Context." *Anthropology and Education Quarterly* 29(1): 68–88.

———. 2000. "The Death of Cyberspace and the Rebirth of CALL." *English Teachers' Journal* 53: 61–67. Available online at www.gse.uci.edu/markw/cyberspace.html (November 2, 2002).

———. 2002. "A Developmental Perspective on Technology in Language Education." *TESOL Quarterly* 36(3): 453–475. Available online at www.gse.uci.edu/markw/developmental.html (November 2, 2002).

———. Forthcoming. "Language, Identity, and the Internet." In B. Kolko, L. Nakamura, and G. Rodman (eds.), *Race in Cyberspace.* New York: Routledge. Available online at www.gse.uci.edu/markw/lang.html (November 30, 2002).

Zhao, Yong. 1996. "Language Learning on the World Wide Web: Toward a Framework of Network Based CALL." *CALICO Journal* 14(1): 37–52.

Computer-Mediated Communication (CMC)

Computer-mediated communication refers to a range of functions in which computers are used to support human communication (Santoro 1995). G. Santoro describes three broad categories of CMC functions. These categories are distinguished by the nature of the human-computer interaction and by the role taken by the computer in mediating the human communication process. Although these categories are listed separately, it is most important to note that they are not mutually exclusive. Generally, these functions are combined to meet educational goals. The three categories are *computer-based conferencing, informatics,* and *computer-assisted instruction.* Computer-based conferencing refers to the computer as an interpersonal communications device and includes e-mail, interactive messaging (e.g., Internet chat, instant messaging), and conferencing systems (e.g., learning management systems such as WebCT®, Blackboard®, and FirstClass®) that may be web-accessed or not. It should be noted that it is becoming harder to distinguish what is web-based or web-accessible. Most conferencing systems, even those that are client-server applications, are becoming accessible via the web. While the primary function of computer conferencing is discussion, the discussion in online instruction is usually text-based and requires a lot of reading and writing on the part of student and instructor. Informatics is the use of the Internet as a giant library with a rapid growth of Internet-accessible resources, including online public-access library catalogs, interactive remote databases, and program/data archive sites. In computer-assisted instruction, the computer functions as a tutor or instructor and is programmed to take a more active role (i.e., it carries out commands, whereas in the two categories described above the computer passively executes human instructions).

In its broadest definition, CMC can mean virtually all instances when a computer is used as a communications device and can support a wide variety of educational formats that can range from simply providing students with e-mail in an otherwise traditional class to actually delivering instruction and supporting student-to-student and student-to-teacher interaction at a distance. For example, an entire course can be web-based, with almost all instructional activity involving the use of a computer. Or an instructor may use e-mail lists to support discussion groups to augment mainly in-person class instruction. Guest lecturers can join any class (online or not) from anywhere via video over the Internet. CMC can facilitate collaboration among students, individual pacing for self-directed learning, and student practice and reflection.

Functions of CMC

Computer conferencing is human communication enabled via the computer. A computer conference can be described as two or more people coming together to talk, using computers as the communications medium and software developed for computer communication. Computer conferencing can be as simple as one person talking to another via e-mail or as complex as people spread around the world communicating in real time. Although people may not be aware of computer conferencing, they may have used it without even being aware. America Online, CompuServe, and Prodigy were among the first commercial information services widely used to enable human contact via computer. In fact, a marketing strategy may include computer conferencing services so customers can meet new people from all over the world without leaving the safety and comfort of home. Talk and chat tools also enable people to interact with others online in real time.

Informatics means the computer as research tool. Informatics can save time and money because people can bring a world library into their homes or workplaces instead of driving to a physical building to conduct research book by book. In informatics the computer, through links to other computers and networks, serves as a virtual library with access to many databases. New additions to this library are continuously being added. A web browser allows one to search for materials throughout this worldwide library. Of course, one needs to be careful regarding quality control, given there are no librarians serving the sorting and authenticating functions on the web.

Besides using search and locator tools, information can also be found on electronic bulletin boards, usenet groups, and mailing lists. Usenet groups are discussion groups and each group is aimed at a particular topic, such as hobbies, cultural information, technical material, or specific interests. By subscribing to these discussion groups, people can read the posted messages, reply to previously discussed topics, and post a message for others to reply to.

Computer-assisted instruction and computer-based training (CAI and CBT) are growing fields rooted in a combination of other disciplines, such as education and computer science. The role of the computer as instructor is evident in the five basic elements common to CAI and CBT. First, the computer directly or indirectly orients the learner to the new insights that will be gained from the program and what is expected of the learner. Second, the computer then presents new information. Interaction is the next component as the student responds to the new information as it occurs throughout the program. Initially this interaction takes the form of prompting and cueing the learner and gradually moves to more complex questioning. Feedback, the fourth factor, is immediately provided to the learner for each response that she inputs. Finally, some form of assessment is incorporated into CAI/CBT programs to test the learner's level of mastery of the content presented. Just as a teacher in a traditional classroom instructs and evaluates students, CAI/CBT programs are developed to orient students to the program content, to present information, to provide ample opportunities for the learners to interact with the material, to acknowledge student input, and to ultimately test for comprehension and application (Zellhofer, Collins, and Berge 1998).

However, CAI and CBT differ from traditional classroom instruction because the learner is able to control her own pace through the material. If one topic is more difficult than another, the learner can take as much time as she needs to grasp the content before moving on to the next topic. CAI and CBT also vary from the traditional classroom because the learner is given the control of choosing the sequencing of the information. CAI and CBT programs can be linear, modular, or hypermedia. In the two latter cases, the learner is able to select the content of most interest to her, complete that segment of the program, then choose the next topic to be covered. Individual pacing and information sequencing are rarely choices offered to students in traditional classrooms, but they are the norm for CAI and CBT.

CAI/CBT, computer conferencing, and informatics are distinctive fields in their own right, but they all fall in the general domain of computer-mediated communication. The wide scope of CMC unifies online communication, information, and instruction. A person can communicate with another person, groups of people, or databases through the infrastructure of system networks designed to send and receive information and data that enable people to communicate online—the true purpose of computer-mediated communication.

Limitations of CMC

Of course, CMC is not the educational silver bullet we sometimes seek. Limitations include the suitability of some learning domains for CMC, technology accessibility issues, technophobia and novice users, a lack of writing skills by some persons, and lack of face-to-face contact.

Although CMC is an enabling medium for human interaction, research, and instruction, it is not practical for all purposes in education and training. First, because it is currently text-based, CMC is weak in addressing motor skills and tactile learners. CMC is a good instructional vehicle for dealing with writing, analysis, problem-solving, interests, and attitudes. Audio and video capabilities are quickly becoming economical for use in CMC and allow CMC to move beyond text-based modalities.

Another limitation of CMC is accessibility to hardware and software. To interact electronically, a computer, modem, telephone line or cable, and communications software are needed. However, if people do not have access to hardware and software at the workplace, home, educational institution, local library, or cyber café, CMC cannot take place. For those who do have the necessary hardware and software at home, there is the problem of constantly upgrading hardware and software and of ensuring that new add-ons and upgrades are compatible with existing hardware and software. And though the prices for new technologies are falling, add-ons and upgrades for better communication can still be expensive. Furthermore, significant bandwidth issues will arise as video and audio formats increasingly support CMC.

Some people are fearful of even touching a keyboard, much less trying to work through a software program. Such fears are a barrier to becoming proficient in using CMC and require individuals to confront their hesitancy to learn the minimum skills for computer use. Ideally, they will become comfortable with using the technology and become confident enough to learn more and develop their new skills. Novice computer users may also experience bewilderment and confusion when confronted with the vast realm of CMC. Fortunately, the hardware and software needed for CMC are becoming easier to use as new versions of software and new systems are developed for the mass consumer market.

Good typing skills are essential in CMC because it is currently a text-based medium. Novice users who have not acquired the necessary keyboarding and writing skills may become frustrated with not being able to communicate expressively with others. In everyday face-to-face contact, people rely on body language, facial expressions, and language inflection to send and interpret messages. These social communication cues are nonexistent in CMC, so writing ability becomes increasingly important.

CMC is a great medium for reaching out to others, but its aim is not to replace personal human contact. CMC will never, and should not, replace human interaction. Human contact is necessary and important, especially when people are confronted at practically every turn with a bombardment of information. In some cases, such as the resolution of conflict, face-to-face interaction is crucial. However, CMC is able to conveniently bring people together who, by any other means, would never have the chance to interact.

Advantages of Using CMC

Although there are limitations, CMC offers myriad benefits that can enhance, improve, and support education and training. As mentioned above, one advantage of CMC is the abundance of online information and instructional content. CMC makes it possible to access vast amounts of information and resources germane to the education and training sectors. Instructors can communicate with their colleagues about online instruction, download useful files for classes from remote databases, and post lessons and tutorials online. Similarly, students can form virtual study groups with classmates and research assignments utilizing Internet resources for class assignments. CMC is a great asset for classes that have a strong "read about, talk about, think about, or write about" component.

CMC in education and training makes the teacher's physical presence and setting invisible and places the responsibility of learning on learners. CMC encourages the use of a variety of instructional methods where the instructor acts more as a facilitator rather than as a didactic authority figure. CMC provokes changes in the roles of instructor and students (see, e.g., Berge 2000).

The learner-centered approach is typical of CMC courses, prompting teachers to think about instructional design in innovative ways. Traditional lesson plans may not be suitable for online instruction. Besides choosing appropriate instructional activities that will cater to the different needs of the students, the instructor must also take into consideration staffing, class schedule, budget, varied learning styles, and the new frontier of the online environment (Berge and Collins 1995a, 1995b; Eastmond and Ziegahn 1995; Nalley 1995). Although some teachers may view this as more work, the benefit is that instructors are again asked to consciously examine a process that may have become habitual in order to ensure effective instruction. There are a variety of issues, such as suitable online content, varied ways of learning, instructional methods, and interactive activities that best address the content, the technology itself, and time, to be considered when designing online instruction.

Interactions in an in-person classroom environment may be limited due to distractions, fear of ridicule, insufficient time for all students to interact, or lack of encouragement. Students of classes delivered through CMC are more responsive to interacting with others because the pressures of the traditional classroom have been removed. Students can take the time to gather their thoughts to best reply to messages, and with this extra reflection time they can be confident of the remarks they submit. Interaction in online courses is often mandated; to be silent in a computer conference is to become invisible.

Online class discussions can be reviewed, or even printed out, to serve as notes and memory refreshers for students. Also, the instructor can use electronic text documents and archives for record-keeping purposes to track all class sessions, related problems, outcomes, and so on. This serves to make all parties more accountable and to avoid he said/she said scenarios. This is a valuable way of monitoring and pacing student progress and improving upon it in the future.

CMC overcomes geographic barriers. People do not have to be physically present at the same location in order to communicate. Whether at the workplace, home, or school, people are able to log on to networks to interact online with someone in the neighborhood or continents away. This is especially beneficial to people with physical disabilities or those in remote locations who would otherwise not have the chance for such interaction. CMC can bring people from around the world together to interact.

Besides eliminating geographic barriers, the dual temporal nature of CMC bridges time barriers. Interactions via CMC can either be synchronous or asynchronous. Synchronous communication is similar to in-person exchanges where all parties participating in the conversation are present (online) and provide immediate input and feedback to messages that appear on the computer screen in real-time. Asynchronous communication means leaving messages to be read by others later. Asynchronous communication (e.g., web-conferencing) allows participants to "go to class" when it is convenient. These synchronous and asynchronous aspects of CMC give communicators great flexibility.

No social group with computer access is excluded from taking part in communicating electronically with others because CMC is able to overcome space, time, and social boundaries that could not be bridged in the past. Social status is established by communication content and style, as well as the free sharing of information, thoughts, ideas and opinions—all physical attributes are rendered invisible.

Finally, CMC addresses demographics and lifestyle changes. Whereas computers were seen as complex technological tools in past years, they are now used by "average" people for entertainment and educational benefits. People's expectations of technology have changed with the coming of the information age.

Implications for Education

The training and industry sectors are experiencing changes in how business is conducted, and these changes will continue to evolve as CMC expands (Berge 2001; Schreiber and Berge 1998). In order to stay competitive, reach a larger portion of the student population, and meet the widest variety of student needs, such as flexible scheduling and the demand for greater selection of course offerings, educational institutions are being pushed to implement CMC in their programs. Online teaching is part of the larger context of technology-mediated learning and an educational framework for transforming students into self-directed, lifelong learners. The main teaching styles used by online teachers are discussion, collaboration, authentic learning activities, and self-reflection/self-assessment. Teachers value these purposes and methods. The online classroom CMC, with its tripartite functions, can permit and promote efficient and effective learning and serve as a catalyst for improving education.

Zane L. Berge

See also Collaboration and Negotiation Tool for Case-Based Instruction; Collaborative Technologies; Communities of Practice; Computer-Assisted Language Learning; Computer-Supported Collaborative Learning; Courseware; Web-Based Course-Management Systems

References

Berge, Zane L. 2000. "New Roles for Learners and Teachers in Online Higher Education." In G. Hart (ed.), *Readings and Resources in Global Online Education* (pp. 3–9). Australia: Whirligig Press.

———. 2001. *Sustaining Distance Training: Integrating Learning Technologies into the Fabric of the Enterprise.* San Francisco: Jossey-Bass.

Berge, Zane, and Mauri Collins. 1995a. "Introduction." In Z. Berge and M. Collins (eds.), *Computer-Mediated Communication and the Online Classroom, Volume 1: Overview and Perspectives* (pp. 1–10). Cresskill, NJ: Hampton Press.

———. 1995b. "Introduction." In Z. Berge and M. Collins (eds.), *Computer-Mediated Communication and the Online Classroom, Volume 3: Distance Learning* (pp. 1–12). Cresskill, NJ: Hampton Press.

Eastmond, D., and L. Ziegahn. 1995. "Instructional Design for the Online Classroom." In Z. Berge and M. Collins (eds.), *Computer-Mediated Communication and the Online Classroom in Distance Education,* vol. 3 (pp. 59–80). Cresskill, NJ: Hampton Press.

Nalley, R. 1995. "Designing Computer-Mediated Conferencing into Instruction." In Z. Berge and M. Collins (eds.), *Computer-Mediated Communication and the Online Classroom in Higher Education,* vol. 2 (pp. 11–23). Cresskill, NJ: Hampton Press.

Santoro, Gerald M. 1995. "What Is Computer-mediated Communication?" In Z. Berge and M. Collins (eds.), *Computer-Mediated Communication and the Online Classroom, Volume 1: Overview and Perspectives* (pp. 11–21). Cresskill, NJ: Hampton Press.

Schreiber, Deborah A., and Zane L. Berge, eds. 1998. *Distance Training: How Innovative Organizations Are Using Technology to Maximize Learning and Meet Business Objectives.* San Francisco: Jossey-Bass.

Zellhofer, Sandy, M. P. Collins, and Zane L. Berge. 1998. "Why Use Computer-mediated Communication?" In Z. L. Berge and M. P. Collins (eds.), *Wired Together: Computer-Mediated Communication in K-12, Volume 2: Case Studies* (pp. 1–16). Cresskill, NJ: Hampton Press.

Computer-Supported Collaborative Learning (CSCL)

Computer-supported collaborative learning means the use of computer-based networks to support students learning together in an effective and efficient manner. CSCL was developed based on the broader concept of computer-supported collaborative work (CSCW). CSCW provides support for group work via shared interfaces on a computer-based network. Both CSCW and CSCL are designed to supplement, rather than replace, face-to-face interactions, and both are designed to support groups working together for a common reason. CSCW and CSCL facilitate communication, information access, and collaborative problem-solving. The main distinction is that CSCW is primarily used in business and industry, whereas CSCL is used in educational settings.

Several major theories covered in other entries in this encyclopedia have contributed to the development of CSCL, including constructivism, sociocultural theory, problem-based learning, situated cognition, active learning, cognitive apprenticeships, and cognitive flexibility theory. All these theories share common themes, including the notion that learning should be active, purposeful, collaborative, and authentic. CSCL uses technology as a tool to support these notions and to function as a collaborator in the learning process by offloading some of the cognitive demands typically placed on learners. When technology is responsible for organizing, storing, and retrieving information and modeling various concepts and procedures, learners have more of an opportunity to focus on in-depth learning that requires higher-level thinking skills such as reflection and analysis.

Examples of CSCL

Collaborative Learning Environments Online (CLEO) supports inquiry and collaboration in science and mathematics via the web. Students use authentic data to solve research questions using real data. The online environment provides tools that enable students to share, analyze, and discuss results with geographically disparate peers and provides a clearinghouse of completed projects that can be used by others. Each project in CLEO supports authentic scientific investigations by highlighting a research question that drives the investigation, information about the procedures and tools used to gather data, authentic data shown in a variety of formats including graphs and charts, and a conclusion that leads to new areas of research.

The International Clean Air Project is one example of an authentic investigation supported by CLEO. In this project students from around the world explore issues related to what citizens can do to reduce air pollution in their community and what health issues related to air pollution should be brought to the attention of all citizens in the community. Through this project students are able to compare and contrast communities, share suggestions for dealing with air pollution, and become aware of issues that influence local and global economics and politics.

Computer Supported Intentional Learning Environments (CSILE) was developed by Marlene Scardamalia and Carl Bereiter at the Ontario Institute for Studies in Education. This collaborative learning environment has graphical, multimedia, and text-based capabilities that enable students to generate nodes with relevant information related to the topic being studied. These nodes then become avenues for other students to analyze, comment upon, and add to, thus building a community-based center of knowledge.

CSILE has been integrated into the K-12 curriculum across a variety of disciplines. One example of its use is the DIG Project in which two upper elementary classes studied ancient civilizations through interdisciplinary, project-based learning. Through the use of CSILE students created their own civilization based on cultural universals they studied in their classrooms such as housing, language, and government. CSILE enabled students to make complex connections among ideas and nodes and facilitated a deep level of collaboration among students working toward a similar goal. Students also had an opportunity to practice their writing skills in an authentic environment, to supplement this writing with graphics, and to respond to the writing of other students. Higher-level thinking skills were developed because students needed to ensure that the cultural universals created for this civilization fit with the group's original goals and to collaboratively decide whether each node or idea was appropriate. Each student had an individual responsibility to the group's final goal.

Kara Dawson

See also Collaborative Technologies; Computer-Mediated Communication; CSILE/Knowledge Forum®

References

Koschmann, Timothy, ed. 1996. *CSCL: Theory and Practice of an Emerging Paradigm.* Hillsdale, NJ: Lawrence Erlbaum Associates.

Cone of Experience

Introduced by Edgar Dale (1946) in his textbook on audiovisual methods in teaching, the Cone of Experience is a visual device meant to summarize Dale's classification system for the varied types of mediated learning experiences. The organizing principle of the Cone was a progression from most concrete experiences (at the bottom of the Cone) to most abstract (at the top). The original labels for Dale's ten categories are: Direct, Purposeful Experiences; Contrived Experiences; Dramatic Participation; Demonstrations; Field Trips; Exhibits; Motion Pictures; Radio—Recordings—Still Pictures; Visual Symbols; and Verbal Symbols. (See Figure 1.)

Dale made minor modifications of the visual in the second edition (1954), changing Dramatic Participation to Dramatized Experiences and adding Television. By the third edition of the textbook, Dale (1969) acknowledged the growing popularity of Jerome Bruner's (1966) cognitive psychology concepts by overlaying Bruner's classification system for modes of learning—enactive, iconic, and symbolic—on top of his own categories. This adaptation of his own schema may have been portentous, perhaps giving implied license to others to make other creative adaptations and interpretations, not always to the credit of Dale's original notion.

Application of the Construct

Dale's textbook in its three editions remained popular for over a quarter-century. Inasmuch as the Cone provided the organizing principle for the book, it became ingrained in the thinking of generations of educational technology students and professors who used the textbook. It stimulated many efforts to extend the original idea by developing its implications for elementary education, secondary education, adult education, corporate training, and even counseling.

As a visible leader in audiovisual education, Dale and his work had a great deal of authority within the field, and the Cone may be regarded as the earliest highly influential conceptual schema in the field. In his landmark work on visual learning, F. M. Dwyer (1978) credits Dale as one of the thinkers who inspired the visual education movement in the 1940s and 1950s

Dale's own claims for this classification system were modest and qualified. He advised against viewing the categories as rigid, inflexible divisions (1946). He insisted that the classifications should not be regarded as any sort of hierarchy or rank order (1946). This addresses one of the most

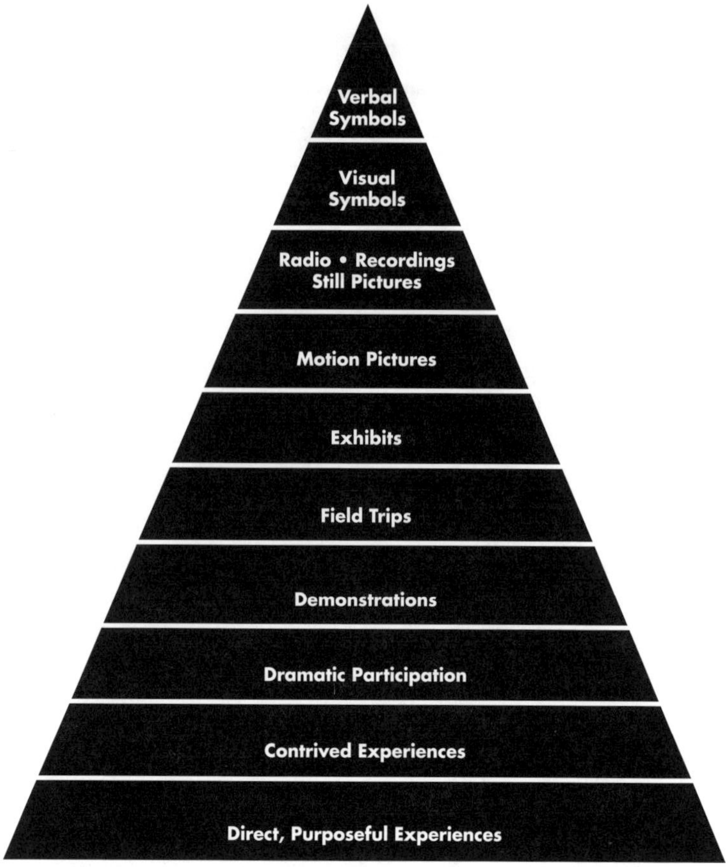

FIGURE 1: Dale's Cone of Experience (first edition)

SOURCE: From *Audio-Visual Methods in Teaching,* 1d. by Dale. © 1969. Reprinted with permission of Wadsworth, a division of Thomson Learning: http://thomsonrights.com.

prevalent misconceptions of the Cone: that the progression from concrete to abstract represented a value judgment about concrete over abstract learning activities. Instead, Dale advocated the use of whatever methods or media were appropriate for the learner and the task, acknowledging that words can be a powerful and efficient means of conveying ideas even for the youngest children. If he had a bias regarding media it was toward rich combinations of concrete and abstract experiences: "Abstractions must be combined, if we are to have rich, full, deep, and broad experience and understanding. In brief, we ought to use all the ways of experiencing that we can" (Dale 1946, 48).

Because many of those who referred to the Cone were advocates for specific media or for audiovisual media in general, they had a tendency to selectively emphasize those parts of Dale's work that supported their claims. Thus by the time of the third edition of *Audiovisual Methods in Teaching* Dale found it necessary to devote six pages of the chapter on the Cone to

"Some Possible Misconceptions" (1969, 128–134). At the core of the misconceptions are the notions that the value of an activity increases with its realism and that the learner's understanding grows by beginning with direct experience and progressing to increasingly abstract activities.

One explanation for the prevalence of other interpretations of the Cone is that Dale did not explicitly draw the distinction between a descriptive construct and a prescriptive theory. He surely intended the Cone to be descriptive—a classification system—and not prescriptive—a road map for lesson planning. He came close to drawing this distinction when he stated in the summary of his chapter on the Cone: "The cone, of course, is merely an aid to understanding this subject . . . something to help explain the relationship of the various types of sensory materials" (1946, 52). The key words are "understand" and "explain," which indicate a descriptive purpose, not a prescriptive one.

Yet Dale himself sometimes fell prey to the urge to extend the descriptive construct to prescriptions, as pointed out by D. P. Subramony (forthcoming). References to "uses" or "implications" of the Cone are scattered throughout the various editions of Dale's textbook (Dale 1946, 1954, 1969). An example found in the third edition states, "When properly understood and used, however, the Cone can be a helpful and practical guide" (1969, 110). With this sort of ambiguity from the author, it is not surprising that many of his followers attempted to use the Cone as a prescriptive guide to lesson planning.

Origins of the Cone's Concepts

Ideas parallel to those expressed by Dale in the Cone of Experience appeared in the literature of education prior to 1946. Paul Saettler (1990), a historian of the field of instructional technology, points to *Exposition and Illustration in Teaching,* published in 1910 by John Adams, "which included the following 'order of merit' concerning concreteness: '(1) the real object, for which anything else is a more or less inefficient substitute; (2) a model of the real object; (3) a diagram dealing with some of the aspects of the object; and (4) a mere verbal description of the object.'" However, a more direct ancestor of the Cone is probably another diagram developed by Hoban, Hoban, and Zisman (1937). They made the conceptual breakthrough of constructing a graph in which visual media are arranged along the *y* axis, while the learner's level of development—from the concrete level of thinking to the abstract level of thinking—is arrayed along the *x* axis. In applying the graph to a particular case, one would locate the learner's current level of conceptual development (concrete to abstract), then trace up to the slope line and then horizontally over to the visual medium that intersects at the same point. For example, an experienced learner with a highly developed (abstract) knowledge of jet propulsion would be expected to be able to learn more about jet propulsion effectively with diagrams and verbal texts.

The categories were: total situation, objects, models, films, stereographs, slides, flat pictures, maps, diagrams, and words. Dale's schema differs mainly in the addition of several classes of media and active learning experiences and the simplification of the schema by showing only the *y* axis—the media, indicating the other dimension (concrete-abstract) by the pyramidal shape of the cone. Although Dale's schema appears to be quite derivative of this, he does not explicitly acknowledge this source, although he makes several references to the work elsewhere in his textbook.

Misappropriation of the Cone

It is important to discuss what the Cone is *not* as well as what it is because of a widespread misrepresentation that has become ubiquitous in recent years. At some point someone conflated Dale's Cone of Experience with a spurious chart that purports to show what percentage of information people remember under different learning conditions. The original version of this chart has been traced to the Socony-Vacuum Oil Company in the 1960s, according to Dwyer (1978). Charles Cyrus, a training specialist at the University of Texas, indicates (letter to Michael H. Callahan, Department of the Navy, November 27, 1963) that the "people generally remember" data were disseminated by the Division of Extension, University of Texas, in the 1950s and may have reached oil company trainers during that period. Cyrus traces the chart back to Paul John Phillips, who brought the data with him after serving as head of the Training Methods branch of the U.S. Army's Ordnance School at the Aberdeen (Maryland) Proving Grounds. Phillips held this position and the rank of lieutenant colonel from 1940 to 1943. He later claimed that during this period his unit conducted research on training that supported these findings, although no documentation of that research is available.

As Dwyer points out, the reported percentages are impossible to interpret or verify without specifying at least the method of measurement, the age of the learners, the type of learning task, and the content being remembered. Despite the lack of credibility, this formulation is widely quoted, usually without attribution, and in recent years has become repeatedly conflated with Dale's Cone, with the percentage statements superimposed on the cone, replacing or supplementing Dale's original categories. The examples are too numerous to document here but are discussed in detail and with citations in Subramony (forthcoming).

The Cone of Experience is essentially a visual metaphor for the idea that learning activities can be placed in broad categories based on the extent to which they convey the concrete referents of real-life experiences. Although it has sometimes been interpreted as advocating the selection of certain media and methods over others, or as a prescriptive formula for selecting instructional media, such was not Dale's stated intent. Dale's own explanations are nebulous enough to support a wide variety of interpretations regarding its applications. Finally, there is the contemporary problem of the conflation of the Cone with the unverified "people generally remember" ordnance school percentages. Nevertheless, the fact that the Cone has been used in so many ways testifies to the robustness and attractiveness of Dale's visual metaphor.

Michael Molenda

See also Bruner, Jerome S.; Instructional Technology; Research on Media and Learning; Visual Literacy

References

Bruner, Jerome S. 1966. *Toward a Theory of Instruction.* Cambridge, MA: Belknap Press of Harvard University Press.

Dale, Edgar. 1946. *Audio-Visual Methods in Teaching.* New York: Dryden Press.

———. 1954. *Audio-Visual Methods in Teaching,* rev. ed. New York: A Holt-Dryden Book, Henry Holt and Company.

———. 1969. *Audiovisual Methods in Teaching,* 3rd ed. New York: Dryden Press and Holt, Rinehart, and Winston.

Dwyer, Francis M. 1978. *Strategies for Improving Visual Learning.* State College, PA: Learning Services.

Hoban, Charles F., C. F. Hoban Jr., and S. B. Zisman. 1937. *Visualizing the Curriculum.* New York: Cordon Company.

Saettler, Paul. 1990. *The Evolution of American Educational Technology.* Englewood, CO: Libraries Unlimited.

Subramony, Deepak P. Forthcoming. "Dale's Cone Revisited: Critically Examining the Misapplication of a Nebulous Theory to Guide Practice." *Educational Technology* 43.

Constructivism

Constructivism represents one set of assumptions about how people learn, that is, how they come to know. Many people think of constructivism as a teaching method, but it is not. Rather, it provides a lens through which we interpret any teaching-learning process. In the constructivist view, learning is not the receipt of information; teaching is not the transmission of information to the student. Rather, constructivism is based on the fundamental assumption that learning is in the activity of the learner. Learning is in the doing, and learning is a process of making sense of the world.

The importance of the constructivist view can perhaps best be appreciated through the contrast to more traditional views of learning. The traditional view that guides most teaching today is what is known as the information processing model, or more generally the "acquisition" view of learning. In this view, knowledge consists of symbols (concepts, procedures, etc.) and the ability to manipulate those symbols. Learning, then, is the acquisition of these symbols, whereas "instruction" involves finding the most efficient means of facilitating this acquisition. For example, it is common in algebra classes for the teacher to stand at the front of the room and provide students with formulas and numbers to plug into the formulas—often providing mnemonics and acronyms to help the student remember what to do next. While this leads to proficiency with calculation, it does not help the student understand the world of math.

The contrast is clear. In the constructivist view, as compared to the acquisition model:

- Knowing about is always contextualized or situated, not abstract.
- Knowing about is constructed through the individual interacting in a situation, not objectively defined.
- Knowing about is a "best interpretation" of a situation, not a truth.

Table 1 provides a further summary of the differences in views. Notice in the table that both views can discuss the teacher as coach—what is important is what it means to coach. In the acquisition model, the coach is transferring expertise with a goal of making the student like the coach. However, in the constructivist view, the coach is there to develop the student's potential; thus the coach is successful when she can converse with the student and respect the student's views (and the student's ability to defend those views).

Also notice that "discovery" is listed under the acquisition model. Although the original notion of discovery learning was very much a constructive activity on the part of the learner, in the most typical use of discovery learning (and in the literal meaning of the term), the teacher guides the student to "discover" what is already known. It is much like turning over a rock and discovering something new. This is different from the constructive, inventive process in the constructivist view. Think, for instance, of the difference in learning that results when a student sits through a lecture on strip mining and its devastating effects on the land as compared to the learning that can come from a project in which students are asked by their state senator to advise him on what he should do about strip mining in the state. In the second activity, the students will be active researchers, learning the benefits and drawbacks of strip mining. They will likely look at environmental, economic, and social impacts and develop a deeper understanding of the systemic nature of real-world issues. They will be able to use the Internet to locate information and contact experts as well as

TABLE 1: Contrasting Views of Learning

Constructivism View	Acquisition View
Learning is:	Learning is:
• organic	• cumulative
• continual reorganization	• discovery (finding what is known)
• invention	
Knowledge is:	Knowledge is:
• a construction	• an acquisition
Coach-apprentice relation is:	Coach-apprentice relation is:
• mutual respect for views	• transfer of coach's expertise
• ability to converse	
Assessment is:	Assessment is:
• ability to use knowledge	• mastery of the content

SOURCE: Created by the authors.

their school's local resources. However, in the information delivery model of the first example, they will memorize a few facts about strip mines and move on to the next topic. They will not learn that they can be independent learners, and they will not learn how to work with messy problems—a critical skill in the work world that is often not developed in school.

The role of the teacher is to support the learning process. Indeed, we talk about creating learning environments and scaffolding learning (providing supports that can later be removed, much like training wheels) because we see the efforts of teaching as being invested in supporting the student's construction of understanding. Of course this does not mean the student can study anything he wants. A central role of the teacher is to engage the student in particular issues, topics, and subject matter (i.e., engage the students with the curriculum). Furthermore, a student's constructed understanding is not acceptable simply because it is what he thinks or believes. Rather, the student must be able to test his understanding against known evidence and against the alternative perspectives. The constructed understanding must be rich and robust; it must be defensible.

Constructivists ascribe to three key principles about learning. First, *learning is situated.* What a student learns depends on the goals of the learner, the task set for the learner, the resources available, the expectations that are set, and so on. In essence, all aspects of the situation impact the nature of what the student learns. It is for this reason that we argue for authentic learning environments—environments that are consistent with the contexts we expect the student to be able to work in after the course is over. It is reasonable to think of learning as creating practice fields for the students.

Second, *learning is goal-driven.* Learning does not occur when an individual sees no benefit. Similarly, learning does not occur when the individual understands everything already. This seems obvious, but it is important. Learning is driven by an individual's need to understand and achieve some end. That goal—the learner's goal, not the goal specified by the teacher or text—determines what is learned. (The third principle, that learning is social, is discussed below.)

The implication, of course, is that it is essential for students to be engaged in the inquiry experience that is set for the class. There are three components to this engagement. The student must: (1) see the problem as important and personally relevant; (2) feel that her action is of value and not just an exercise; and (3) have decisionmaking responsibility.

What people do as "teachers" is seek to engage the student in establishing goals consistent with the outcomes we desire. Therefore, the learning problem must be presented in a context, with supporting discussion and documentation, that brings the student to see the value of working on the problem for its own sake—not simply because it is a course assignment. Without this engagement, even the richest problems offer no more learning potential than a textbook.

Once the student sees the value of the inquiry, the next hurdle is maintaining the student's interest by emphasizing that this is not just a classroom exercise. There must be some way for the student to demonstrate the understandings that have developed. Thus the outcomes must include a performance (a letter, presentation, report, video, product, etc.) that allows students to demonstrate their understanding outside the classroom. This could be for parents, but it could also involve a report delivered to Congress, the mayor, the principal, or some other community figure.

Finally, students must feel they have ownership of the inquiry process and the decisionmaking. The teacher plays the role of coach, but if the process is dictated and only particular outcomes are acceptable, students will reject the inquiry as an exercise. It is not the answer they come to but rather their ability to defend the answer (i.e., the critical thinking) that is important. An unacceptable answer or outcome should only be one that the student cannot defend in relation to alternative perspectives. Consider again the strip-mining example. In the science classroom where this problem was used, the teacher may have expected that students would develop an understanding that strip mines are destructive and should be closed down. What really happened, though, was that students saw the positive economic impact in the form of high-paying jobs for people in their town—perhaps even in their own family—in an area where good jobs were hard to find. Therefore, they concluded that the benefits of strip mining outweighed the consequences. This was sound logic based on well-researched facts; it was an acceptable answer.

The third principle is that *learning is social.* Interacting with others is one of the most efficient ways to test one's views. It provides an opportunity to articulate the views and to hear evidence from others that supports or contradicts your views as well as to hear alternative positions. It is the ability to reconcile these differences that is critical. The implication for this is that the constructivist learning environment emphasizes the collaborative aspect of learning. Collaboration can help to ensure that students share responsibility and seek to support each other. There is a considerable literature on the development of collaborative teams to guide this work. But in the end each individual must be able to defend the outcome of the work, providing rationale and answering questions. Thus while there is collaborative work, there also continues to be individual responsibility.

The constructivist framework has a long history. One researcher (von Glasersfeld 1989) attributes the first constructivist theory to Italian philosopher Giambattista Vico in the early eighteenth century: "One of Vico's basic ideas was that epistemic agents can know nothing but the cognitive structures they themselves have put together . . . 'to know' means *to know how to make*" (von Glasersfeld 1989, 123).

John Dewey was perhaps the greatest proponent of situated learning and learning by doing. Dewey reacted against the traditional educational framework of memorization and recitation and argued that "education is not preparation for life, it is life itself." Most important, learning was organized around the individual rather than around subject-matter topics and predetermined organizations of domains. Dewey emphasized perturbations of the individual's understanding as the stimulus for learning (Roschelle 1992). In essence, the learner's interest in an issue had to be aroused, and learning was then organized around the learner's active effort to resolve that issue.

Jerome Bruner brought the constructivist view back to the forefront in the mid-1960s. He popularized discovery learning, an approach in which teachers support students as they seek to understand issues that are personally and socially relevant. In Bruner's framework, the active struggling by the learner with issues *is* learning. In his best known curriculum effort, a study of anthropology for fifth-graders, Bruner began the curriculum with the unknown—the study of baboon communities and the Nestik Eskimos—as a means of stimulating the child's curiosity. Students were asked to relate the use of tools, language, social organization, and so on to their familiar culture (family, school, etc.).

Arguably, Lauren Resnick's (1987) presidential address to the American Educational Research Association was a primary stimulus for the current resurgence of interest in constructivism. In that address, she contrasted the way we learn out of school to the way we learn in school, emphasizing that outside of school learning is situated and serves as a tool and that it is collaborative. Others (Brown, Collins, and Duguid 1989) expanded upon these ideas and argued that learning and doing are one and the same.

Constructivist Teaching Methods

The constructivist view argues that "teaching" methods should revolve around creating inquiry-based environments that will engage the students in using the concepts, principles, and procedures. All learning involves making sense of the world (i.e., it is inquiry-driven). Thus the course curriculum and the teacher's efforts with the student must be geared toward engaging the students' inquiry in such a way that they are involved in authentic uses of the relevant concepts, principles, and procedures. The students must come to own the problem and the inquiry.

There are many inquiry-based methods to choose from. Problem-based learning involves creating problems that will engage the students (and engage them with the relevant issues), then having students work as teams in cycles of collaborative analysis and self-directed learning. That is, the team is given a problem to solve such as "Should the United States change its immigration policy?" The team members consider the problem, brainstorm, analyze, and define learning issues. They then gather evidence and do analyses related to those learning issues. This cycle is repeated as they progress in their thinking—and their understanding of the issues. Project-based learning involves a similar process, though the cycles of collaboration and self-directed learning may not be so well defined.

Service learning and experiential learning are two alternative strategies that also involve inquiry environments where students are learning through doing. These are less structured than problem-based learning, and the problems are more emergent based on the particular situation. As such, students naturally take ownership of the problems. Here, part of the teacher's role is somewhat different, focusing on helping to define a problem and then supporting the inquiry process.

The Jasper Woodbury series, focusing on mathematics, is perhaps one of the best-known sets of inquiry materials developed to support constructivist teaching strategies. A Jasper adventure is presented on videodisc. The adventure presents a lot of data situated in natural settings; then a problem is posed, and the student must use that data in working on the problem. For example, one problem had to do with developing a business plan for running a dunking machine at the fall festival. Students had to determine how many students would participate and each of several different ticket prices (using survey data) and then take into account the cost, including evaluating alternative strategies for acquiring needed resources (e.g., water).

One of the most effective strategies for supporting inquiry is to let the inquiry evolve from the students' own interests in the subject matter. It has been demonstrated (Bereiter, Scardamalia, Cassells, and Hewitt 1997; Cohen and Scardamalia, 1998; Scardamalia et al., 1992; Scardamalia and Bereiter, 1991) that even young students have interesting questions that are relevant about subject areas. They have developed a distributed, computer-based discussion tool (in this case, CSILE) and methodology to support students' natural collaborative inquiry in the classroom.

In the constructivist view, the teacher plays a critical role in facilitating inquiry-based learning environments. First, the teacher engages the student in inquiry in the domain. There are many ways to do this, but all are built on two key foundations: thinking of linkages between the subject matter and the real world—ways in which the students can be authentically engaged in using the information; and engaging students in the issues so they take ownership. Second, the teacher serves as a coach. The expertise the coach uses is as an expert inquirer—not expertise as a subject-matter expert. The coach needs to: monitor the students' discussion so she understands their thinking; ask questions the students should be asking themselves given their current state of understanding and thinking (i.e., reflect expertise in inquiry); and promote reflection, which is a critical activity for students and one that the teacher generally must facilitate.

Reflection should occur regularly throughout the inquiry process. Summarizing or synthesizing the conversation or work thus far is one strategy for reflecting. It helps to highlight what we consider important. As the end of an inquiry it is important to pause and ask: What did I learn? Where do I need to improve my understanding? Where else does all of this apply? Without this reflective activity, we rapidly forget the key learning issues. Reflection serves as a mean for connecting experiences, tying the experiences to the students' lives outside the classroom and expanding student understanding. It enhances student metacognition and communication abilities; as an added bonus, reflection also provides the teacher with a way of gaining an understanding of where individual students are in their own development.

The Constructivist Learning Experience

Constructivist learning environments are collaborative: Students learn from each other through sharing different experiences, refining understandings, and questioning each other. These environments are also resource-rich. The learning activities that happen in constructivist environments require students to have access to a wide variety of materials, from books and magazines to experts. One key to a successful curriculum can be the use of technology as a resource.

Technology can support the constructivist environment not only as an informational resource but also as a communication and thinking resource. A classroom with an Internet connection can communicate with people worldwide, thus broadening the social context for learning. Furthermore, computers offer tools that help students collect and organize information in ways that are meaningful to them and that can prompt later reflection. Spreadsheets and databases allow students or teams to create sets of relevant information that can be easily accessed. Presentation software and word processors allow students to create professional-level output. Tools such as dynamic geometry software and three-dimensional modeling software allow students to visually see math and explore their evolving understanding of it or create their own version of the solar system that reacts to gravitational pull. Technology can also help students engage in rich problem-solving situations because of the availability of resources. These tools go far in the effort to create rich learning experiences for students by enabling simulations and modeling authentic learning environments (Jonassen 2000; Barab, Hay, and Duffy 1998).

Thomas Duffy
Chandra Orrill

See also Bruner, Jerome S.; Cognitive Psychology; Collaborative Technologies; Communities of Practice; CSILE/Knowledge Forum; Jasper Woodbury; Learner-Centered Environment; Mindtools; Problem-Based Learning; Situated Cognition; Vygotsky, Lev

References

Barab, Sasha, and Thomas M. Duffy. 2000. "From Practice Fields to Communities of Practice." In D. Jonassen and S. Land (eds.), *Theoretical Foundations of Learning Environments.* Mahwah, NJ: Lawrence Erlbaum Associates.

Barab, Sasha A., Kenneth E. Hay, and Thomas M. Duffy. 1998. "Grounded Constructions and How Technology Can Help." *TechTrends* 43(2): 15–23.

Bereiter, C., M. Scardamalia, C. Cassells, and J. Hewitt. 1997. "Postmodernism, Knowledge Building, and Elementary Science." *Elementary School Journal* 97(4): 329–340.

Brown, John Seely, Allan Collins, and Paul Duguid. 1989. "Situated Cognition and the Culture of Learning." *Educational Researcher* 18: 32–42.

Cohen, A., and M. Scardamalia. 1998. "Discourse about Ideas: Monitoring and Regulation in Face-to-Face and Computer-Mediated Environments." *Interactive Learning Environments* 6(1–2): 93–113.

Duffy, Thomas M., and Donald J. Cunningham. 1996. "Constructivism: Implications for the Design and Delivery of Instruction." In David H. Jonassen (ed.), *Handbook of Research for Educational Communications and Technology* (pp. 170–198). New York: Simon and Schuster Macmillan.

Duffy, Thomas M., and David H. Jonassen. 1992. *Constructivism and the Technology of Instruction.* Hillsdale, NJ: Erlbaum.

Hewitt, Jim, and Marlene Scardamalia. 1998. "Design Principles for Distributed Knowledge Building Processes. CSILE (Computer-Supported Intentional Learning Environments)." *Educational Psychology Review* 10(1): 75–96.

Jonassen, David H. 2000. *Computers as Mindtools for Schools: Engaging Critical Thinking,* 2nd ed. Upper Saddle River, NJ: Merrill.

Resnick, Lauren B. 1987. "Learning in School and Out." *Educational Researcher* 16: 13–20.
Roschelle, Jeremy. 1992. "Reflections on Dewey and Technology for Situated Learning." Paper presented at the annual meeting of the American Educational Research Association, San Francisco.
Savery, John, and Thomas Duffy. 1996. "Problem-Based Learning: An Instructional Model and Its Constructivist Framework." In Brent Wilson (ed.), *Constructivist Learning Environments: Case Studies in Instructional Design* (pp. 135–148). Englewood Cliffs, NJ: Educational Technology Publications.
Scardamalia, M., and C. Bereiter. 1991. "Higher Levels of Agency for Children in Knowledge Building: A Challenge for the Design of New Knowledge Media." *Journal of the Learning Sciences* 1(1): 37–68.
Scardamalia, M., et al. 1992. "Educational Applications of a Networked Communal Database." *Interactive Learning Environments* 2(1): 45–71.
von Glasersfeld, Ernst. 1989. "Cognition, the Construction of Knowledge, and Teaching." *Synthese* 80: 121–140.

Copyright

Copyright provides legal protection for intellectual property by bestowing a bundle of rights on the copyright owner. New technologies seem to challenge copyright's ability to apply to them, but the law, while retaining its basic tenets, has proven capable of accommodating the new challenges. Although Internet services like Napster (where users download music) challenged copyright's boundaries, copyright law appears to be winning—at least in court.

As with patents, Congress's power to legislate copyright laws derives from the U.S. Constitution, which gives Congress the power "to promote the progress of science and useful arts, by securing for limited times to authors and inventors the exclusive right to their respective writings and discoveries" (art. I, sec. 8, cl. 8). In other words, the Constitution tries to encourage the creation of new works by promising an economic reward to copyright owners. The Constitution sees the creation of new works as a means of advancing the public welfare.

The current Copyright Act, often called the Copyright Revision Act of 1976, actually took effect on January 1, 1978. It replaced the Copyright Act of 1909. However, the 1909 act still plays a role in some court cases. For example, in 1998 Martin Luther King Jr.'s 1963 "I Have a Dream" speech formed the basis of an infringement case brought by King's estate against CBS. A CBS video of the speech had been used in an A&E video. The King estate claimed ownership of the speech; CBS claimed the speech had entered the public domain. To resolve the case, the court looked to the 1909 act, which was in effect at the time. The court found that the speech had been freely distributed with no copyright notice and published in a newsletter with no copyright. Thus, under the strict copyright notice requirements of the 1909 act, the court found the speech had entered the public domain.

What Can Be Copyrighted?

Five sections of the Copyright Act (secs. 102, 103, 104, 104A, and 105) address the question of what constitutes a copyrightable work. Section 102 provides the basic criteria for copyrightability. A work must be an original work of authorship fixed in a tangible medium of expression from which it can be communicated, either directly or with the aid of a machine or device. The eight categories into which such a work may fall are: literary works; musical works, including any accompanying words; dramatic works, including any accompanying music; pantomimes and choreographic works; pictorial, graphic, and sculptural works; motion pictures and other audiovisual works; sound recordings; and architectural works. It also clearly states that copyright does not "extend to any idea, procedures, process, system, method of operation, concept, principle, or discovery." This includes titles, names, short phrases, mere listings of ingredients, standard calendars, and rulers, as well as works such as improvisational speeches that have not been fixed in a tangible form such as a recording.

Section 103 includes compilations and derivative works. The catch here is that the compilation or derivative work's copyright "extends only to the material contributed by the author." It does not affect the copyright of the preexisting work.

Sections 104 and 104A address the national origin of the work. Section 104 sets forth the criteria for published works, such as one or more of the authors is a U.S. national or lives in the United States or is a national of a country with which the United States has a treaty. Section 104A restores copyright protection to certain foreign works that entered the public domain in the United States before 1996 for the remainder of the copyright term the work would have been granted if it had not entered the public domain.

Section 105 simply says works of the U.S. government are not copyrightable. This places many items in the public domain. However, government works created by an independent contractor may indeed be copyrightable by that contractor depending on the wording of the contract. Postage stamps, although created by the U.S. Postal Service, are indeed copyrightable.

Who's the Copyright Owner?

Sections 201 and 101 describe the copyright owner. As soon as a work is created in a fixed form, the copyright initially rests with the author, who may then decide to transfer some or all of the rights granted by copyright to others (see the discussion of section 106 below). In the case of a work made for hire, the employer is the copyright owner. Section 101 defines a work made for hire as one "prepared by an employee within the scope of his or her employment" or a work specially ordered or commissioned for use (e.g., a contribution to a collective work, a translation, a test, an atlas, answer materials for a test, bibliographies, an index, and a sound recording). The parties must have expressly agreed to the work-made-for-hire designation in a signed written instrument.

Section 106

Section 106 spells out the copyright owner's bundle of rights. It grants a copyright owner the exclusive rights to do and to authorize any of the following: (1) to reproduce the work in copies or phonorecords; (2) to prepare derivative works based upon the copyright work; (3) to distribute copies or phonorecords of the work to the public by sale or other transfer of ownership, or by rental, lease, or lending; (4) to publicly perform literary, musical, dramatic, and choreographic works, pantomimes, and motion pictures and other audiovisual works; (5) to publicly display literary, musical, dramatic, and choreographic works, pantomimes, and pictorial, graphic, or sculptural works, including the individual images of a motion picture or other audiovisual work; and (6) to publicly perform sound recordings by means of a digital audio transmission. In addition, section 106A grants certain authors of visual-arts works the rights of attribution and integrity. However, all the rights granted by section 106 are limited by exemptions and compulsory licenses laid out in sections 107 through 121 (discussed below).

For How Long?

How long do the copyright owner's exclusive rights last? Chapter 3 of the Copyright Act sets the duration of copyright. Section 302, amended in 1998 by the Sonny Bono Copyright Extension Act, establishes the basic copyright term for works created on or after January 1, 1978, as the life of the author plus seventy years. For joint works (i.e., two or more authors) the term is the life of the last surviving author plus seventy years. Anonymous and pseudonymous works and works made for hire receive copyright protection for ninety-five years from the year of first publication or 120 years from the year of its creation, whichever expires first.

Limitations on Section 106

Having granted copyright owners certain exclusive rights in sections 106 and 106A, Congress immediately added sections 107–122, placing "limitations on exclusive rights." Section 107 restates the judicial doctrine of fair use, incorporating it for the first time into the copyright act itself. This section allows the reproduction of a copyrighted work in, for example, copies or phonorecords "for purposes such as criticism, comment, news reporting, teaching (including multiple copies for classroom use), scholarship, or research." Section 107 identifies four criteria for determining if a particular use of a copyrighted work is indeed a fair use:

- *Purpose and character of the use:* Is this for a nonprofit educational purpose or commercial in nature?
- *Nature of the copyrighted work:* Is it factual or fictional?
- *Amount and substantiality:* Is it a small or large portion of the copyright work as whole?
- *Effect of the use upon the potential market for or value of the copyrighted work:* Is the original work's market devalued by the reproduction?

Section 107 also clearly states that unpublished works may indeed fall under fair use based upon a consideration of the four criteria.

Section 108 turns to reproduction by libraries and archives, granting certain exemptions hedged about by detailed criteria. Thus the nine subsections in this section generate more subsections. The first subsection allows a library or archives, or any of its employees acting within the scope of their employment, to reproduce one copy or phonorecord of a work (except as noted in the following two subsections) or to distribute that copy or phonorecord if it is not intended for commercial use, the library is open to the public, and it includes a copyright notice. The second subsection allows up to three copies of an *unpublished* work solely for preservation and security purposes. If the copy is in digital format, it may not be made available in that format outside the library or archives' premises. The third subsection allows three copies or phonorecords of a *published* work to replace a damaged, deteriorating, lost, or stolen copy or phonorecord or if the work's existing format has become obsolete. But these copies may be made only if a reasonable effort has been made to find a replacement at a fair price and if any copy in digital format is not made publicly available outside the library or archives' premises. In terms of this subsection, a format is considered obsolete if the machine or device necessary to render a work stored in that format is either no longer manufactured or is no longer reasonably commercially available. The next four subsections are concerned with patron copying and interlibrary loan copy-related issues such as a copy of an article or an entire work. Subsection 8 is an attempt to mitigate the effect of the life-plus-seventy-years duration of a copyright granted to an author (see discussion above):

> During the last 20 years of any term of copyright of a published work, a library or archives, including a nonprofit educational institution that

functions as such, may reproduce, distribute, display, or perform in facsimile or digital form a copy or phonorecord of such work, or portions thereof, for purposes of preservation, scholarship, or research.

However, the library or archives may not reproduce, distribute, display, or perform a work if (1) the work is subject to normal commercial exploitation; (2) copies or phonorecords can be obtained at a reasonable price; or (3) the copyright owner posts a notice that either of the prior two conditions applies. The last subsection clearly states that Section 108 does not apply to a musical, pictorial, graphic, or sculptural work or a motion picture or other audiovisual work unless it is one dealing with news.

Section 109 deals with the transfer of a particular copy or phonorecord. It says if a person owns a phonorecord or computer program, she may not rent, lease, or lend that copy. But this does not apply to the rental, lease, or lending of a phonorecord for nonprofit purposes by a nonprofit library or education institution. "The transfer of possession of a lawfully made copy of a computer program by a nonprofit educational institution to another nonprofit educational institution or to faculty, staff, and students does not constitute rental, lease, or lending." It goes on to indicate that nonprofit libraries may lend computer programs if each copy has a copyright warning affixed to its packaging.

Section 110 identifies performances and displays that are not considered copyright infringements. The first subsection is referenced in discussions about the legality of showing videotapes marked "for home use only." Instructors or students of nonprofit educational institutions may perform or display a lawful copy of a work "in the court of face-to-face teaching activities . . . in a classroom or similar place devoted to instruction." The general consensus explains this subsection as allowing the showing of videotapes in class for instructional purposes only. "For home use only" videotapes may not be shown as rewards (e.g., for good attendance, or simply to baby-sit a class).

Section 117 allows the owner of a copy of a computer program to make a copy of that program for archival purposes. If the program is no longer owned by that person, the archival copy must be destroyed. This section also allows a copy to be made when it is "an essential step in the utilization of the computer program in conjunction with a machine and that it is used in no other manner."

Section 121, also known as the Chafee Amendment, after Senator John H. Chafee, who introduced the measure, was added to the Copyright Act on September 16, 1996. It addresses the reproduction of copyrighted works for blind or other disabled people. This refers to individuals who are eligible to receive books and other publications in specialized formats under the 1931 act to provide books for the adult blind. According to this section, "It is not an infringement of copyright for an authorized entity to reproduce or to distribute copies or phonorecords of a previously published, nondramatic literary work if such copies or phonorecords are reproduced or distributed in specialized formats exclusively for use by blind or other persons with disabilities." These copies or phonorecords may be reproduced or distributed only in a specialized format exclusively used by blind or other disabled people. These specialized formats are "Braille, audio, or digital text which is exclusively for use by blind or other persons with disabilities." They must include a notice stating that "further reproduction or distribution in a format other than a specialized format is an infringement" and include a copyright notice. However, the section does "not apply to standardized, secure, or norm-referenced tests and related testing materials, or to computer programs, except the portions . . . in conventional human language (including descriptions of pictorial works) and displayed to users in the ordinary course of using the computer programs."

Section 111 focuses on secondary transmissions, especially by cable systems; section 112, ephemeral recordings; section 113, pictorial, graphic, and sculptural works; section 114, transmission of sound recordings; section 115, compulsory licenses for making and distributing phonorecords; and section 116, public performances by means of coin-operated phonorecord players (e.g., jukeboxes). Section 118 centers on the use of certain works in connection with noncommercial broadcasting; section 119 on the secondary transmissions of superstations and network stations for private home viewing; and section 122 on secondary transmissions by satellite carriers within local markets. Section 120 turns to architectural works. These sections have limited impact on the average classroom.

Copyright Notice

Chapter 4 focuses on the details of copyright notice, deposit, and registration. Before March 1, 1989, failure to include the proper copyright on publicly distributed copies and phonorecords jeopardized the copyright in a work. What changed on that date? U.S. adherence to the Berne Convention Implementation Act of 1988 became effective. The Berne Convention does not require a notice, but use of the notice is strongly recommended because it clearly informs people that the work is copyrighted, by whom, and in what year. It also prevents a plea of innocent infringement (i.e., the infringer claims he or she did not realize the work was copyrighted).

The legal formality of copyright registration creates a public record of a particular copyright. Although registration is not required for protection, it establishes a public record; it *is* required before an infringement suit may be filed; it establishes evidence of the copyright's validity; it makes available statutory damages and attorney's fees in court actions; and it allows registration with the U.S. Customs Service to protect against the importation of infringing copies. Two copies of works published in the United States must be submitted for deposit in the Library of Congress.

Remedies

Chapter 5 enumerates the remedies for copyright infringement. It might also be called "crime and punishment." Remedies include injunctions; impounding and disposition of infringing articles; awards of damages and profits; and awards of costs and attorney's fees. Statutory damages can range from $750 to $30,000 for one work, although at the court's discretion it may range from $200 to $150,000. Section 504(c)(2) offers some protection for educators and librarians "in any case where an infringer believed and had reasonable grounds for believing that his or her use of the copyrighted work was a fair use under section 107" and was acting within the scope of his or her employment.

Esther Sinofsky

See also Educational Fair Use

References

Sinofsky, Esther R. 1994. *A Copyright Primer for Educational and Industrial Media Producers,* 2nd ed. Washington, DC: AECT/Copyright Information Services.

U.S. Congress. Senate. 1976. *An Act for the General Revision of the Copyright Law, Title 17 of the United States Code.* Pub. L. 94–553, 94th Cong., S. 22. U.S. Copyright Office, *Circular 92* (www.loc.gov/copyright).

U.S. Library of Congress. Copyright Office. 2000. *Circular 1: Copyright Basics.* Washington, DC: Government Printing Office (www.loc.gov/copyright).

Courseware

The origins of the term "courseware" are obscure, although it is obviously derived from the term "software." The latter has been attributed to John W. Tukey, a statistics professor at Princeton University, who in the January 1958 issue of *American Mathematical Monthly* wrote: "Today the 'software' comprising the carefully planned interpretive routines, compilers, and other aspects of automative programming are at least as important to the modern electronic calculator as its 'hardware' of tubes, transistors, wires, tapes and the like."

Credible claimants to the creation of the term "courseware" include programmers at IBM who in 1960 released the IBM 1500 computer complete with a courseware authoring system called Coursewriter, as well as Donald Bitzer and his colleagues at the University of Illinois who in the 1960s began developing a system (Programmed Logic for Automatic Teaching Operations, or PLATO) to automate instruction. Regardless of the origins, the term generally refers to any educational materials available in digital format. Some people extend the term to refer to nonelectronic materials such as print-based textbooks, audiotapes, videotapes, and the like, but the more common usage is restricted to computer-based software formats that today are most widely distributed via CD-ROM or accessible through the World Wide Web or corporate intranets.

Although the word "course" in "courseware" may be interpreted to refer to educational software that is part of a formal course of study in the sense it is used in U.S. higher education, this is not a widespread distinction. Courseware comes in many forms, ranging from the *Sesame Street* CD-ROM a parent might buy a child to assist in early language development to sophisticated multimedia simulations that airlines utilize to train pilots.

In the early days of personal computers, there was a widespread belief that teachers should develop their own courseware, and a generation of teachers studied the BASIC (Beginner's All-Purpose Symbolic Instruction Code) or primitive courseware authoring systems such as PILOT (Programmed Inquiry, Learning, or Teaching) so that they could do so. Another wave of teacher-produced courseware resulted from the introduction of HyperCard by Apple Computer in 1987. Today, teachers all over the globe are creating web-based courseware with hypertext markup language (HTML) editors and webpage production software such as DreamWeaver from Macromedia®.

Courseware encompasses both content and pedagogy. With respect to content, it would be difficult to find subject matter for which some form of interactive courseware has not been developed by someone somewhere, often an individual teacher, student, or programmer. Not surprisingly, most commercial courseware is focused on content related to information technology or some formal educational subject such as mathematics. Commercial courseware is typically developed by a team of specialists including subject-matter experts, instructional designers, graphic artists, and programmers.

With respect to pedagogy or instructional design, courseware exists in a variety of formats, including tutorials, drill-and-practice programs, simulations, and interactive learning environments. The earliest forms of courseware were heavily influenced by the behavioral psychology of B. F. Skinner (1968). These programs were essentially automated forms of programmed instruction. They presented information to the student in small segments, required the student to make overt responses to the information as stimulus, and provided feedback to the student along with differential branching to other segments of instruction or to drill-and-practice routines. Although this basic behavioral model continues to dominate commercial courseware such as integrated learning systems (Bailey 1993), the nature of the interactions between learners and computers in today's most innovative courseware, such as constructivist learning environments (Wilson 1996), is based upon advances in cognitive psychology (Salomon, Perkins, and Globerson 1991) and constructivist pedagogy (Papert 1993). In these types of courseware environments, learners are less passive and may even contribute to the creation of interactive learning materials (Jonassen and Reeves 1996).

Courseware has recently become the focus of large investments from the U.S. Department of Defense, industry, and universities through the SCORM (Sharable Courseware/Content Object Reference Model) initiative of the Advanced Distributed Learning (ADL) organization (www.adlnet.org). The purpose of the ADL initiative is "to ensure access to high-quality education and training materials that can be tailored to individual learner needs and made available whenever and wherever they are required." Also referred to as "e-learning" and "online learning," distributed learning courseware is being "tagged" using extensible markup language (XML) to define all of the courseware elements, structure, and external references necessary to move courseware from one learning management system environment to another. As a result, courseware is supposed to become "accessible, interoperable, durable, reusable, adaptable and affordable." The ADL website claims that such sharable courseware will reduce the cost of instruction by 30–60 percent and the time of instruction by 20–40 percent as well as increase the effectiveness of instruction by 30 percent and student knowledge and performance by 10–30 percent. These findings are generally in agreement with other reviews of the literature (Kulik and Kulik 1991; Coley, Cradler, and Engel 1997).

Thomas C. Reeves

See also Computer-Assisted Instruction; Instructional Design; LEGO/Logo; PLATO; Web-Based Course Management Systems; Web-Based Instruction

References

Bailey, Gerald D., ed. 1993. *Computer-based Integrated Learning Systems.* Englewood Cliffs, NJ: Educational Technology.

Coley, Richard J., J. Cradler, and Penelope K. Engel. 1997. *Computers and Classrooms: The Status of Technology in U.S. Schools.* Policy Information Report. Princeton, NJ: Educational Testing Service.

Jonassen, David H., and Thomas C. Reeves. 1996. "Learning with Technology: Using Computers as Cognitive Tools." In D. H. Jonassen (ed.), *Handbook of Research for Educational Communications and Technology* (pp. 693–719). New York: Macmillan.

Kulik, C.-L., and Joanna A. Kulik. 1991. "Effectiveness of Computer-Based Instruction: An Updated Analysis." *Computers and Human Behavior* 7(1/2): 75–94.

Papert, Seymour. 1993. *The Children's Machine: Rethinking School in the Age of the Computer.* New York: Basic Books.

Salomon, Gavriel, David N. Perkins, and Tamar Globerson. 1991. "Partners in Cognition: Extending Human Intelligence with Intelligent Technologies." *Educational Researcher* 20(3): 2–9.

Skinner, Burrhus F. 1968. *The Technology of Teaching.* New York: Appelton-Century-Crofts.

Tukey, John W. 1958. "The Teaching of Concrete Mathematics." *American Mathematical Monthly* 65(1): 1–9.

Wilson, Brent G., ed. 1996. *Constructivist Learning Environments: Case Studies in Instructional Design.* Englewood Cliffs, NJ: Educational Technology.

Crossroads Project

The American Studies Crossroads Project (Crossroads) was one of the first websites and research projects devoted to technology and the humanities and one of the first to act as a comprehensive resource to an academic professional society, the American Studies Association (ASA). Founded in 1993 by Randy Bass, associate professor of English at Georgetown University, Crossroads's central research focus is the relationship between technology and pedagogy, highlighting the ways that the introduction of technology into classrooms leads to questions about teaching practices.

Crossroads serves the American studies community and a range of allied fields such as literature, history, art history, ethnic studies, and women's studies. The site provides a variety of web-based resources, including lists of links for national and international programs and centers in American studies, syllabi libraries, and a database of 5,000-plus links to American studies sites classified by subject and annotated for undergraduate users. In collaboration with the ASA executive director's office, the project publishes an online version of the ASA *Newsletter* and provides a listing of dissertations in American studies as well as a variety of information related to the annual meeting. In addition to these resources, the site also provides links and information for various constituent groups of the ASA and is beginning to provide these groups with online tools to represent themselves and their work online.

From the beginning, Crossroads sought to encourage faculty not only to integrate new technologies into their classrooms but also to ask tough questions about the implications of these new technologies for student learning. Crossroads engaged in a variety of research projects situated in faculty practice in classroom situations. Early project questions focused on, among other topics, the ways that electronic access to the cultural record created new ways of making knowledge, as well as how new technologies created opportunities for new forms of narrative. In asking these questions, faculty discovered that the introduction of new technologies brought forward questions about technology as well as questions fundamental to research and teaching practice in the disciplines.

These questions led to the design of the Visible Knowledge Project, a five-year project that seeks to deepen these inquiries through a focus on student learning. Using national networks of institutions and faculty involved in Crossroads and related projects, the Visible Knowledge Project provides for faculty to engage in course innovations involving technology, examine the evidence of student learning as a result of those innovations, and plan for further pedagogical change. Building off Crossroads, the Visible Knowledge Project site includes resources such as an online glossary, online project posters, and a set of tools to facilitate local discussions around technology and pedagogy. In addition to the Visible Knowledge Project, the Crossroads Online Institute continues the emphasis on helping faculty work through complex questions by providing an online faculty development seminar focused on the design of online or hybrid learning modules. The modules will be placed in a digital library along with faculty reflections and evidence of student learning.

Michael Coventry

See also *Who Built America?*

References

American Studies Crossroads Project (crossroads.georgetown.edu).

Bass, Randy. 1995. "Crossroads: The Release Version." American Studies Association *Newsletter* (September).

———. 1996a. "American Studies Crossroads Project Supports Change and Innovation Through Campus-Based Programs" American Studies Association *Newsletter* (March).

———. 1996b. "Getting Beyond Linking. Or, How Crossroads Are Always Arrivals at Beginnings." American Studies Association *Newsletter* (December).

Bass, Randy, Teresa Derrickson, Bret Eynon, and Mark Sample. 1998. *Intentional Media: The Crossroads Conversations on Learning and Technology in the American Culture and History Classroom.* Special issue of *Works and Days* 16(1–2) (spring/fall).

Guernsey, Lisa. 1998. "On-Line 'Expo' Points Scholars to New American-Studies Resources." *Chronicle of Higher Education* (May 1): A36.

Visible Knowledge Project (crossroads.georgetown.edu/vkp).

CSILE/Knowledge Forum®

The origins of CSILE (Computer Supported Intentional Learning Environments) are in research on knowledge-building, on the nature of expertise, and on the sociocultural dynamics of innovation. CSILE was designed to (1) make advanced knowledge processes accessible to all participants, including children; (2) foster the creation and continual improvement of public artifacts or *community knowledge* (Scardamalia

2002); and (3) provide a community space for carrying out this knowledge-building work collaboratively. It has evolved to the second-generation knowledge-building environment, known as Knowledge Forum.

The term "knowledge-building" (as a definable educational enterprise) originated with CSILE and accompanying scholarly works and represents an integrated framework for knowledge-building pedagogies, practices, and environments (Bereiter 2002; Bereiter and Scardamalia, 1989, 1993; Scardamalia et al. 1989; Scardamalia 2002). CSILE was first prototyped in a university course in 1983. By 1986 a fully functioning networked version was in daily use in an elementary school. In 1995 it was reengineered, with its affordances for knowledge building substantially enhanced, and published as Knowledge Forum (www.knowledgeforum.com) by Learning in Motion. Currently in version 4, with version 5 under development, Knowledge Forum continues to evolve in response to research findings and new opportunities. Thus, for example, while CSILE was built prior to the World Wide Web, Knowledge Forum now offers browser as well as client versions that can link Knowledge Forum classrooms to one another via the Internet. Current developments are taking advantage of the potential of wireless technologies to allow synchronization of online and offline knowledge-building. Knowledge Forum is used in education (grade 1 to graduate), health care, community, and business contexts in the Americas, Asia, Australia, Europe, and New Zealand. Knowledge Forum's cross-sector, cross-age, cross-cultural framework reflects the theoretical idea that the socio-cognitive and cultural processes underlying knowledge acquisition and knowledge creation are fundamentally the same (cf. Piaget 1971; Popper and Eccles 1977) and so must apply to knowledge-builders of all sectors, ages, and cultures.

From the earliest days of educational computing, leadership was defined through model-school projects that demonstrate what classrooms enhanced with information and communication technology should look like. Typically, these classrooms exemplify discriminating consumership and creative use of off-the-shelf technology. Creating model-school projects was the idea behind the Apple Classrooms of Tomorrow program (Dwyer 1993). However, Apple's first venture into advanced educational software (Bowen 1990) did not come out of their model classrooms. It came from CSILE, which Apple released in 1993 under the name Collaborative Learning Product. Apple's press release announced:

> **APPLE INTRODUCES GROUND-BREAKING PRODUCT. . . .**
> During a meeting of key education press at Apple headquarters today, the company introduced . . . Collaborative Learning Product, an integrated, research-based product and the first collaborative learning offering available for the K-12 education market.

Apple distinguished this product from bulletin-board services and e-mail, citing its affordances for inquiry-based work and knowledge construction, its basis in research, and its ability to address the skill requirements identified by a U.S. Department of Labor commission on achieving necessary skills for the workplace of the twenty-first century. It cited the following skills: "the ability to organize resources, to work with others, to learn a variety of technologies, as well as the ability to acquire, understand, and evaluate information."

As the Apple press release claimed, CSILE represented a new generation of educational technology that specifically addressed the educational challenges of the twenty-first century. CSILE was not management, planning, or productivity software retooled for children; it was technology specifically designed to support knowledge creation. It was not a collection of tools; it was an environment to support the kinds of inquiry, information search, and creative work with ideas that go on in knowledge-building organizations of all kinds. There have since been a number of experimental efforts to construct knowledge-building tools and environments. However, Knowledge Forum is the only product continuously improved over the years based on research results arising from active and diverse user communities and reflecting knowledge-building theory, principles, and practices. Knowledge Forum development is not driven by technology but uses advances in technology to continually enhance and unfold the knowledge-building agenda.

The heart of CSILE/Knowledge Forum is a multimedia community knowledge space. In the form of notes, participants contribute theories, working models, plans, evidence, reference material, and so forth to this shared space. The software provides knowledge-building supports in the creation of these notes as well as in the ways they are displayed, linked, and made objects of further work. Revisions, elaborations, and reorganizations over time provide a record of group advances, like the accumulation of research advances in a scholarly discipline.

Multiple Perspectives, Multiple Literacies, and Teamwork

User-created graphical views constitute a higher level of organization and conceptually useful workspaces. A given note may appear in multiple views. As an example, four views provide different ways of conceptualizing the same group of notes.

The notes, represented by small icons, are produced by grade 1–3 students contributing information and graphics concerning their favorite dinosaurs. The notes appear on a blank white background, and show no particular organization. However, from the note titles children discovered classmates who had the same favorite dinosaur (triceratops, brontosaurus, etc.). Several students had produced graphic rather than text notes, and others wanted to link their notes to these graphics. So students added these graphics to the background of a new "dino types" view. Knowledge Forum's keyword searches were used to collect all relevant notes (e.g., all notes with the keyword "triceratops") and move them to the appropriate picture. The result was a new view. At about the same time, students in a university course were provided with access rights to this grade 1–3 dinosaur knowledge space. The university students noted, in reading these same notes, that they contained references to geological time. The university students created a new "geological time" view and entered a geological-timeline graphic from the Internet as a background. Student notes were searched again, now for periods of time (e.g., Jurassic), and the new collection was added at the appropriate point to the geological timeline. Students who had not yet identified the time when their dinosaur roamed the earth quickly extended their research so their note would appear in this new view. There was yet another sequence with these same notes. A biologist was invited to join the knowledge-building collaborative efforts. She signed in from afar and created the "food chain" view that referenced students' dinosaurs as plant or meat eaters. The work illustrates the following knowledge-building affordances:

- Notes and views support teamwork and collaborative design. Notes are multimedia objects that can be coauthored. View backgrounds are not simply white backdrops or static bulletin boards (although they can be used that way); they are collaborative design environments with a built-in graphics program that makes it easy for users to design their own backgrounds.
- Views provide different perspectives on information. As the work of the biologist presented suggests, this networked technology is used to expand perspectives rather than to solidify roles in which students ask questions and experts answer them.
- Multimedia and other supports provide a way in for all participants to a common discourse. Notes and views support a range of multimedia objects, from text to video. As an example, some of these grade 1 students were not writers at the beginning of the year but did represent their ideas graphically. Easy assignment of keywords (touching a word in a note with a movable key icon is all it takes) made it possible for these young students to keyword their notes. These notes were then available, via keyword searches, for incorporation into new conceptual views.
- Notes and views can be individually or group authored. They can also be entered into private or public spaces; the default option is that they are contributed to a public forum. Accordingly,

the environment encourages openness in knowledge work while enhancing both individual and group processes.
- Emergent ideas and goals are supported. Knowledge Forum represents an open environment, without predetermined boundaries or structures around ideas or activities. Through collective responsibility for public knowledge spaces, with input from varied sources of expertise, the environment favors the emergence of big ideas and deep principles.

Creating Connections and Public Knowledge

Notes are situated in build-on structures that result in visible links to parent notes, with both notes and links modifiable.

- *Flexible build-ons:* The first five panes show the varied forms that a build-on structure can take. The note complex has been rearranged a number of times to give greater meaning to the concepts of frequency and amplitude. This flexibility can be contrasted with the downward branching of these same notes in threaded discourse. Threaded discourse now dominates the Internet, despite the fact that in many ways it defeats knowledge-building. Thus, for example, popular applications such as WebCT® and Blackboard® were built without supports for linking ideas across threads or for placing them in new and varied contexts. Linear discourse forms and isolated "conferences" entrap ideas. Knowledge Forum's client version supports the full range of connections elaborated below; these more varied forms are being incorporated into the browser version (a process slowed by the limited interactivity allowed by current webpage technology).

Idea connectedness is further facilitated through the following means:

- *Annotation, citation, and reference links:* Annotations and reference links can be added to any note. References include pointers back to source notes, so ideas can be viewed in both original and new contexts. Links to views and subviews can be embedded within either notes or views. Deep embedding of ideas facilitates deep processing of information.
- *Interconnected views:* Views can be interlinked: Views reference other views, and different levels of access to views can be used to indicate the centrality or distance of particular views to the current work of the community.

- *Multifaceted indices:* Author-assigned indices (keywords, scaffolds, problem fields, titles) and automatically assigned indices (author, date, semantic field) make notes available through a variety of search parameters; a note-sorter allows the notes resulting from a search to be sorted for viewing or transported into another view. Citation, commentary, and notification enhance engagement by drawing all relevant authors back into the discourses that involve their ideas.

Advanced Knowledge Processes

Software typically includes palettes for text, graphics, and other productions. CSILE/Knowledge Forum introduced palettes for high-level knowledge processes.

Scaffolding knowledge processes. Scaffold supports are Knowledge Forum's most imitated feature. They were originally designed to provide procedural facilitation (Scardamalia and Bereiter 1983) for fostering expertise in writing and were titled "thinking types." Others refer to them as prompts and have used them to serve purposes similar to templates or fill-in-the-blanks forms. Knowledge Forum scaffolds can serve these purposes, but the following uses represent the real goals underlying their design:

- Scaffolds give ideas defined roles in such processes as theory refinement (e.g., "This theory cannot explain . . .") and constructive criticism. The opportunistic rather than mandated use of scaffold supports helps students embed these forms of discourse in their everyday work with ideas; "forms of discourse become forms of thinking" (Pontecorvo 1993, 191).
- The supports contained within any particular scaffold can be used opportunistically and flexibly in any order. Once selected, the term is entered into the main body of the text, at the point assigned by the author. The scaffold support then additionally serves as a searchable parameter. Scaffolds can be yoked to views, so different discourse forms can be encouraged in different views. It is also easy to toggle between scaffolds, so that multiple scaffolds can be used within any view or hidden when not wanted.
- Scaffolds are customizable. They can be easily added or modified to support the discourse needs of a particular community.

Reference and contribute. Knowledge Forum replaces the common "say it in your own words" norm with the more mature "contribute-and-reference" norm. Cited material is automatically quoted (made visually distinct), with automatic links back to original sources and automatic compilation of a bibliography. Users are especially drawn to this feature, because they see their work referenced rather than copied.

Problems of understanding. A problem field at the header of the note encourages a shift from topic- to problem-based inquiry. Knowledge Forum's "problem space" encourages the identification of problems of understanding to guide inquiry. "Problem" searches produce an overview of the problems that others are working on, so it is easy to contribute to those problem spaces or start a new field of inquiry.

Rise-Above and Improvable Ideas

"Rise-above" notes play a pivotal role in idea improvement. The idea, based on the philosophical concept of dialectic, is that the most constructive way of dealing with divergent or opposing ideas is not to decide on a winner or a compromise position but rather to create a new idea that preserves the value of the competing ideas while "rising above" their incompatibilities. In the simplest cases a rise-above may be simply a summary or distillation; in the most compelling cases, the rise-above presents a new idea that all the participants can recognize as an advance over their previous ideas.

A student's high-level summary of knowledge advances over a period of several months. This student packaged the set of notes that led to the discovery reported here; his older notes are now accessible only through this rise-above note. Rise-above notes are also used to synthesize ideas, create historical accounts and archives, reduce redundancy, and in other ways impose order on ideas.

The rise-above idea can be applied to views rather than notes. The linked views (e.g., eye, circulation) were created first, and this higher-order "human body" view then served to integrate these separate views and to support a new discourse on how different parts of the body work together. As this suggests, notes and views operate as a form of zoom in/zoom out, encouraging users to think in terms of relationships.

Endless improvability of ideas is further supported by the following:

- Ability to create increasingly high-order conceptual frameworks. It is always possible to reformulate problems at more complex levels, create a rise-above note that encompasses previous rise-above notes, or create a more inclusive view-of-views.
- Review and revision. Notes and views can be revised at any time, unlike most discussion environments that disallow changes after a note is posted.

- Published notes and views. Processes of peer review and new forms of publication engage students in group editorial processes. Published works appear in a different visual form, and searches can be restricted to the published layer of a database.

Individual and Group Portfolios: Evolution of Ideas

The entries in a database, taken collectively, provide process accounts of the contributions of each participant, as well as accounts of how those different inputs were combined to lead to the collective achievements of the group. Portfolios are simply new views created by an author, a group, or a class to highlight different aspects of this work. The evolution of ideas can be studied through searches and analytic tools. Thus, for example, the work that led to a new theory can be analyzed. Additionally, the process of constructing portfolios affords deeper student reflection about their own knowledge and puts self- and peer-assessment into evaluative accounts of student learning

Ideas and Artifacts as Objects of Discourse

Anything that can be represented digitally can be transformed into an object of discourse, with the full range of knowledge-building activities applicable. There are four different objects, along with transparent overlays, markup, and build-ons.

Embedded and Transformative Assessment

Knowledge-builders monitor their work and engage in self-assessment rather than being totally dependent on external evaluations. Individual and group portfolios help this process. Research tools work in the background of Knowledge Forum to automatically record activity patterns such as reading, building-on, referencing, and creating views. Results from these analytic tools can then be fed back into the work as it proceeds, rather than waiting until the end of a unit of work to provide feedback, when it is too late to make adjustments.

The Knowledge Society Network and the Virtual Suite of Possibilities

Through Knowledge Forum's flexible database access and linking structures, knowledge-building discourse may be confined to a single classroom or distributed across the world. As this suggests, the work of local communities can be enhanced through the worldwide network of communities using Knowledge Forum. Sophisticated semantic analysis tools can be used for matchmaking (locating groups working in similar semantic fields) and common problems. Searches of the Knowledge Society Network access the published layers of all the Knowledge Forum databases linked to the worldwide network, with these published layers operating, in effect, as websites. Thus the work of a local community "rises up" to become an object of discourse within an extended discourse community. Adults frequently find that the work of even the youngest students helps them advance their own understanding—even if only their understanding of how students think.

There are virtual tours of a database that the Knowledge Society Network enables. This tour begins with an overview of the science work of a grade 1 class, September to June. By following links it is possible to get a more in-depth view of specific units (e.g., the unit on leaves). The tour is presented from multiple perspectives (e.g., links to the students' and teacher's perspectives). Other perspectives include the curriculum, research, assessment, parents', and state-of-the-art perspectives. Access to all or parts of a database, with or without a guided tour, can be granted to visitors or telementors through the Internet. Participants can build on, comment, and in other ways create reference links to this work in the Knowledge Society Network. This network also supports virtual workshops, practica, seminars, and other events surrounding a knowledge base. Some of the most successful instances of collaborative knowledge-building have involved school students, teachers, researchers, graduate students, curriculum, and subject-matter experts coming together to tackle a problem of understanding.

Toward a Knowledge Society

From the start, the CSILE/Knowledge Forum initiative has aimed at revolutionary change: from a focus on carrying out tasks and activities to a focus on the continual improvement of ideas; from an emphasis on individual learning and achievement to the building of knowledge that has social value; from a predominantly teacher-directed discourse to distributed knowledge-building discourse. In line with the magnitude of the intended change there has developed, along with the CSILE/Knowledge Forum technology, knowledge-building pedagogy, practices, and principles (Scardamalia 2002; Scardamalia and Bereiter, forthcoming). Results indicate significant advances in textual, graphical, and computer literacy, as well as in depth of inquiry, collaboration, and a host of mature knowledge processes (Scardamalia, Bereiter, and Lamon 1994). More generally, use in grade 1 through tertiary education, in health care, in public organizations, and in workplaces suggest that Knowledge Forum not only enhances learning but also enables the creative work with ideas that set students on a course of knowledge creation—a course that helps to drive lifelong learning and innovation. A worldwide community of educational innovators—the Knowledge Society Network—has begun to take shape (Scardamalia and Bereiter 1996), supported by the Institute for Knowledge Innovation and Technology (www.ikit.org).

Marlene Scardamalia

See also Active Learning; Computer Supported Collaborative Learning; Constructivism; Open-ended Learning Environments

References

Bereiter, Carl. 2002. *Education and Mind in the Knowledge Age.* Mahwah, NJ: Lawrence Erlbaum Associates.

Bereiter, Carl, and Marlene Scardamalia. 1989. "Intentional Learning as a Goal of Instruction." In L. B. Resnick (ed.), *Knowing, Learning, and Instruction: Essays in Honor of Robert Glaser* (pp. 361–392). Hillsdale, NJ: Lawrence Erlbaum Associates.

———. 1993. *Surpassing Ourselves: An Inquiry into the Nature and Implications of Expertise.* Chicago and La Salle, IL: Open Court.

Bowen, Barbara, ed. 1990. *Design for Learning: Research-Based Design of Technology for Learning.* Cupertino, CA: External Research, Apple Computer Inc..

Dwyer, David. 1994. "Apple Classrooms of Tomorrow: What We've Learned." *Educational Leadership* 51.

Piaget, Jean. 1971. *Biology and Knowledge: An Essay on the Relations between Organic Regulations and Cognitive Processes.* Chicago: University of Chicago Press.

Pontecorvo, Clotide. 1993. "Forms of Discourse and Shared Thinking." *Cognition and Instruction* 11(3/4): 189–196.

Popper, Karl, and John Eccles. 1977. *The Self and Its Brain.* Berlin, Germany: Springer-Verlag.

SCANS (U.S. Department of Labor, Secretary's Commission on Achieving Necessary Skills). 1991. *What Work Requires of Schools: A SCANS Report for America 2000.* Washington, DC: U.S. Department of Labor.

Scardamalia, M. 2002. "Collective Cognitive Responsibility for the Advancement of Knowledge." In B. Smith (ed.), *Liberal Education in a Knowledge Society* (pp. 76–98). Chicago: Open Court.

Scardamalia, M., and C. Bereiter. 1983. "The Development of Evaluative, Diagnostic, and Remedial Capabilities in Children's Composing." In M. Martlew (ed.), *The Psychology of Written Language: Developmental and Educational Perspectives* (pp. 67–95). London: John Wiley and Sons.

———. 1996. "Engaging Students in a Knowledge Society." *Educational Leadership* 54(3): 6–10.

———. Forthcoming. "Knowledge Building." In *Encyclopedia of Education,* 2nd ed. New York: Macmillan Reference, USA.

Scardamalia, M., C. Bereiter, and M. Lamon. 1994. "The CSILE Project: Trying to Bring the Classroom into World 3." In K. McGilly (ed.), *Classroom Lessons: Integrating Cognitive Theory and Classroom Practice* (pp. 201–228). Cambridge, MA: MIT Press.

Scardamalia, M., et al. 1989. "Computer Supported Intentional Learning Environments." *Journal of Educational Computing Research* 5: 51–68

Curriculum Integration

At the beginning of the twenty-first century there is a national movement to change the way we teach students. Teachers in our nation's schools are charged with teaching students who will spend all of their lives in a technology-driven society. The ability to use and, more important, integrate technology into the curriculum is a necessity in the teaching profession. This has given rise to the topic of curriculum integration.

Curriculum integration is difficult to define because of the many ways it can be accomplished. First, "curriculum" is defined as a set of courses constituting an area of specialization or a field of study; "integration" is defined as bringing together different parts to combine into a whole; "technology integration" refers to the combination of all technology parts, such as hardware and software, together with subject-related content to enhance student learning and accomplish curriculum goals. Thus the term "curriculum integration" includes the effective integration of technology throughout the curriculum to help students meet the standards and outcomes of each lesson, unit, or activity. Curriculum integration encompasses computer literacy, information literacy, and integration literacy.

A person must possess a certain level of computer literacy, which is the understanding of computers and technology and their uses. For many years, a majority of educators believed that if they used technology in the classroom, then they were integrating technology into their instruction. In other words, if a teacher uses any type of technology as a productivity tool or during instruction, then that was curriculum integration. These terms, "using" and "integrating," are continuously interchanged. Yet clearly they do not mean the same thing. However, a person must know how to use technology prior to being able to appropriately integrate technology into classroom curriculum.

The second literacy one must possess is information literacy. Information literacy is knowing how to find, analyze, and use information. Information literacy is the ability to gather information from multiple sources, select relevant material, and organize the information into a form that will allow the user to make decisions or take specific actions.

Although computer and information literacy are very important for educators, today's educators also must integrate technology as a tool to facilitate learning. Educators must be able to assess technology resources and plan classroom activities using any and all available technologies. These skills are part of integration literacy, which is the ability to use computers and other technologies combined with a variety of teaching and learning strategies to enhance students' learning. Integration literacy means that teachers can determine how to match appropriate technology to learning goals, objectives, and outcomes (Shelly et al. 2002). Effective curriculum integration includes understanding how to integrate technology into the classroom curriculum successfully. This relies on a solid foundation of computer and information literacy.

To have effective curriculum integration, teachers must find methods and strategies that maximize the delivery and design of learning experiences. Educational technologists have become the advocates in integrating computers and other technologies into content areas. For years, schools have focused their efforts on getting technology or computer labs into the schools. Computer labs clearly provide solutions to some educational dilemmas and are an excellent addition to any school. Research shows that computers and related technologies, however, are more effective when integrated into subject content and placed in the classroom at the point of instruction (Web-Based Education Commission 2000). Point of instruction means having the technology in the classroom at the teachers' and students' fingertips.

In order to accomplish this, the role of the teacher must change to become that of a facilitator. The teacher's role changes from being the "sage on the stage" to being the "guide on the side." As teachers plan authentic learning experiences that incorporate a variety of tools and technologies, they need to be prepared to guide students through the learning experience. This requires a good foundation in computer literacy, information literacy, and integration literacy. Initially, teachers may be uncomfortable with the role of facilitator; however, as students adjust and learn to be more responsible for their learning, they will be more motivated and become better problem-solvers. When teachers are facilitators, they enable students to develop higher-order thinking and problem-solving skills. In addition, teachers help students develop the technology and social skills necessary to be successful in the workplace or in higher education.

Curriculum integration is evident when there is a seamless transition from one learning element to another using multiple technologies. An educator has always had the job of making teaching look easy. Integrating educational technology into the curriculum suggests determining which electronic tools and the appropriate methods for implementing them for any given classroom situations and learning condition. The definition of integrating technology in the curriculum means moving beyond trivial uses of instructional technology to uses that advance student learning of the technology in the curriculum that prepares students for the workforce of tomorrow. We can teach teachers to use and integrate technology, but until they experience hands-on authentic integration successes in their classrooms, we are just standing still. Standing still is not integrating, because integrating is like dancing: When you dance, the parties and equipment come together in perfect motion—that is curriculum integration.

Planning and Evaluation

To obtain effective curriculum integration, an educator must plan and think through the lessons. Lessons should be broken up into individual lesson plans, and those should be developed with an instructional planning tool or instructional model like the ASSURE model (Heinich et al. 2002). The ASSURE model, a six-step procedural guide for planning and delivering instruction that integrates technologies and media, can assist teachers with this process. The six-steps are: (1) analyze the learners; (2) state objectives; (3) select methods, media, and materials; (4) utilize media and materials; (5) require learner participation; and (6) evaluation and revise.

The tools used can be books, computers, multimedia, overhead projectors, televisions, videos, digital cameras, PDAs, photographs, or other tools used to accomplish learning goals. Technology is a tool, and if the tool is not appropriate for the instructional goals, then it should not be used. The instructional goals, the needs of the student, and the availability of the media for a specific set of objectives must be the driving force that causes a teacher to choose a particular media. The instructional goals should be the focus, not the technology. When integrating technology, it should always be viewed as a tool that assists the teacher in meeting the goals of the curriculum and the learning needs of the students. The teacher becomes a mentor and colearner, who is actively engaged in enabling students to access, analyze, apply, and create information electronically. When this happens, so does curriculum integration.

Glenda Gunter
Donna Baumbach

See also ADDIE; Analysis; Apple Classrooms of Tomorrow Project; Information Literacy; Instructional Design; Instructional Technology; Technology in K-12 Schools; Teachers: Preparation and Training

References

Heinich, Robert, et al. 2001. *Instructional Media and Technologies for Learning.* Upper Saddle River, NJ: Prentice-Hall.

Shelly, Gary, et al. 2002. *Teachers Discovering Computers—Integrating Technology in the Classroom,* 2nd ed. Boston: Course Technology.

Web-Based Education Commission. 2000. *The Power of the Internet for Learning: The President and the Congress of the United States.* Available online at interact.hpcnet.org/webcommission/index.htm.

Cyberculture and Related Studies

"Cyberculture" is a general, fluid, and contested term referring to the intersections among three relatively recent phenomena: digital technologies that make up cyberspace or new media; the social interactions and cultural manifestations that take place within or are made possible by new media; and the multiple contexts that surround new media. Often used interchangeably with the terms "Internet culture" or "digital culture," cyberculture first surfaced in the early 1990s and gained popularity throughout the decade, directly coinciding with the widespread adoption and whirlwind popularization of the Internet. It is a term used by the popular press, academic scholars, and aficionados of the Internet.

Defined traditionally, the technologies that comprise cyberculture include the computers, servers, and wires that constitute the Internet, as well as the various applications that run on the Internet, including e-mail, mailing lists, listservs, bulletin boards, Internet relay chat, instant messaging, multiple-user domains, and, of course, the World Wide Web. Cyberculture technologies also include more recent developments like Webcams, blogs, networked games, chat rooms, and peer-to-peer networks. In more general terms, cyberculture technologies refer to digital technologies that are interactive and/or networked, including, for example, cell phones, PDAs, and stand-alone computer games, as well as DVDs, CD-ROMs, and software applications like word processors and spreadsheets.

In addition to the digital technologies that give it shape, cyberculture is defined by the ways in which humans use it. Cyberculture is found in the social interactions between users frequenting the same mailing list or chat room. Cyberculture is maintained by the collective behaviors and social norms agreed upon by users, enforced by list moderators, or stipulated in frequently asked questions. Cyberculture is apparent in the playful exercises and creative experiments found in Internet theater and cyberart, as well as in the wired attitude of writers of online sites like Salon, Suck, and Slashdot.

And finally, cyberculture includes the multiple and complex contexts from which the Internet emerges. For example, cyberculture can be ideological and geographical, as in the case of the ARPANET, developed by the U.S. military and firmly directed by Cold War politics, or in the case of Radio B92, the web-based radio station that delivered alternative news from a war-torn Belgrade in the late 1990s. Furthermore, cyberculture is both a product and a reflection of economic contexts, ranging from the free-market libertarianism popular in the early to mid-1990s to the dot-com daze of the late 1990s. Cyberculture also functions this way in relationship to political contexts, as witnessed, for instance, in the child-porn scare of the mid-1990s that led in part to the Communications Decency Act, as well as the most recent curbs on online privacy resulting from the post–September 11 climate and the emphasis on homeland security. Cyberculture includes and is given shape in the popular media via depictions of the Internet in film, television, and print, advertisements for the Internet, and rhetoric about cyberspace.

At the heart of cyberculture is an interest and concern with how the Internet and its related technologies are restructuring almost every aspect of lived experience among residents of developed and, increasingly, less developed countries. In a relatively short period, these technologies have affected the way we understand time, space, and the world in general. For example, national boundaries, previously self-contained, are quickly becoming eclipsed by the constant and instantaneous flow of information and capital. In addition, new media technologies have altered the ways we live our everyday lives. For many, a typical day now includes a succession of interactions with new media technologies, including everything from online banking to reading the newspaper on the Internet. The Internet has also offered us new possibilities and challenges, including the use of virtual learning environments and growing cases of Internet-aided plagiarism. In order to understand how these technologies have transformed our interactions within our changing society, it is imperative that we examine the role new media plays in the way our culture is being reorganized and understood.

Although the rapid integration of new media technologies into our society has confronted us with an expansive amount of questions and issues related to the various ways they have modified our lives, several key areas of concern have arisen. Among these, three specific discussions necessitate further consideration. The first is the so-called digital divide, which reflects concerns about the widening gap between those who have and do not have easy access to the Internet, as well as the ways in which such access is tied to education, ethnicity, and income, among other factors. The second includes the issues surrounding use of new media technologies as tools for communication and community-building. And finally, there are the questions involving the effects of new media technologies on our understanding of the body and identity.

The technologies associated with cyberculture have increasingly become essential tools for economics, politics, and education; questions of access, knowledge, and use surrounding new media technology are important in considering who will have power, and who will be powerless, in our increasingly technological society. Some studies, including "Falling through the Net," a 1999 report commissioned by the U.S. Department of Commerce, argue that there is a digital divide forming in the United States and that ensuring all Americans have the skills and tools to participate in the emerging digital society is essential to guarantee the nation's participatory democracy and economic success. Even though the number of Americans connected to the nation's information infrastructure continues to rise, the digital divide still exists and, in many cases, is growing. The digital divide has created additional barriers for individuals and groups that have traditionally faced various boundaries in America; as the 1999 "Falling through the Net" report notes, minorities, low-income persons, the less educated, and children of single-parent households, particularly when they reside in rural areas or central cities, are among the groups that lack access to information resources.

The digital divide is of specific interest to educators. Educators are increasingly expected to assist in bridging the digital divide. In its study, the Department of Commerce suggests that a primary solution for making the technology available to those left behind by the information revolution is through public access points, referred to as community access centers (CACs), which include schools and libraries. The study intends the CACs to assist in providing public and free access to the Internet along with the training needed to survive in our emerging digital society. Although this recommendation has helped schools gain financial assistance for the procurement of new technologies, it has also created new demands for the educator, such as additional training in how to teach the essential skills needed for securing a job in the digital workplace.

Access to the Internet and its related new media technologies has become increasingly essential due to its accelerated use as a medium of communication and community-building. Communication has played a key role in both the development and initial popularization of the Internet. ARPANET, the Internet's predecessor, was conceived of by the U.S. Department of Defense in 1969 in an attempt to create a communications network that would survive a nuclear attack. In the 1980s the use of the Internet spread beyond the military to academics and scientists, when they began using the technology for sharing research along with various collaborative projects. It was also during this time that many Internet users discovered its power for creating new ways of connecting with each other, engaging with applications such as e-mail, usenet groups, and MUDs. In the early 1990s the introduction of the World Wide Web brought the information superhighway to the masses; while surfing the web became the popular face of the Internet, new communication applications continue to flourish in the guise of e-mail, chat rooms, and instant messaging.

In addition to debate regarding access and the digital divide, the widespread use of the Internet for the purpose of communication has placed questions concerning the relationships between community and cyberspace in a central position among those interested in cyberculture. Within this discussion, the earliest arguments focused on the perceived decline of face-to-face community in contemporary culture. Several writers, including Howard Rheingold, contended that online communities were providing members with an arena for connecting with others in our modern society where the spaces traditionally associated with communities were disappearing. Conversely, others argued community could never exist in virtual space, and new media technologies were one of the major reasons, rather than a solution, for the failing sense of community in everyday life. While this debate lingers, more and more people are expressing either that they have found a true feeling of community online or that they have supplemented and strengthened real-life communities using new media technologies.

While virtual technologies have provided us with new opportunities for the creation of collective communities, the same technologies have also given us a set of tools for the expression of individual identities. One of the most popular uses of technology in this way is the creation of the personal webpage. People are employing text, images, and links in order to narrate their personality and interests to the web-viewing public. Whereas in real life we often are able to present only a limited view of ourselves to those we encounter—usually contingent on the situation we are in—the webpage allows people to explore and present as many aspects of their identity as desired. Several other technologies are being used in this way to varying degrees, including the inclusion of a tag line with a favorite quote attached to an electronic message, perhaps installing a webcam to one's computer in order to broadcast aspects of one's daily life over the web. While many value new media technologies as a novel way of expressing who they really are, others are interested in the anonymity afforded by the Internet. Many visitors of chat rooms, MUDs, and online gaming environments have explored the multiplicity of personas available to them as disembodied Internet users. This practice has been valorized by several writers as a way of deconstructing the limiting categories of gender, race, and class that surround and influence our offline lives. Others, however, fear that the naive passing that occurs in virtual environments works only to reinforce already existing prejudices and stereotypes.

With respect to the field of education in general and higher education in particular, the study of cyberculture, or cyberculture studies, has become a growth industry. Stretching back to the early and mid-1990s with the work of such thinkers as Julian Dibbell (1993), Howard Rheingold (1993), and Sherry Turkle (1995), and gaining steam during the mid- to late 1990s with the influential anthologies edited by Steve Jones (1995, 1997), the field of cyberculture studies (alternatively called Internet studies or new media studies) is currently undergoing institutionalization thanks in part to the publication of *New Media and Society,* the first conference of the Association of Internet Researchers in 2000, and select universities offering degree programs on the topic.

As witnessed in the countless monographs and anthologies published by academic and popular presses, and the increasing number of papers and panels presented at scholarly conferences from across the disciplines and around the world, the field of cyberculture studies is a diverse and growing field of inquiry. It is possible, however, to trace the major works of scholarship on cyberculture by establishing three stages or generations. The first stage, *popular cyberculture,* is marked by its journalistic origins and characterized by its descriptive nature, the tendency to divide the debate between utopian and dystopian arguments, and the use of the Internet-as-frontier metaphor. The second stage, *cyberculture studies,* focuses largely on virtual communities and online identities and benefits from an influx of academic scholars. The third stage, *critical cyberculture studies,* expands the notion of cyberculture to include four areas of study—online interactions, digital discourses, access and denial to the Internet, and interface design of cyberspace—and explores the intersections and interdependencies among any and all four domains.

Our disciplinary lineage begins with popular cyberculture, a collection of essays, columns, and books written by particularly wired journalists and early adopters. Starting in the early 1990s, these cultural critics began filing stories on the Internet, cyberspace, and the information superhighway for major U.S. newspapers and magazines. Significantly, what began as an occasional column in a newspaper's technology section soon became feature articles appearing on the front page, in the business section, and in lifestyle supplements, as well as within the new media/cyberspace beat of many mainstream magazines. Between 1993 and 1994, for example, *Time* magazine published two cover stories on the Internet, and *Newsweek* released the cover story "Men, Women, and Computers." Moreover, in 1994 the second editions of the popular how-to books *The Internet for Dummies* and *The Whole Internet* became best-sellers.

The popular cyberculture writings were generally descriptive. Usually required to follow the term "Internet" with the parenthetical phrase "the global computer network system," these journalists had the unenviable task of introducing nontechnical readers to the largely technical, pre–World Wide Web version of cyberspace. Accordingly, much of this work included lengthy descriptions, explanations, and applications of early Internet technologies such as file transfer protocol, gopher, lynx, UNIX configurations, telnet, and usenet.

In addition to being overly descriptive, early popular cyberculture often suffered from an unproductive dichotomy: Early popular cyberculture often took the form of dystopian rants or utopian raves. From one side, cultural critics blamed the Internet for deteriorating literacy, political and economic alienation, and social fragmentation. Often branded neo-Luddites, these critics feared a society based on mediation rather than face-to-face communication and wondered aloud whether our children would find themselves more at home in front of a computer screen rather than at school, on the playground, and around the dinner table. From the other side, a vocal group of writers, investors, and politicians loosely referred to as technofuturists declared cyberspace a new frontier of civilization, a digital domain that could and would bring down big business, foster democratic participation, and end economic and social inequities. Although finding friendly platforms within major U.S. newspapers and popular magazines, their primary pulpit was a new line of technozines—glossy, visually impairing magazines with names like *Mondo 2000, bOing bOing,* and *Wired.*

Finally, in addition to its descriptive nature and rhetorical dualisms, early popular cyberculturalists employed the frontier as the reigning metaphor. For example, in the now canonical 1990 essay "Across the Electronic Frontier," Mitchell Kapor and John Perry Barlow (1990) described the Internet in the following terms: "In its present condition, cyberspace is a frontier region, populated by the few hardy technologists who can tolerate the austerity of its savage computer interfaces, incompatible communication protocols, proprietary barricades, cultural and legal ambiguities, and general lack of useful maps or metaphors."

If early cyberculture scholarship can be characterized by its descriptive nature, binary dualism, and frontier metaphors, the second stage, cyberculture studies, rests upon the twin pillars of virtual communities and online identities. One of the earliest and certainly the most referenced articulators of virtual communities is Howard Rheingold (1993), who defines a virtual community as: "A group of people who may or may not meet one another face-to-face, and who exchange words and ideas through the mediation of computer bulletin boards and networks. . . . We do everything people do when people get together, but we do it with words on computer screens, leaving our bodies behind" (58).

If Rheingold's *The Virtual Community* is the first pillar of cyberculture studies, the second is Sherry Turkle's *Life on the Screen: Identity in the Age of the Internet* (1995). Turkle addresses the idea of online identities by exploring ethnographically a number of virtual environments, including multiuser domains, or MUDs. She finds that while some people use cyberspace to repress an otherwise less-than-functional "real" or offline life, most use the digital domain to exercise a more true identity or a multiplicity of identities. According to Turkle (1995, 263), "Virtuality need not be a prison. It can be the raft, the ladder, the transitional space, the moratorium, that is discarded after reaching greater freedom. We don't have to reject life on the screen, but we don't have to treat it as an alternate life either."

By the mid-1990s, cyberculture studies was well under way, focused primarily on virtual communities and online identities. Furthermore, as a result of the enthusiasm found in the work of Rheingold and Turkle, cyberculture was often articulated as a site of empowerment, an online space reserved for construction, creativity, and community. Fortunately, however, this simplification was matched by the richness found in the nascent field's welcoming of interdisciplinarity. With the growing popularity of user-friendly Internet service providers such as AOL and CompuServe and the widespread adoption of Netscape, the great Internet rush was on. Significantly, the introduction of the web was not only a *technological* breakthrough but also a *user* breakthrough. Replacing tricky file transfer protocol and burdensome gopher with a simple, point-and-click graphical interface, the web helped to foster a less technical, more mainstream Internet populace. Coupled with these technological breakthroughs were academic considerations. In addition to a concerted effort on the part of university administrators to get faculty wired, scholarly conferences, papers, archives, and discussions came online, leading all but the most technophobic academics to the Internet.

As expected, new scholars brought new methods and theories. For example, while some sociologists approach virtual communities as "social networks," others employ the sociological traditions of interactionism and collective action dilemma theory. Within anthropology, scholars began formulating a new subfield, cyborg anthropology, devoted to exploring the intersections between individuals, society, and networked computers. Researchers from a related field, ethnography, took their cue from Turkle and began to study what users do within diverse online environments, ranging from online lesbian bars and usenet newsgroups to web-based "telegardens" and online cities.

At the same time, linguists began to study the writing styles, netiquettes, and (inter)textual codes used within online environments. Similarly, feminist and women's studies researchers have used textual analysis and feminist theory to locate, construct, and deconstruct gender within cyberspace. Furthermore, a collection of community activists and scholars began to explore the intersection of real and virtual communities in the form of community networks, including the Public Electronic Network in Santa Monica, California, the Blacksburg Electronic Village in Blacksburg, Virginia, and the Seattle Community Network in Seattle, Washington.

Although cyberculture studies strays in many directions, a product in part of the massive influx of scholars from across the disciplines, an increasingly well-traveled path is taking shape and represents a third stage of work: critical cyberculture studies. Influenced heavily by recent scholarship in critical theory and cultural studies, critical cyberculture studies deals primarily with the social construction of cyberspace and focuses especially on issues of cultural difference, consumerism, and power. Whereas early proponents proclaimed the Internet as a virtual space free of gender, race, and class, critical cyberculture studies scholars place issues of cultural difference in the center and explore the ways in which policy, code, and rhetoric reveal and reflect cultural differences. Similarly, wary of discussions of the digital divide framed by corporate computer interests suggesting that the appearance of hardware will ensure access, contemporary scholars locate the debate within a matrix of political, cultural, and economic considerations.

In some ways, critical cyberculture studies represents an intervention into the dot-com daze that practically overcame the Internet in the late

1990s. While acknowledging the contribution of companies like AOL to broaden the spectrum and diversity of the Internet's population, critical cyberculture studies encourages us to think about the differences between communication and consumption, e-mail and e-commerce, and active participation and passive reception.

Perhaps more than anything, critical cyberculture studies seeks to position cyberspace and its attendant technologies in as large and broad a context as possible. For example, contemporary scholars are beginning to look at the Internet within a historical context, comparing its social, technological, and economic developments to those found in the history of other once-new communication technologies like film, radio, and television. In this light, new media's "newness" becomes less important in favor of a more holistic sequence of media history, one that encourages critical perspectives and affords a broader understanding of media convergence.

David Silver
Donald Snyder

See also Communities of Practice; Community Technology Centers; Digital Divide; MOOs and MUDs; Netiquette; Virtual Reality; Weblog

References

Association of Internet Researchers (www.aoir.org).

Bell, David, and Barbara M. Kennedy, eds. 2000. *The Cybercultures Reader.* London: Routledge.

Dibbell, Julian. 1993. "A Rape in Cyberspace; or How an Evil Clown, a Haitian Trickster Spirit, Two Wizards, and a Cast of Dozens Turned a Database into a Society." *Village Voice* (December 21): 36–42.

Falling through the Net: Defining the Digital Divide. 1999. Available online at www.ntia.doc.gov/ntiahome/fttn99/contents.html.

Gauntlett, David, ed. 2000. *Web Studies: Rewiring Media Studies for the Digital Age.* London and New York: Arnold Publishers and Oxford University Press.

Internet Studies Center (www.isc.umn.edu).

Jones, Steven G., ed. 1995. *CyberSociety: Computer-Mediated Communication and Community.* Thousand Oaks, CA: Sage.

———. 1997. *Virtual Culture: Identity and Communication in Cybersociety.* London: Sage.

Kapor, Mitchell, and John Perry Barlow. 1990. *Across the Electronic Frontier.* Available online at www.eff.org//Misc/EFF/electronic_frontier.eff.

Kolko, Beth, Lisa Nakamura, and Gilbert Rodman, eds. 2000. *Race in Cyberspace.* New York and London: Routledge.

Resource Center for Cyberculture Studies (www.com.washington.edu/rccs).

Rheingold, Howard. 1993. *The Virtual Community: Homesteading on the Electronic Frontier.* Reading, MA: Addison Wesley.

Turkle, Sherry. 1995. *Life on the Screen: Identity in the Age of the Internet.* New York: Simon and Schuster.

D

Diffusion of Innovations

"*Diffusion* is the processes by which an innovation (a new idea) is communicated through certain channels over time among the members of a social system" (Rogers 1995, 5). There are four factors that interact to influence the diffusion of an innovation: (1) the innovation (e.g., educational technology); (2) the communication channel (e.g., media, word of mouth); (3) time; and (4) the social system (e.g., school, university). Everett M. Rogers has done the most to synthesize all of the significant findings and theories related to diffusion. His book *Diffusion of Innovations,* first published in 1960 and now in its fourth edition (1995), has become the standard reference in this area. It provides a theoretical framework for models of diffusion and offers a critical examination of the research evidence.

Diffusion theory is valuable to the field of educational technology for several reasons (Surry 1997). First, the causes of educational technology's diffusion problems remain unclear; some blame the teacher while others blame the institutions. Second, educational technology is an innovation-based field in which the various products and methods used represent innovations in the form, organization, and delivery of instruction. Third, by studying diffusion theory, it is possible to come up with a model for adoption and diffusion of educational technology. Diffusion theory provides important guidance for those interested in seeing educational technology adopted within all levels of the educational system. There are five concepts that are central to diffusion of innovations theory: (1) the innovation-decision process; (2) attributes of successful innovations; (3) adopter categories; (4) characteristics of successful change agents; and (5) consequences of innovations.

Innovation-Decision Process

The innovation-decision process may be considered from the perspective of an individual or an organization. Potential adopters have to first know about an innovation, be persuaded as to the merits of such an innovation, decide to adopt or reject the innovation, implement the new idea, and finally confirm their decision whether or not to continue with the original decision.

This process is predictable regardless of the innovation and provides a framework for those working toward diffusing educational technology in the classroom. The type of innovation-decision has an important influence on the rate of adoption of an innovation. Organizational adoption of innovations (e.g., school or university adoption) tends to fall into one of three categories:

1. *Optional* innovation-decisions are choices to adopt or reject an innovation that are made by an individual or an organization. A school or a university may choose to use a certain educational technology idea or teaching method. Teachers may as well have their own initiatives to teach using certain educational technology techniques or methods.
2. *Collective* innovation-decisions are choices to adopt or reject an innovation that are made by consensus among members of an organization. A school board or a university committee may collectively decide to standardize certain educational technology tools and methods. Such a decision is normally made through agreement by the majority members of the organization.
3. *Authority* innovation-decisions are choices to adopt or reject an innovation that are made by one or more individuals in an organization who possess power, status, or technical expertise. An educational policy may require all employees to adopt and use certain educational technology innovations regardless of the opinion of the teachers.

Attributes of Successful Innovations

For an innovation to be readily and easily adopted, it is usually evaluated with respect to five attributes. First, an innovation must have relative advantage to the potential adopter. For example, an educator should clearly see the benefits of educational technology in enhancing the teaching process over traditional teaching methods. Second, the innovation must fit and be *compatible* with existing working patterns and experiences. Adopting an innovation should be consistent with an organization's existing values, past experiences, and needs. Third, a new innovation such as educational technology should allow for experimentation (trialability) by potential adopters before they make their decision to adopt or reject the idea. Fourth, the positive results of an innovation should be visible to others. Teachers need to see the results of educational technology use by their colleagues and examine its effect on others' classes in order to decide whether or not to use it in their own classes. Fifth, an innovation that is perceived as difficult to use and complicated will be adopted more slowly than others. These attributes and many others have been found to play an important role in several technology-related adoption studies.

Adopter Categories

Adopters of an innovation are usually categorized based on the degree to which an individual is relatively earlier in adopting new ideas than the others (innovativeness). Typically, there are five main adopter categories:

1. *Innovators:* individuals who are obsessed with new ideas and believe that there is always a better way to do things. For example, innovators in the field of educational technology may invent new ideas, but most of the time they improve the methods by which already invented tools are used. Innovators find ways of overcoming obstacles to their plans, love challenges, and usually never stop until they achieve their goals.
2. *Early Adopters:* people who are the first to adopt an innovation in an organization. They tend to have a higher level of formal education and a higher social status. They are the individuals who are considered by many as "the individuals to check with" before using a new idea. Such people serve as a role model for speeding the diffusion of an innovation since they are not too far ahead of the average individual in innovativeness.
3. *Early Majority:* individuals whose innovation-decision period is relatively longer than that of the innovator and the early adopter. They follow with the intention and willingness to adopt an innovation, but they seldom lead.
4. *Late Majority:* those who adopt new ideas just after the average member in an organization. The adoption for the late majority is mostly due to increasing network pressure from peers. For example, as more teachers adopt educational technology for their teaching and learning activities, people in the late majority feel that they are left behind and realize the increasing pres-

sure from their students and from the organization to catch up with the rest.

5. *Laggards:* the last people in an organization to adopt an innovation. Their decisions are often made based on what was practiced in the past, and their interaction is normally with those who have relatively traditional values. An example of this category would be traditional teachers who have served for a very long time and don't see a compelling reason to adopt new educational technologies. From their perspective, what they have been doing traditionally works just fine, and there is no need to learn new technologies and methods to change the teaching and learning processes.

Individuals falling into these adopter categories for any given innovation tend to follow the normal curve with a few innovators (2.5 percent), early majority (13.5 percent), and laggards (16 percent), and many early and late majority categories (68 percent to 34 percent each).

Characteristics of Successful Change Agents

A *change agent* is an individual who influences the innovation-decision of others. Change agents play a key role in the innovation-decision process and can be opinion leaders or peers of the potential adopter. Typically, change agents desire to secure the adoption of the innovation. However, change agents may assert their power of persuasion to slow or stop the diffusion of innovations they perceive as harmful to potential adopters (Rogers 1995). Teachers, students, and institutions are all seen to be effective change agents in the diffusion process of educational technology. Successful change agents have a strong client orientation as opposed to an orientation toward the innovation. They tend to understand their clients and have credibility, possibly because of similarities to the client. Teachers are ideal change agents because of their homophily with other teachers.

Near-peer influence, institutional support, and colearning with students are methods that can provide the channels for change agents to diffuse educational technology use in schools and universities. Students with technology skills have a powerful influence to assist and persuade teachers to adopt technology in their instruction. Finally, institutional support is essential to creating an environment conducive to the adoption of educational technology use among teachers and students alike.

Consequences of Innovations

The adoption or rejection of an innovation leads to certain consequences that may affect an individual or an organization. There are desirable or undesirable consequences, direct or indirect, and anticipated or unanticipated. It is usually hoped that an innovation will lead to desirable, direct, and anticipated consequences. However, consequences do not always occur as intended, leading to undesirable, indirect, and unanticipated consequences. Although it is possible to describe and establish categories for consequences of innovations, it is difficult to generalize or predict when and how consequences will happen.

Diffusion Theory

Educational technology is an innovation-based field that follows an unpredictable path guided by highly dedicated and motivated individuals. These individuals share a common vision and work to bring that vision into reality. Motivated teachers, students, and administrators usually work hard to make educational technology better understood, successful, and diffused faster in an institution. Rogers's 1995 work mostly analyzed distinct innovations whose characteristics were not changed by the adopters during the diffusion process. However, educational technology cannot be described as static or unchangeable innovation. Adopters of educational technology tend to reinvent and mold the innovation in accordance with their needs and perceptions. This may cause the innovation to be unmanageable unless a centralized organizational structure is developed and implemented to maintain stability and ensure continuous growth of the innovation.

Possible Strategies for Diffusion

For educational technology to be widely adopted as a tool for teaching and learning, several strategies should be considered. As mentioned earlier, most members in a social system or organization fall in the early and late majority categories. This category of potential adopters should be targeted the most. They should be recognized as a distinct group within the institution and made a part of the planning and policymaking process. Attempts to "convert" them to a point of view are likely to be fruitless, and imposing the new technology on them would be disastrous. Diffusion of the innovation to the late majority and laggards is more likely to occur through this early majority involvement since the vertical lines of communication between the three groups are more direct than with the innovators and early adopters.

The problems of diffusing educational technology may vary widely from one institution or social setting to another. Technical and pedagogical support such as infrastructure, expertise, and training are some of the problems normally cited by educators. Training programs will help introduce educational technology to teachers and students and teach them how technology can assist them in teaching and learning tasks. The changing nature of technology makes the case that change is making technology simpler and easier to use. This will encourage educators to explore emerging technologies and to integrate technology into their teaching.

Additional barriers to the effective diffusion of educational technology are related to insufficient support by institutions. There is a big role to be played by technical personnel in institutions to provide support for teachers and students alike. Institutions should develop a scheme for appropriate recognition or acknowledgment of the extra effort invested in developing new approaches to teaching. Teachers may complain about the time-consuming nature of developing technology-based materials and approaches given the scarcity of time available to commit to such tasks.

As for students, educational technology can be used as a way to add excitement to the learning process. Methods can be used to produce education-entertainment programs such as educational CD-ROMs and games that combine teaching with entertainment. Today, students in schools and universities are highly skilled in using computers, the Internet, and other related technologies. Students have been acting as change agents among themselves, and there is a need to extend their skills to reach the educators in order to encourage the adoption of educational technology and its integration into the curriculum.

Ziad Akir

See also Communication Theory; Entertainment-Education; Instructional Communications; School Reform

References

Rogers, Everett M. 1995. *Diffusion of Innovations.* New York: The Free Press.

Surry, Daniel W. 1997. "Diffusion Theory and Instructional Technology." Available online at intro.base.org/docs/diffusion.

Digital Divide

The fundamental premise of the so-called digital divide is that factors such as income, race, gender, age, location, and education affect access to technology and the equity of technology experiences overall. It is a multifaceted problem and involves unequal opportunities regarding information technologies, in the United States and globally. Put simply, the digital divide describes the differences between the technology/information "haves" and "have-nots." Yet the daily use of technology in society expands, so an individual's fluency with technology and his ability to use technology as a functional tool in daily life will need to be considered. The many obstacles include educational experiences, workplace opportunities, access to information, and communications opportunities.

Some may argue that the digital divide is closing (e.g., the number of computers in classrooms has mushroomed in recent years). But mere numbers do not always translate into access and learning opportunities for all students. Learning activities and experiences need to be evaluated.

Much of the early research on the digital divide focused on the number of computers in schools. In 1992, results showed that schools with a high percentage of poor and minority students had fewer computers available to students, with the average ratio being 19.2 students to one computer. Strides have been made in giving poor and minority students more access to technology. In one report (Coley, Cradler, and Engel 1998), the penetration of computers in U.S. schools was examined using data from a 1997 survey. The report stated that 98 percent of all schools own a computer, with the typical number of computers per school being between twenty-one and fifty. The average ratio was 5.7 students to one computer. However, schools with a high percentage of minority and Title I students have an above-average student-to-computer ratio. Schools containing 90 percent or more minority students have an average ratio of 17.4 students to one computer, and as the percentage of Title I students increase, so does the ratio of students to computers.

Students in K-12 and higher education school systems who are impacted by the digital divide may be at a disadvantage without the equivalent learning experiences. This could occur in a variety of ways. These students might not have the exposure to much of the technology that the workplace uses. They also might not have the opportunities to learn how to acquire, organize, and evaluate instructional resources. Finally, these students often engage in different learning activities than students not affected by the digital divide. Underserved students are frequently asked to complete drill-and-practice tasks rather than research-oriented learning exercises.

The digital divide also influences an individual's ability to have access to information and opportunities in the workplace. Research has shown that there is a disparity among various ethnic groups with respect to computer ownership. A Department of Commerce (1999) report found that African Americans and Hispanics are far behind Asian Pacific Islanders and whites when it comes to owning a computer. The gaps could partly be explained by income disparities.

One can examine the digital divide from another perspective by looking at the use of technology to function in society. In a Department of Labor (1991) report, industry leaders stated that it was important for employees of the future to be fluent in the use of technology. With information technology evolving at such a rapid pace, individuals without opportunities for ongoing training will quickly fall behind better prepared individuals. Mere access to technology will not help these individuals; training will. An examination of how to work effectively within various communities must occur in our society. The training of various minority groups, the poor, the elderly, rural residents, and others is fundamental to providing the skills necessary to function in an increasingly technologically advanced society. Some organizations and agencies are working to improve technology access and fluency for individuals affected by the digital divide. For example, community technology centers are working with many rural and urban populations to provide additional access and training. As technology continues to penetrate all aspects of society, all individuals will need to be fluent in the use of technology.

Finally, opportunities to communicate with computer technology is another issue deserving attention. The ability to use communications tools available on the Internet is another place that training is needed. However, there are other avenues to investigate. The content of the Internet is considered to be a stumbling block for many individuals affected by the digital divide. A Children's Partnership (2000) study reported that much of the material on the Internet does not provide a service for many Americans. The study findings show there are barriers such as literacy levels, language, lack of cultural diversity, and lack of local information. Appropriate content could certainly be a reason more poor and minority households are not using the Internet. All organizations, institutions, industries, and educators that use the Internet need to be aware of this problem with online content and encourage the creation of projects and groups that can work to alleviate obstacles for all individuals using the Internet.

Colleen Swain

See also Community Technology Centers; Cyberculture and Related Studies

References

Children's Partnership. 2000. Online content for low-income and underserved Americans. Available online at www.childrenspartnership.org/pub/low_income/low_income.pdf.

Coley, Richard, John Cradler, and Penelope Engel. 1996. *Computers and Classrooms: The Status of Technology in U.S. Schools.* Princeton, NJ: Educational Testing Service.

Digital Divide Network. Available online at www.digitaldividenetwork.org.

Lazarus, Wendy, and Francisco Mora. 2000. *Online Content for Low-Income and Underserved Americans: The Digital Divide's New Frontier.* Available online at www.childrenspartnership.org.

SCANS (U.S. Department of Labor, Secretary's Commission on Achieving Necessary Skills). 1991. *What Work Requires of Schools: A SCANS Report for America 2000.* Washington, DC: U.S. Department of Labor.

U.S. Department of Commerce. 2000a. *Falling through the Net: Defining the Digital Divide* (third in a series). Available online at www.ntia.doc.gov/ntiahome/fttn99/FTTN.pdf.

———. 2000b. *Falling through the Net: Toward Digital Inclusion* (fourth in a series). Available online at search.ntia.doc.gov/pdf/fttn00.pdf.

Digital Video

Digital video is a technology used to record and display moving images as binary data. Because video in digital format is transmitted using binary code (a system that communicates information in a series of discrete signals represented as 1's or 0's), it offers several advantages over analog video, a format that transmits information in a continuous range of electromagnetic waves. Producers and consumers of this technology benefit from its efficient use of transmission and storage space, its high-quality, robust signal, its compatibility with other information in digital format, and its user-friendly viewing and editing interfaces. These advantages contribute to making digital video an increasingly popular means for delivering education and training content via CD-ROM and the World Wide Web. The recent technological advances involving the production of digital video equipment, compression schemes, and network communication modes suggest it will remain the standard for some time to come.

Analog versus Digital Video Formats

Analog, the signal format most common before the rise of the digital signal format, is used with various technologies, including telephone communications, broadcast radio, and television. Analog appliances record or transmit information as a continuous electrical signal composed of electromagnetic waves. In analog video cameras, waves are created by a charged coupled device. This device captures visual information and transforms it into an electrical signal that can be broadcast directly or recorded on magnetic tape. When analog video is sent from its source through the air, wire, or other network conduit, the electromagnetic signals weaken. During transmission, signals run the risk of picking up vibrations from other sources that can result in errors or noise. To compensate for the weakened strength of signals at a distance, amplification is performed at various points during transit. The amplification of a weakened and sometimes error-filled signal often results in reduced signal quality and distortion.

By contrast, digital data travels as a series of discrete, high- and low-voltage pulses representing binary data. In digital video cameras, the charged couple device transmits high-voltage pulses that represent a 1 and low-voltage pulses that represent a 0. Simple and discrete in form, these pulses can travel great distances with little deterioration in signal quality. When signal boosting is necessary, pulses can be replicated identically through signal regeneration. Digital signals do not pick up noise or vibrations in the way that analog signals do.

Advantages of Digital Video

The basic differences between the analog and digital data formats account for several advantages of video recorded and transmitted in digital format.

Transmission and Storage Space

First, digital video makes more efficient use of storage and transmission space because it is composed of discrete pulses rather than continuous waves. In its natural, uncompressed state digital video requires less recording and transmission space. When it is compressed (a process that uses mathematical algorithms to eliminate extraneous data) it requires even less. As a result, digital video requires less network bandwidth for transmission and less storage space for recording data. Extra space in the network conduit is made available for multiple signals. Interactivity, or two-way communications, often results from using extra space. When storing digital video, increased efficiency results in better use of storage units such as cassettes, disks, and hard-drive space. Efficiency also results in products (digital videocassettes, cameras, and other equipment) that are much smaller.

Superior Quality

Because digital data can be stored efficiently, large amounts of data can be captured in less space. This results in the second advantage of digital video: superior quality. Digital video reproduces images that have greater accuracy to their original source. Betacam, an expensive professional analog video, is the only analog standard that rivals the quality of digital video. Higher-quality reproduction of moving images results from increased resolution, greater color rendition (the ability to produce accurate colors), and increased frame rate (the number of still images captured per second). For example, the mini-DV format captures 500 lines of horizontal resolution, a nearly 25 percent improvement over pictures recorded in the S-VHS or Hi–8 analog formats.

Robust Signal

Another advantage of digital video is its robust signal. This signal is free of problems encountered by an analog signal when it is transmitted over distances or copied. As a result, it is possible to dub multiple generations, or additional copies from other copies of the original source information, without signal loss. Reproduction of digital video is less expensive because the process does not require as much time or as many resources.

Random Access

One more significant advantage of digital video is its support for random access of stored or transmitted information. This means that information recorded in digital format can be and usually is edited in a nonlinear format. Unlike linear editing, where moving forward and backward through the medium is required before editing can occur, nonlinear editing allows instant access anywhere in the data. This instant access saves time and allows greater creative control over material. For example, changing or switching one shot with another in nonlinear editing is a difficult proposition; reediting of the entire piece is required to change even one frame. But editing video in nonlinear format is easy and takes little time. Nonlinear editing also makes it easy to link portions or clips of video together with transitions like dissolves and fades. It also makes it easier to modify the data and add special effects such as picture-in-picture and colored filters. Although it is possible to digitize analog video and edit it using nonlinear editing, this process requires special equipment and results in the degradation of video quality. Nonlinear editing systems work easily and fluidly with digital video, requiring no special equipment and preserving video quality.

The Production Process

Content

The creation of digital video content involves several steps. First, content is recorded in a digital or analog video signal. This signal is stored on a tape. Popular analog tape formats include VHS, S-VHS, Hi–8, and Betacam. Popular digital video formats include Sony's DVCAM, HDCAM, and DigiBeta and Panasonic's DVCPRO. The method of storing data on the tape is the main difference among the different tape formats. The source and the quality of the initial video content that is recorded will vary considerably. Source content may range from previously edited, broadcast-quality data to raw, unedited images. Source content might be recorded directly through the camera to the tape or prerecorded. The techniques for shooting video vary depending on the purpose and use for the footage. Video that will be viewed in conventional ways (e.g., on a television or monitor) will likely be filmed in different ways.

Capturing

Second, video content is "captured" using a video capture card—an internal device within a computer that allows individuals to input video as binary data. Video capture cards range in quality based on cost. Cards that support high video resolution, color rendition, and frame rates are more expensive than those with less resolution, no color rendition, and smaller frame rates.

Users have many options when selecting nonlinear editing systems because they vary in sophistication, quality, and cost. Low-end systems, the type commonly used by amateurs or hobbyists, can be constructed from integrating several basic computer components. These include a personal computer with a fast processor, video capture card, sound card, and SCSI (small computer systems interface) controller. In order to control this hardware and edit the video, a simple video editing software program must be installed and configured. Apple's iMac computer, marketed as a preconfigured, low-end, easy-to-use system, represents one popular option.

Middle-range systems, the kind commonly used by low-budget production companies and schools, consist of the basic hardware and other equipment such as video decks, mixing boards, disk arrays, and recording equipment. More powerful software programs such as Premiere and Dazzle and Final Cut Pro are the most commonly used software options.

High-end, commercial, nonlinear editing systems, the kind used by large-scale production houses producing movies and TV programs, consist of basic hardware components modified to work specifically with proprietary nonlinear editing software. Such systems are available from major multimedia companies, including Media 100 and Avid.

Editing

Third, video content is edited and saved as an uncompressed file. This step in the production process marks the greatest advantage of digital over analog video. Digital editing software programs allow nonlinear interaction with digital content. Editing video using a digital editing software program allows instant access to specific images that may be located anywhere in the content. This process enables greater creative control over the editing process. Transitions, titles, and special effects can be easily inserted, and audio can be mixed seamlessly and simultaneously. Although the output of digital video is a linear sequence (e.g., a series of images edited together to tell a story), constructing it in a nonlinear process facilitates creativity and spontaneous decisionmaking about content. Once edited, the video image is saved as an uncompressed file. Several standard file formats are commonly used for uncompressed files. These include Apple's Quicktime (.MOV, .MOOV) and Video for Windows (.AVI).

Compression and Distribution

Because uncompressed files are often incredibly large (often exceeding two gigabytes for four-minute video clips), the content is compressed for distribution and then delivered to the viewer. Data compression transforms a data into a code that is smaller than the original code. Methods of coding data vary significantly and are evolving. Perhaps the most significant difference in data-compression processes exists between the lossy compression and the lossless compression schemes. Lossy compression, the kind used in video and audio compression, discards some information during the compression process. Because video and audio are not an exact medium (meaning they can still communicate their message with some loss of information), the human ear and eye do not notice the loss of some data provided it is the right type of data. Lossless compression, the kind used when exact information is required (like when sending text information), does not discard data. Compression performance is measured by the size of the output stream compared to the size of the input stream.

Regardless of the compression process used for coding, a compressed file must be decoded before it can be viewed. Decompression is performed by software installed on the output computer. This software is referred to as a player or plug-in. This software may be preinstalled on a user's computer operating system or web browser. If not, it must be packaged with digital video content.

The standard compressions for coding and decoding video content are called codecs. The codec selected for distribution of digital video will vary based on the distribution medium and file format determined to be appropriate. Each codec is created for specific purposes. Professional video developers make decisions about appropriate codecs based on their knowledge of these codecs and the technological limitations of their intended audience.

The delivery of digital video is undergoing an even more dramatic evolution than other aspects of digital video production. The most important development in the delivery of digital video content has been the development of streaming media. Streaming media is prerecorded or live content transferred to users on demand. Formats created by RealNetworks, Windows, Apple, and nonprofit groups such as the Moving Pictures Expert Group are competing to become the standard in streaming media formats. Streaming media resides on a server or web server and can be accessed through links on. HTML files situated throughout the Internet or through direct addressing. Streaming media turns digital video into fluid content that resembles other conventional media technologies such as broadcast and cable TV.

Technological Advances

Continual technological advances resulting in improved compression rates, network bandwidth, software usability, and equipment costs have contributed to the growing preference for digital video. These advances influence every aspect of digital video production, from the recording of video content to its capture, production, and output on disk, tape, CD, or the web. As a result, users at every skill level, from beginners to professionals, are experiencing increased accessibility to the medium and the advantages it offers in the way of quality, usability, and compatibility.

Progress in the hardware industry is steadily improving the performance of video capture cards, computer ports, processors, and other devices associated with digital video. The improvement of the information transfer rates (the speed of sending information from a user to a device that controls it) in cameras, microprocessors, and media drives (special hard disks used to store video data) continually speed up the amount of time required to produce video. In large commercial video production houses, where time is money, speed increases have shortened production times, resulting in increased profitability. Amateurs have also appreciated the ability to create small video clips without large investments of time. It is becoming more common for individuals to create and edit their own home videos on personal computers, an activity previously limited to professionals with expensive equipment.

This progress has been accompanied by the creation of more efficient codecs. By using various codecs to compress digital video, large, high-quality files can be reduced for storage and transmission. The development of more sophisticated compression algorithms has led to both a decrease in the file space required for storing digital video and the power of hardware for playing it. Software programs have made it easier to determine appropriate codecs for video depending on the method by which they will be stored and transferred.

The popularity of digital video is also related to the physical expansion of global and local network infrastructures. The network expansion that promises the most dramatic impact on the use of digital video by the masses (specifically for entertainment purposes) is the placement of broadband networks in residential neighborhoods. Here, narrowband access through telephone lines is being replaced by broadband access enabled by cable and digital subscriber lines. These expansions could help to overcome bottlenecks (caused when information travels from the broadband Internet to narrowband residences) and thereby enable video use. Networks of all sizes have had difficulty keeping up with growing demands for bandwidth. If network bandwidth fails to meet the growing need for digital video, its increasing popularity as a medium for education and training on the Internet will likely wane.

Groups working to support the increasing use of digital video have begun to install dedicated wide area networks. One such network (Internet2) has been in use by colleges and universities since February 1999. This network boasts huge transfer rates and is becoming more and more common for businesses and corporations to install on their own internal dedicated high-speed networks (intranets). Corporate intranets are already providing education and training materials, including videos, for employees. In fact, many of the best examples of digital video training materials are being developed on intranets. Although the growth of these networks means support is provided for the applications of digital video, it is unfortunate that these resources are not accessible to the public. The removal of private resources from the Internet and access promise to be issues in the future for digital video.

Software for recording and editing and playing digital video is also improving dramatically. The increased usability of nonlinear digital editing software programs with graphical user interfaces has made complex editing processes simple and enabled greater creative control over the medium. Nonlinear editing programs have lowered the technical skill required to create digital video. As a result, more individuals have begun using these programs, resulting in greater sales and declining product costs. Plug-ins/players are also becoming easier to use and more affordable (free, in most cases). As a result, access to digital video on the Internet or CD-ROM is easier and faster and becoming more affordable for the general population.

Clare R. Kilbane

See also Interactive Television

References

Alesso, H. Peter. 2000. *e-Video: Producing Internet Video as Broadband Technologies Converge.* Boston: Addison-Wesley.

Distance Education

Distance education has gained increasing attention from the public and would appear to be a modern-day phenomenon. It has, however, been around a long time but has only recently become known as distance education and distance learning. As the terminology has evolved, so have the components that define this type of instruction.

The most obvious is the physical distance that separates teachers and learners. Beyond that, practitioners in the field would include the use of mechanical or electronic means for delivery of content; interaction between teachers and learners (with a requirement that any participant can initiate interaction with the others); and the influence of a formal educational organization such as a training department or a university (sometimes referred to as a "contract") that delineates the roles of the participants, the expectations, and the anticipated outcomes. (This final component is what distinguishes distance education from informal self-study or information dissemination in the form of a radio news program, for example.)

Examples of distance education are as varied as the technologies that provide content and as divergent as the needs of the student population. Low-tech options for learning at a distance include print-based correspondence study, videotaped classes that are distributed through the mail or private delivery services, and live audioconferences during which students call in to a telephone bridging service to engage in a real-time discussion with the instructor and other students. The familiarity of broadcast TV has long played a role in distance education, with lavishly produced telecourses (often featuring nationally recognized experts in the field acting as commentators, speakers, or hosts) aired on public stations with faculty support and college credit provided by individual postsecondary institutions. Online and computer-based instruction has garnered the greatest attention in the distance learning field, with course management software systems providing easy-to-use templates for the development of web-based coursework. Instructionally, many of these options closely resemble their historical antecedent of correspondence study, albeit with greater flexibility for participant interaction and the potential to integrate a variety of instructional media.

Overall, the purposes of distance education eventually boil down to three guiding principles. The first is the decision to provide educational opportunities for unserved or underserved populations. This would include those students for whom traditional coursework is unrealistic due to geographic factors or to competing priorities of family or employment, as well as individuals who may be homebound or otherwise unable to travel. In some instances, distance delivery of instruction can break through cultural barriers to educational opportunity, as in the case of religious strictures limiting face-to-face interactions among men and women, for example.

The second purpose for distance delivery of instruction is to save money. This is particularly true in the corporate/industrial training sector, where students may have previously been flown to a central location for classes. By providing courses directly to the student's workplace, the company not only saves travel costs but also does not lose the productivity of that employee during this period. Even though many delivery systems are relatively expensive compared to traditional in-person courses, the savings over time have proven to be a desirable incentive for its implementation. Few, if any, traditional K-12 or postsecondary schools have undertaken distance education programs as a way to cut costs, however, and instead find such endeavors to be significantly more expensive than continuing to offer courses in a face-to-face configuration.

Finally, distance education can help traditional schools offer a much wider variety of courses to students. In a rural school district in South Dakota, employing a teacher of Japanese (if one could even be found in the area) for the few students interested in studying this language would be cost-prohibitive. However, by utilizing the economies of scale possible when students across a large geographic area take Japanese classes together, these kinds of curricular options become highly desirable in remote locations. When one considers the vast distances separating Alaskan villages or ranches in the Australian outback, utilizing all of the available technologies to provide a well-rounded education is not a luxury.

History of Distance Education

Although distance education may appear to be a recent phenomenon, it has a long history in the United States and abroad. Correspondence instruction was already popular in Europe when it began in this country in the early 1870s, although it was considered "informal" education intended for the working class and those unable to attend traditional "formal" classes. Interest in correspondence study was fueled, in part, by the industrial revolution that created a need for trained workers. In 1881, William Rainey Harper, a teacher of Hebrew, began teaching courses for Sunday school teachers via correspondence and shortly thereafter, as the president of the University of Chicago, developed the first distance education program within a university.

When radio gained popularity early in the twentieth century and was available in the typical American home, it was only natural that universities and trade schools would utilize this new medium to deliver instruction. By 1919, thirty-six universities had begun to incorporate radio programming into their correspondence courses. Unfortunately, postsecondary institutions failed, at that time, to recognize the potential of distance learning and relegated it to a lower status within their organizations, essentially opening the door to a wide variety of entrepreneurial efforts that were often fly-by-night scams or barely disguised diploma mills. In this mood, distance education programs at accredited institutions languished in a state of limbo until the 1950s.

The next upswing for distance education was generated by a heightened interest in the sciences and technology and can be traced to two significant events, both occurring in 1958: the launching of the first geostationary satellite, and the National Defense Education Act, which was in direct response to the Soviet Union's launch of *Sputnik.* Federal funds poured into TV-studio construction projects, and universities were once again interested in distance education (albeit primarily in its high-tech incarnation as educational television). By 1960 there were more than fifty educational television stations in the United States. One of the most fascinating and imaginative projects was the Midwest Program on Airborne Television Instruction, in which coursework was transmitted from one of several DC-6 aircraft that flew over a six-state area. This initiative began in 1961 and lasted an amazing six years. However, the cost of maintaining all of these projects took its toll, and as the novelty began to wear off viewership declined, partly because the quality of the courses was questionable at best.

Throughout the 1980s and early 1990s, televised distance learning programs adopting satellite delivery began a slow but steady rise as costs dropped and more schools and businesses could realistically consider utilizing these technologies. Two major initiatives helped to revitalize interest in distance learning. The first was the founding of the Annenberg/Corporation for Public Broadcasting grants program to fund the development and production of high-quality courses that incorporated video programs, texts, study guides, test banks, and related teaching materials. Examples of these courses included well-respected series like *Eyes on the Prize* and *The Mechanical Universe.* The second initiative, aimed primarily at the K-12 environment, was the Federal Star Schools Program Assistance Act, a project started in 1988 and designed to encourage statewide partnerships for distance education programming. By 1994 students who had participated in coursework funded by this program numbered in the millions. However, even with these exemplary programs, resistance to distance education lingered in sometimes subtle ways. Universities restricted the number of credits earned "at a distance" that could be applied to a degree, for example, and the stigma of these courses as supposedly less rigorous and of lower quality remained.

The most recent renaissance of distance education activity in the United States has, once again, been sparked by advances in technology. The widespread availability of the Internet, coupled with decreasing costs for increasing computing power, ignited the creativity of trainers and educators who are recognizing the enormous potential of distance instruction. While private-sector corporations explore computer networking and online learning as a solution to expensive face-to-face training, traditional colleges and universities are suddenly competing with institutions that offer complete degree programs online, wooing students who might prefer the convenience of taking classes from home.

In 1995 there were approximately 16 million Internet users worldwide; by June 2000, that number had climbed to well more than 332 million (Sly 2000). In the United States alone, close to 1 million students enroll in distance-delivered courses through postsecondary institutions each year.

Delivery Systems

One component of the definition of distance education refers to the use of a delivery system, some kind of "mediating interface" that ties the participants together and provides opportunities for interaction. In the traditional classroom, delivery systems may consist of the human voice, gestures, images projected within the room, and other sensory stimuli. If learners are separated geographically, however, some kind of device or technology is required to enable ideas to be exchanged, and the variety of delivery systems grows richer every year. If something can be considered as a method of transporting video, audio, or data, it has probably been used for distance education. Although there are many different ways of providing these connections, they are typically categorized using one of three structures: tangible/virtual, synchronous/asynchronous, or distributed/centralized.

Tangible/virtual delivery systems refer to the actual medium of distribution; like the artist who expresses herself with clay or watercolor, these systems provide a method not only for fixing content in a permanent form but also for disseminating that content beyond the immediate surroundings. Tangible delivery systems are those in which a physical element is used to contain ideas (e.g., text materials printed on paper or moving images on videotape) or to allow for human interaction (copper telephone lines or fiber-optic data connections). What makes a delivery system tangible is its physical presence.

Virtual delivery systems are those in which ideas are distributed through a medium that is volatile and impermanent. Satellite TV signals travel through the air and may be converted to video images, but their delivery medium remains ethereal. Other virtual delivery systems might include radio programming, microwave TV distribution, and wireless data networking.

Whether tangible or virtual, delivery systems are subject to constraints in bandwidth (a way to measure the complexity or richness of a given signal). More robust signals are those that carry large amounts of information, such as broadcast-quality video with its wide color palette and audio accompaniment. Bandwidth is often compared to a garden hose: The larger the hose, the greater its capacity and distribution capability, and large bandwidth delivery systems are those that can carry a rich array of sensory stimulation. Unfortunately, the larger the bandwidth, the more expensive it becomes. Inexpensive delivery systems, like standard telephone lines, carry a limited amount of information, whereas large bandwidth systems, such as fiber-optic cable or microwave TV signals, can transport a heavy load of video, audio, and data signals concurrently but are significantly more expensive as a result.

One of the more frequently used methods for categorizing delivery systems depends on time. Synchronous systems are those in which instruction is delivered and received in real time, whereas asynchronous systems are those in which there is a noticeable time delay between the sending of a signal and its receipt by the learner.

Synchronous delivery of instruction, such as live satellite TV programming or an online chat system, requires all learners to participate at the same time. They may be geographically separated, but instruction occurs according to a predetermined schedule during which they may simply receive programming (e.g., by watching a televised lecture/demonstration) and/or interact with the instructor and possibly other learners in a real-time format. Although some students prefer to engage in coursework on their own schedule rather than according to a preset timetable, for others the opportunity to interact in a live conversation adds an element of personal involvement that may otherwise be limited and can provide a sense of connectedness not found in many time-delayed delivery systems. In addition, coursework delivered as a synchronous module can often be captured for use later as an asynchronous review of material (e.g., taping a satellite-delivered program).

Asynchronous delivery systems—those in which instruction is received significantly later than it was sent—offer distant learners the opportunity to study at times that are convenient for them. Such systems might utilize e-mail, pretaped video programs, online discussion boards, or something as familiar as printed modular texts that are designed to be self-paced. Many traditional face-to-face courses also utilize these elements as a supplement to the live in-class activities as a way of extending the classroom beyond the constraints of real-time schedules. The major disadvantage to such delivery systems appears to be the potentially isolating effects of such coursework if not carefully planned to incorporate activities that require interaction among students and instructors.

Delivery systems are also categorized by whether the interactions are centralized or distributed, that is, whether communication is one-to-one, one-to-many, or many-to-many. One-to-one interactions occur in most correspondence courses, whether utilizing postal delivery of materials or the online environment. Although instructors might interact with many students, they will communicate with each student privately, and students will not interact with (or probably even be aware of) other students taking the same course. This is especially true in self-paced programs in which each learner is at a different point in her work.

A satellite video broadcast is considered a one-to-many delivery system, with the instructor at an "originating location" (typically a television studio/classroom) where the video program is sent up to a satellite that then amplifies and distributes the signal across the earth's surface in a pattern known as the satellite's "footprint." A good analogy for this type of delivery system is a showerhead. A narrow stream of water (the television signal) travels to a device that sprays the water (signal) over a large surface. The program can then be received by anyone inside the footprint who has the appropriate antenna or dish. Because a satellite's footprint typically covers almost one-third of the earth's surface, the potential audience for satellite-delivered instruction is enormous and would be limited only by instructional or administrative constraints. Microwave delivery of instruction using the frequency range designated for educational use (Instructional Television Fixed Service) works in a similar one-to-many way. Signals are sent out from a tower to a radius of twenty to thirty miles and can be picked up with specialized antennae anywhere within range of the signal.

A configuration in which any participant has the option of initiating interactions with one or more other participants would be classified as a many-to-many delivery system. The use of online chats is a good example, where no one individual is the "originator" based on their location or available technologies, and anyone who chooses to participate can initiate discussion with others. Although communications organized in this way

TABLE 1: Characteristics of Example Delivery Configurations

	Tangible	Virtual	Central	Distrib.	Synch.	Asynch.
Print-Based Correspondence	x		x			x
Prerecorded Media (e.g., tapes)	x		x			x
1-Way Audio (e.g., radio)		x	x		x	
2-Way Audio	x	x		x	x	
1-Way Video	x	x	x		x	
2-Way Audio, 1-Way Video	x	x		x	x	
2-Way Audio & Video	x	x		x	x	
Online Discussion	x	x		x	x	x

SOURCE: Created by the authors.

have been utilized successfully for many years in telephone audioconferences, its popularity is a relatively recent phenomenon in education as instructional paradigms began moving away from primarily teacher-centered models while advances in technology were enabling nonlinear communication structures. (See Table 1.)

There are countless other means for categorizing and describing delivery systems, but the critical point is that these categories can be helpful when determining how best to distribute instruction to learners at a distance. For example, a needs assessment of a potential student clientele reveals that most of them work full-time and would be unable to attend classes regularly, so a model that utilizes asynchronous components (e.g., printed materials and videotaped lecture/demonstrations) would likely be successful. Or if coursework will require a high degree of discussion and student participation, a system utilizing both synchronous and asynchronous elements in a many-to-many configuration (e.g., audioconferencing and online discussion) would be desirable. Also no successful distance education program utilizes only one type of delivery system, and most programs incorporate at least three types of interaction and content delivery. For those systems that are completely or predominantly organized around messages going from teacher to student, additional technologies must be integrated to accommodate the need for student-to-teacher and student-to-student interactions. The more options for learning that are presented, the more likely it is that the students' various learning needs will be met and their obstacles to education hurdled.

Theories

Many of the earliest distance education programs, especially those originating in Europe, developed out of a theoretical perspective that emphasized the industrialization of education. These programs (described by the German educational theorist Otto Peters) included components such as mass production of standardized course materials that required a high level of quality control, long-term planning that focused on careful decisionmaking, organization of programs based on their overall purposes and goals, and centralization of course administration. This theoretical model exhibits characteristics that most educators would find admirable, but it has never found strong acceptance in the United States. This is probably due to the decentralized nature of the U.S. school systems and structures that emphasize the autonomy of the individual teacher.

A more accepted theory (at least in the United States) that has guided many distance education programs focuses on the independence of the learner and a separation of teaching from the end result of learning. Charles Wedemeyer, a proponent of independent study and the freedom of individuals to pursue learning at times and places of their own choosing, wrote extensively of the need for students to take on responsibility for their own learning. This theoretical approach emphasizes the need for educational programs to encourage students to engage in learning at their own pace, stopping and starting as they choose, and to incorporate whatever media are available to provide a wide array of options. Wedemeyer recognized distance education as a less restrictive form of education that was more likely to accommodate these tenets of individualism, and many of his ideas are gaining acceptance in traditional learning environments as the roles of learner and teacher shift to accommodate constructivist thought.

Börje Holmberg posited a theory of distance education with an emphasis on interaction and communication. His idea of learning as a "guided didactic conversation" was intended as an explanatory framework for how distance education can support student motivation and facilitate a sense of connectedness to the learning group, something not easily accomplished when learners are physically removed from one another. To critics of distance education who charge that it can be an isolating experience devoid of human contact, this theory responds that the interactions incorporated into such teaching are purposeful and considered critical to the student's success. Unlike many face-to-face courses, distance education course design includes planned interactions, never assuming that meaningful communication will occur simply because the opportunity exists. This theory has been embraced in many distance education programs worldwide, leading many new distance teachers to marvel at how well they get to know their students, even though they may never meet in person.

These three theories (industrialization, learner autonomy, and interaction) represent a sampling of theoretical bases upon which planning, decisionmaking, and research have been based. Other theories that have been used to guide distance education practice include the adoption of innovations, instructional equivalency, communications, and systems theory; as an interdisciplinary field that incorporates aspects of many domains, one of the strengths of distance education remains its diversity of thought.

Implementation

The perspective of distance education as belonging to a somewhat marginalized category of instructional activity has a long history, and this has reinforced the idea that, by definition, such teaching is somehow less rigorous, less scholarly, and certainly less respectable than traditional, face-to-face models. Fortunately, the overwhelming majority of programs delivered at a distance prove this to be a patently false assumption, although a silver lining has emerged from such biases. Calls for the careful evaluation of distance education programs require a clear delineation of what constitutes a rigorous course, how the courses under consideration are to be evaluated against these criteria, and ultimately how the results compare to our familiar models of teaching and learning. It comes as no surprise to practitioners in this field that nearly every well-controlled comparison of distance education courses with their classroom-based counterparts has returned the verdict of "no significant difference" in learning gains. The long-term benefit of these evaluative activities, however, rests in the furthering of our understanding of how best to design instruction for any type of environment or delivery system and apply these standards to all of our teaching efforts.

Issues of copyright and intellectual property rights are of particular concern to teachers who are developing instructional materials for distance delivery. Copyright law and the accompanying fair use guidelines are sometimes difficult to interpret for specific cases, and many institutions are justifiably worried about inadvertent infringements. With current technologies, reproducing, altering, and distributing someone else's work has become nearly effortless, and it has never been easier to break the law as a result of ignorance. Those working in distance education recognize the need to provide access to educational resources but must also strive to observe the property rights of those whose works they (and their students) rely upon, and when in doubt they should obtain written permission to utilize these materials. Ideally, the terminology utilized in legal decisions of copyright compliance may eventually reflect current practice (e.g., if a class is conducted in a real-time format with all participants able to see and hear one another, can it be considered face-to-face?), but until then educators must temper demands for access with respect for the law and the rights of creative individuals.

Intellectual property rights are being renegotiated at many postsecondary institutions as a direct result of distance education. When an instructor designs and develops an online course, the question of who that course belongs to can become a delicate matter, especially if the instructor chooses to take a teaching position elsewhere and hopes to reuse those materials. In programs that utilize satellite delivery of lecture/demonstration sessions, the question of videotape ownership can prove controversial. Instructors may be uncomfortable with the idea that students (or anyone) could purchase an entire set of such tapes, potentially devaluing the marketability of that instructor's writings or consulting opportunities. Additionally, the possibility that an institution might choose to simply replay the tapes during later semesters, rather than hiring the live instructor, is a significant disincentive to teaching in such programs, and it is in the best interests of the institution and its faculty to clarify policies regarding such arrangements.

This represents only a small fraction of the issues that must be resolved before distance education becomes an unremarkable and everyday option for learning. Other topics currently of concern to professionals in the field include technology access for underprivileged students; assisting faculty and trainers with appropriate professional development and technical support; ensuring that distance-delivered materials are accessible to all students (including those with disabilities); reconciling accreditation guidelines that require minimum amounts of "seat time" with asynchronous learning initiatives; guiding students as they take on greater responsibility for their learning; and designing research on distance education practices to help guide planning and decisionmaking. Efforts to improve instruction at a distance will ideally be applied in many learning environments and lead to the eventual integration of these practices into the broader categories of exemplary teaching and training.

Susan M. Zvacek

See also Adult Learners; Copyright; Courseware; Instructional Communications; Interactive Television; Open University; Television and Learning; Virtual Universities; Web-Based Course Management Systems; Western Governors University

References

Holmberg, B. (1983). "Guided Didactic Conversation in Distance Education." In D. Stewart, D. Keegan, and B. Holmberg (eds.), *Distance Education: International Perspectives* (pp. 114–122). London: Routledge.

Keegan, Desmond. 1990. *Foundations of Distance Education,* 2nd ed. London: Routledge.

Moore, Michael G., and Greg Kearsley. 1996. *Distance Education: A Systems View.* Belmont, CA: Wadsworth.

Schlosser, Charles, and Mary L. Anderson. 1994. *Distance Education: Review of the Literature.* Washington, DC: Association for Educational Communications and Technology.

Simonson, Michael, Sharon Smaldino, Michael Albright, and Susan Zvacek. 2000. *Teaching and Learning at a Distance: Foundations of Distance Education.* Upper Saddle River, NJ: Prentice-Hall.

Sly, Liz. 2000. "Mongolia's Nomads Roam World on Internet: New Technology Opening Trade, Education Opportunities for Remote, Low-Income Regions." *Chicago Tribune* (July 16).

E

Educational Fair Use

The term "educational fair use" refers to a set of negotiated guidelines explaining the application of fair use in educational situations. Despite the common myth in education circles, educational use and fair use are nowhere near synonymous. Just because a teacher or student wants to reproduce or show a copyrighted work for classroom use does not mean that the teacher or student may actually do so. Section 106 of the Copyright Revision Act of 1976, which took effect on January 1, 1978, confers upon copyright owners the exclusive right to reproduce, prepare derivative works of, distribute, perform, and display their copyrighted works. Section 107 immediately creates a limitation on the exclusive rights granted in Section 106: fair use.

> Section 107. Limitations on exclusive rights: Fair use
>
> Notwithstanding the provisions of sections 106 and 106A, the fair use of a copyrighted work, including such use by reproduction in copies or phonorecords or by any other means specified by that section, for purposes such as criticism, comment, news reporting, teaching (including multiple copies for classroom use), scholarship, or research, is not an infringement of copyright. In determining whether the use made of a work in any particular case is a fair use the factors to be considered shall include—
>
> 1. The purpose and character of the use, including whether such use is of a commercial nature or is for nonprofit educational purposes;

2. The nature of the copyrighted work;
3. The amount and substantiality of the portion used in relation to the copyrighted work as a whole; and
4. The effect of the use upon the potential market for or value of the copyrighted work.

The fact that a work is unpublished shall not itself bar a finding of fair use if such finding is made upon consideration of all the above factors.

Section 107 is the congressional restatement of a judicial doctrine first introduced in 1802 by Lord Ellenborough in *Cary v. Kearsley:* "A man may fairly adopt part of the work of another: he may so make use of another's labours for the promotion of science, and the benefit of the public: but having done so, the question will be, Was the matter so taken *used fairly*" (emphasis added). In *Wilkins v. Aiken* (1810), Lord Eldon referred to fair quotation and legitimate use. The phrase "fair use" seems to have first appeared in *Lewis v. Fullarton* (1839). In 1869 it first gained recognition as a legal doctrine in *Lawrence v. Dana.* Until incorporated into section 107, it remained a judicial equitable rule of reason. Although never actually mentioning fair use, Justice Joseph Story's 1841 decision in *Folsom v. Marsh* laid out the basic criteria for fair use.

It was only through intensive lobbying efforts that teaching, scholarship, and research was included as possible fair-use purposes in section 107. Even then, no one was quite sure how to apply it to those situations. Committees formed to further explicate fair use apropos print, music, and off-air recording. In March 1976, the committee focusing on print sent its agreed-upon guidelines to Congress. (See Appendix Figure 1.)

The music guidelines followed a month later in April (see Appendix Figure 2). The off-air recording guidelines were announced in 1981 (see Appendix Figure 3). The educational multimedia guidelines are a product of the late 1990s (see Appendix Figure 4). Unlike the original three guidelines, this last set of guidelines has not as yet gained the same widespread acceptance.

Two points to keep in mind about all the guidelines: First, although they have been read into the *Congressional Record,* this does not give them the status of law. It does, however, supply the courts with legislative intent. Second, the intent that the guidelines be considered the minimum and not the maximum standards of educational fair use seems to be endangered. The various standards are increasingly cited as if set in stone.

Esther Sinofsky

See also Copyright

References

Cary v. Kearsley, 4 Esp. 168, 102 Eng. Rep. 679 (K.B., N.P. 1802).

Folsom v. Marsh, 9 F. Cas. 342 (C.C.D. Mass. 1841).

Lawrence v. Dana, 15 F. Cas 26 (C.C.D. Mass. 1869).
Lewis v. Fullarton, 2 Beav. 6, 48 Eng. Rep. 1080 (Rolls Ct. 1839).
U.S. Congress. *General Revisions of Copyright Law.* U.S. Code, Title 17 (1976), 90 Stat. 2541. Copyright Office. Available online at www.loc.gov/copyright.
Wilkins v. Aiken, 17 Ves. 422, 34 Eng. Rep. 163 (Ch. 1810).

Appendix

FIGURE 1: Agreement on Guidelines for Classroom Copying in Not-for-Profit Educational Institutions with Respect to Books and Periodicals

The purpose of the following guidelines is to state the minimum and not the maximum standards of educational fair use under Section 107 of H.R. 2223. The parties agree that the conditions determining the extent of permissible copying for educational purposes may change in the future; that certain types of copying permitted under these guidelines may not be permissible in the future; and conversely that in the future other types of copying not permitted under these guidelines may be permissible under revised guidelines.

Moreover, the following statement of guidelines is not intended to limit the types of copying permitted under the standards of fair use under judicial decision and which are stated in Section 107 of the Copyright Revision Bill. There may be instances in which copying which does not fall within the guidelines stated below may nonetheless be permitted under the criteria of fair use.

Single copying for Teachers	* For personal scholarly research; use in teaching or preparation to teach * Covers a chapter, article, short story, short essay, short poem, chart, cartoon, or diagram
Multiple Copies for Classroom Use	* Limited to one copy per pupil * Must meet tests of brevity, spontaneity, and cumulative effect * Must include copyright notice * Poetry: not to exceed 250 words * Prose: complete article or story of 2,500 words or less; excerpt of not more than 1,000 words or 10% or work * Illustration: one per book or periodical issue * Excludes "special" works such as picture books which are limited to two published pages * Must be requested by the teacher * Limited to 9 instances during class term * Cannot replace or substitute for anthologies and the like * Does not apply to consumables such as workbooks, answer sheets, and test booklets * Cannot replace purchase of items

FIGURE 2: Guidelines for Educational Uses of Music*

The purpose of the following guidelines is to state the minimum and not the maximum standards of educational fair use under Section 107 of HR 2223. The parties agree that the conditions determining the extent of permissible copying for educational purposes may change in the future; that certain types of copying permitted under these guidelines may not be permissible in the future; and conversely that in the future other types of copying not permitted under these guidelines may be permissible under revised guidelines.

Moreover, the following statement of guidelines is not intended to limit the types of copying permitted under the standards of fair use under judicial decision and which are stated in Section 107 of the Copyright Revision Bill. There may be instances in which copying which does not fall within the guidelines stated below may nonetheless be permitted under the criteria of fair use.

Guidelines for Educational Uses of Music

Permissible	***Prohibited***
* Emergency replacement for imminent performance provided replacement copies purchased soon after	* Replacement or substitution for anthologies and the like
* For academic uses other than performance, not more than 10% of whole work as long as not performable unit and only one copy per pupil	* Excludes consumables such as workbooks and answer sheets
* Editing or simplifying as long as fundamental character of work not distorted	* Substitution for actual purchase of sheet music
* Single copy of a student's performance for evaluation or rehearsal purposes	* Copying for performance except in emergency as noted
* Single copy for constructing aural exercises or examinations	

*Please see complete version for more information.

FIGURE 3: Guidelines for Off-Air Recording of Broadcast Programming for Educational Purposes*

In March 1979, Congressman Robert Kastenmeier, Chairman of the House Subcommittee on Courts, Civil Liberties and Administration of Justice, appointed a Negotiating Committee consisting of representatives of education organizations, copyright proprietors, and creative guilds and unions. The following guidelines reflect the Negotiating Committee's consensus as to the application of "fair use" to the recording, retention, and use of television broadcast programs for educational purposes. They specify periods of retention and use of such off-air recordings in classrooms and similar places devoted to instruction and for homebound instruction. The purpose of establishing these guidelines is to provide standards for both owners and users of copyrighted television programs.

Guidelines for Off-Air Recording of Broadcast Programming for Educational Purposes

* Applies to nonprofit educational institutions only
* Applies to programs broadcast for general reception
* May only be retained for 45 consecutive calendar days
* May be used once in teaching during first 10 consecutive school days within the 45 calendar day period
* Must be requested and used by individual teachers; cannot tape in anticipation of requests
* Teacher may only request off-air taping of a program once, not each time it is broadcast
* After first 10 consecutive school days, may use only for evaluation purpose until end of 45-day period
* May show only a part, but may not alter original content or create anthologies
* Must include copyright notice
* Educational institutions must implement appropriate control procedures for monitoring compliance with the guidelines

*Please see complete version for more information.

FIGURE 4: Fair Use Guidelines For Educational Multimedia

Table of Contents:

1. Introduction

1.1 Preamble

Fair use is a legal principle that provides certain limitations on the exclusive rights of copyright holders. The purpose of these guidelines is to provide guidance on the application of fair use principles by educators, scholars and students who develop multimedia projects using portions of copyrighted works under fair use rather than by seeking authorization for non-commercial educational uses. These guidelines apply only to fair use in the context of copyright and to no other rights.

There is no simple test to determine what is fair use. Section 107 of the Copyright Act sets forth the four fair use factors which should be considered in each instance, based on particular facts of a given case, to determine whether a use is a "fair use": (1) the purpose and character of use, including whether such use is of a commercial nature or is for nonprofit educational purposes, (2) the nature of the copyrighted work, (3) the amount and substantiality of the portion used in relation to the copyrighted work as a whole, and (4) the effect of the use upon the potential market for or value of the copyrighted work.

While only the courts can authoritatively determine whether a particular use is fair use, these guidelines represent the endorsers' consensus of conditions under which fair use should generally apply and examples of when permission is required. Uses that exceed these guidelines may or may not be fair use. The endorsers also agree that the more one exceeds these guidelines, the greater the risk that fair use does not apply.

The limitations and conditions set forth in these guidelines do not apply to works in the public domain — such as U.S. Government works or works on which copyright has expired for which there are no copyright restrictions — or to works for which the individual or institution has obtained permission for the particular use. Also, license agreements may govern the uses of some works and users should refer to the applicable license terms for guidance.

The participants who developed these guidelines met for an extended period of time and the result represents their collective understanding in this complex area. Because digital technology is in a dynamic phase, there may come a time when it is necessary to review the guidelines. Nothing in these guidelines shall be construed to apply to the fair use privilege in any context outside of educational and scholarly uses of educational multimedia projects. These guidelines do not cover noneducational or commercial digitization or use at any time, even by non-profit educational institutions. These guidelines are not intended to cover fair use of copyrighted works in other educational contexts such as digital images or archives, distance education, or electronic reserves, which may be addressed in other fair use guidelines.

This Preamble is an integral part of these guidelines and should be included whenever the guidelines are reprinted or adopted by organizations and educational institutions. Users are encouraged to reproduce and distribute these guidelines freely without permission; no copyright protection of these guidelines is claimed by any person or entity.

1.2 Background

These guidelines clarify the application of fair use of copyrighted works as teaching methods are adapted to new learning environments. Educators have traditionally brought copyrighted books, videos, slides, sound recordings and other media into the classroom, along with accompanying projection and playback equipment. Multimedia creators integrated these individual instructional resources with their own original works in a meaningful way, providing compact educational tools that allow great flexibility in teaching and learning. Material is stored so that it may be retrieved in a nonlinear fashion, depending on the needs or interests of learners. Educators can use multimedia projects to respond spontaneously to students' questions by referring quickly to relevant portions. In addition, students can use multimedia projects to pursue independent study according to their needs or at a pace appropriate to their capabilities. Educators and students want guidance about the application of fair use principles when creating their own multimedia projects to meet specific instructional objectives.

1.3 Applicability of These Guidelines (Certain basic terms are identified in bold and defined in this section.)

These guidelines apply to the use, without permission, of portions of lawfully acquired copyrighted works in educational multimedia projects which are created by educators or students as part of a systematic learning activity by nonprofit educational institutions. Educational multimedia projects created under these guidelines incorporate students' or educators' original material, such as course notes or commentary, together with various copyrighted media formats including but not limited to, motion media, music, text material, graphics, illustrations, photographs and digital software which are combined into an integrated presentation. Educational institutions are defined as nonprofit organizations whose primary focus is supporting research and instructional activities of educators and students for noncommercial purposes.

For the purposes of these guidelines, educators include faculty, teachers, instructors and others who engage in scholarly, research and instructional activities for educational institutions. The copyrighted works used under these guidelines are lawfully acquired if obtained by the institution or individual through lawful means such as purchase, gift or license agreement but not pirated copies. Educational multimedia projects which incorporate portions of copyrighted works under these guidelines may be used only for educational purposes in systematic learning activities including use in connection with non-commercial curriculum-based learning and teaching activities by educators to students enrolled in courses at nonprofit educational institutions or otherwise permitted under Section 3. While these guidelines refer to the creation and use of educational multimedia projects, readers are advised that in some instances other fair use guidelines such as those for off-air taping may be relevant.

Overview of Educational Fair Use for Multimedia*

Preparation of Educational Multimedia Projects under These Guidelines

- Must include proper attribution and citation
- Students may use portions of lawfully acquired copyrighted works for multimedia projects
- Teachers may use portions of lawfully acquired copyrighted works to create multimedia teaching tools

Permitted Educational Uses for Multimedia Projects Under These Guidelines

- Students may perform and display in class for which created or keep as examples in personal portfolio
- Teachers may perform and display in face-to-face instruction, assign for directed self-study, use in distance education in certain circumstances
- Teachers may perform and display own projects at workshops and conferences
- Teachers may keep in personal portfolios

Limitations

- Teachers may use creations in class up to 2 years after created
- May use up to 10% or 3 minutes, whichever is less, of motion media work
- May use up to 10% or 1,000 words, whichever is less, of text material
- May use up to 10% but no more than 30 seconds of music and lyrics from a musical work
- May use no more than 5 images by an artist or photographer or not more than 10% or 15 images, whichever is less, from a published collective work
- May use up to 10% or 2,500 fields or cell entries, whichever is less, from a copyrighted database or data table
- Teachers may make a second copy to be placed on reserve or for preservation purposes

Examples of When Permission Is Required

- For non-educational or commercial purposes
- For replicating or duplicating beyond limitations
- For electronic networking

Important Reminders

- Need to be cautious when downloading material from the Internet since many works are copyrighted
- Credit the sources and display their copyright notice and ownership information
- Include acknowledgment that multimedia project prepared according to the guidelines
- Does not apply to computer programs
- Guidelines do not preempt or supersede licenses or contractual obligations

---

*Please see complete version for more information.

Educational Systems Design (ESD)

Educational technology is a broad field that encompasses both "hard" and "soft" technologies. Educational systems design is a "soft" educational technology that is grounded in systems philosophy, systems theory, and design theory and enhances the success of efforts to change our education systems. ESD provides a knowledge base that improves the chances of successful implementation of a change in education, improves the chances of long-term survival of the changes, and, most important, improves the chances that the changes will meet the needs of the social systems and individuals they serve.

Technology can be used to help educators do better what they are already doing, or it can be used to help them do things very differently—and potentially much better—than without it. ESD is an important soft technology for helping educators succeed at the latter.

Because most readers are familiar with instructional systems design (ISD), and because there are many similarities between ESD and ISD, it may be helpful to understand ESD by comparing and contrasting it to ISD. ESD is broader in scope than ISD, for ISD is a subset of ESD, but both encompass knowledge bases for "process" and "product" (means and ends). Furthermore, ESD has a foundation in systems thinking and design thinking, and it focuses on holistic transformation rather than piecemeal change. These four issues are discussed below.

The Scope of ESD

The major difference is that ESD is concerned with the entire education system, whereas ISD is only concerned with one part of it: the instructional subsystem. There are four subsystems in any educational or training system (Banathy 1991):

1. The learning experience subsystem, in which the learner processes information from his environment to produce new or modified cognitive structures;

2. The instructional subsystem, in which instructional designers and teachers use information about learning needs (gained through analysis activities), as well as administrative and governance input, to produce environments or opportunities for learners to learn;
3. The administrative subsystem, in which administrators use information about instructional needs, as well as governance input, to make decisions about resource allocation, including use of leadership; and
4. The governance subsystem, in which "owners" use their goals and values to produce policies and in other ways provide direction and resources for the educational enterprise in order to meet their needs (which usually include those of learners, teachers, and administrators).

ISD provides the knowledge base about designing the instructional subsystem, whereas ESD provides the knowledge base about designing the complete education system.

Process and Product Knowledge in ESD

Knowledge about the *process* of ISD is generally referred to as ISD models (such as those reviewed in Gustafson and Branch 1997), which focus on the activities in which people should engage in order to design a new instructional system. By contrast, knowledge about the *products* of ISD is generally referred to as instructional theories (such as those compiled by Reigeluth 1983, 1999), which focus on what the new instructional system should be like—what instructional methods it should use.

In a similar way, knowledge about the process of ESD is generally referred to as ESD models, or systemic change models, and focuses on the activities in which people should engage in order to design a fundamentally different educational or training enterprise or to fundamentally transform an existing one (see, e.g., Duffy, Rogerson, and Blick 1999; Jenlink et al. 1996, 1998). Knowledge about the products of ESD is generally referred to as comprehensive design and focuses on what features a fundamentally different kind of educational or training system should have in order to meet fundamentally different learning needs of the system it serves (see the vision offered by Reigeluth and Garfinkle 1994).

Foundations of ESD

A critical feature of ESD is its foundation in systems thinking and design thinking (design philosophy and design theory) (see, e.g., Banathy 1996, 155–171). Systems thinking, which includes systems philosophy, systems theory, and systems methodology (see, e.g., Ackoff 1981; Checkland 1981;

Churchman 1979; Hutchins 1996), is the centerpiece of the new sciences, which also include chaos theory, nonlinear science, and the science of complexity (Bohm 1983; Davis 1989; Wheatley 1992). Systems thinking focuses on understanding the mutually interdependent relationships (1) between a system and the larger systems of which they are a part (suprasystems); (2) between a system and its peer systems (other systems that are parts of the same suprasystem); and (3) among the many functions and components that comprise the new system. It also focuses on understanding the complex dynamics that govern the behavior of systems and strongly influence the effects of any changes made to a system (Senge 1990). Particularly relevant to ESD are the systems concepts of holism, self-organization, emergence, coevolution, and dynamical systems (Capra 1988; Michaels 1994).

Design thinking (Ackoff 1981; Checkland 1981; Cross 1984; Nadler 1981; Warfield 1990) informs process knowledge about how to design education systems. The fundamental activities of analysis, synthesis, and evaluation occur continuously throughout the design process. Simultaneity, recursion, and transcendence are important concepts, as are idealized design, values-based design, holistic design, continuous design, and ethical principles of design (e.g., stakeholders or users should design their own systems).

Piecemeal versus Systemic Change

To further understand the nature of ESD, it is helpful to think in terms of two basic kinds of change: piecemeal and systemic. Piecemeal change leaves the structure of a system unchanged. It often involves finding better ways to meet the same needs. In contrast, systemic change entails transforming the structure of a system, usually in response to new needs. For example, one may find that students' characteristics (such as their entering knowledge, learning styles, interests, and motivations) are more diverse than they used to be and that they have very different goals (such as college, vocational school, or immediate employment). To respond to these changed needs, an educator may decide to use customized, team-based, problem-based learning with continuous progress and to use advanced technology in new ways. Piecemeal change usually changes one part of a system in a way that is still compatible with the rest of the system, whereas systemic change entails such a fundamental change that it requires fundamental changes throughout the system, because the other parts of the system would not be compatible with the change.

Thus ESD is concerned with creating a new paradigm of education or training, as opposed to making changes within the existing paradigm. It recognizes that fundamental change in one aspect of a system requires fundamental changes in other aspects for it to be successful. In public education, it must eventually pervade all levels of the system: classroom, building, district, community, state government, and federal government. Similarly, in corporate training, it must eventually pervade all levels of the corporation. In this way, it can encompass not only the nature of the learning experiences and the instructional system but also the administrative and governance systems.

Such an approach to improvement is radical, not to mention difficult and risky. Thus it is important to address the rationale for such a radical change.

When and Why Is ESD Needed?

Systems thinkers know that when a human-activity system (or societal system) changes in fundamental ways its subsystems must change in equally fundamental ways to survive. This is because each subsystem must meet one or more needs of its suprasystem in order for the suprasystem to continue to support it (Hutchins 1996). So if the suprasystems for education systems are undergoing systemic changes, only then do education systems need to undergo systemic change or risk becoming obsolete.

So are education's suprasystems changing dramatically? In the industrial age we needed minimally educated people (dropouts) who would be willing and able to endure the tedium of work on assembly lines. However, those line jobs have been rapidly disappearing. Just as the percentage of the workforce in agriculture dropped dramatically in the early stages of the industrial age, so the percentage in manufacturing has been declining dramatically over the past few decades. Even in manufacturing companies, a majority of the jobs today entail manipulating information rather than materials (Reich 1991). Just as the industrial age represented a predominance of manual labor with machinery, so the information age represents a predominance of "knowledge work" with information technology. This means that more people must be educated to higher levels, with a greater emphasis on preparing students to think, solve problems, assume responsibility and initiative, and work well in teams. These are dramatically different educational needs from those that our industrial age schools were designed to meet. While preparation for work is but one mission of schools, the civic and character-building missions have seen similarly dramatic changes in this age of terrorism, global conflict, drugs and violence, and corporate malfeasance. Thus it is evident that education's suprasystems are indeed changing dramatically.

The Knowledge Base of ESD

Given the need for ESD, what is its knowledge base like? It was mentioned earlier that it includes both process and product knowledge. Product knowledge is concerned with what a new system of education should be like. This part of the knowledge base investigates the changing educational needs of the suprasystem (the community or organization to be served) and individual "clients" and explores the educational implications of those changes in needs. Process knowledge is concerned with how to design a new education system or how to help an existing one transform itself. This part of the knowledge base investigates the obstacles to systemic change and the activities that are most likely to lead to a successful change effort. These are discussed below.

Product Knowledge

Product knowledge is composed of (1) visions of different kinds of education systems and (2) linkages between each kind of education system and the characteristics and needs of the larger systems it serves. Consider, for example, the following analysis.

One thing educators know for certain is that different people learn at different rates. Yet our industrial age education systems present a fixed amount of content to a group of learners in a fixed amount of time, making it like a race to see who receives the A's and who flunks out. By holding time constant, we force attainment of standards to vary. Our current systems are not designed for learning; they are designed for sorting, which was appropriate for industrial age needs. However, our information age society, with a predominance of knowledge work, requires all students to learn—that no child be left behind. If we are ever to meet this new educational need, we must stop holding time constant and instead hold attainments constant by allowing every student as much time as needed to master them.

However, switching from a time-based system to an attainment-based system means that we must switch from group-based progress to personalized progress. This in turn requires changing the role of the teacher to that of a coach or manager, rather than that of dispenser of knowledge to groups of learners who rotate from one teacher to another at the ring of a bell like so many little widgets on an assembly line. This requires that learning occur primarily from sources other than the teacher. Thus technology and other resources (including peers) must be used in dramatically different ways. Rather than focusing on technology integration—using technology to enhance what is currently being done in classrooms—we should focus on technology transformation—using technology to do things that were not possible before (Reigeluth and Joseph 2002). Furthermore, our assessment systems need to change from norm-referenced (comparing students with each other) to criterion-referenced (comparing student performance to a standard). One might envision an "inventory of attainments" replacing our current report cards and transcripts.

TABLE 1: Key Markers That Distinguish Industrial Age and Information Age Organizations

Industrial Age	Information Age
Standardization	Customization
Bureaucratic organization	Team-based organization
Centralized control	Autonomy with accountability
Adversarial relationships	Cooperative relationships
Autocratic decisionmaking	Shared decisionmaking
Compliance	Initiative
Conformity	Diversity
One-way communications	Networking
Compartmentalization	Synthesis and holism
Parts oriented	Process oriented
Planned obsolescence	Total quality
CEO or boss as "king"	Customer as "king"

SOURCE: Charles M. Reigeluth, ed. (1999), *Instructional Design Theories and Models, Volume 2: A New Paradigm of Instructional Theory* (Mahwah, NJ: Lawrence Erlbaum Associates).

To build product knowledge by investigating broader societal changes and exploring their educational implications, it is also helpful to look at the work of people like Daniel Bell (1973), Alvin Toffler (1980), and Robert Reich (1991), who have identified several massive changes that our society has undergone, from the agrarian age to the industrial age and now into what some call the information age. According to these and other scholars, there are "key markers" that characterize the emerging differences between information age organizations and their industrial age counterparts (see Table 1).

These key markers may provide insights as to core ideas that should guide the design of our new education systems (Banathy 1991). For example, changing from a system in which a group of thirty students must learn the same thing at the same time and rate, to a system in which each student takes as long as necessary to master a standard of attainment, clearly entails customization. Similarly, our current systems of education are also based on conformity and compliance. Students are typically expected to sit down, be quiet, and do what they are told. Their learning is directed by the teacher. But employers now want people who will take initiative to solve problems and who will bring diversity—especially diverse perspectives—to the workplace. Both of these enhance the ability of a team to solve problems and keep ahead of the competition. Communities and families also need people who will take initiative and honor diversity. The systems that education serves need diversity and initiative to be fostered by our education systems in the information age. Understanding these key markers can play a valuable role in building a sound product knowledge base in ESD.

So what product knowledge do we currently have? The New American Schools Development Corporation assumed as its mission to foster the development of "break the mold schools" as called for by President George H. W. Bush in 1990. This resulted in the development and implementation of seven comprehensive school designs, such as ATLAS, Co-NET, Modern Red Schoolhouse, and Roots and Wings. However, these designs have largely not arisen through a careful examination of the changing educational needs of society and a fundamental rethinking of teaching and learning processes to meet those new needs and thus have largely not "broken the mold" of the current school system. Furthermore, they have often not been implemented as intended and have not produced the desired improvements in student learning (Berends, Bodilly, and Kirby 2002). Nevertheless, these designs do represent "product" knowledge—guidance about what school systems should be like. Other more promising, though less thoroughly worked out, guidance can be found in a vision of brain-based learning systems (Caine and Caine 1997) and the vision of an information age education system (Reigeluth and Garfinkle 1994).

Process Knowledge

Process knowledge can focus on either how to design a new education system or how to help an existing one transform itself. This part of the knowledge base investigates the obstacles to systemic change and the activities that are most likely to lead to a successful change effort. The result is a set of principles and methods for systemic change (see Caine and Caine 1997; Duffy, Rogerson, and Blick 1999; and Jenlink et al. 1996).

There are many approaches to systemic change, none of which have been well-developed, researched, and validated. One approach is for "experts" to develop a comprehensive new system (product) and to have the leadership in a school district purchase the design, complete with implementation plans and assistance (process). This "invented elsewhere" approach has not been very successful for many reasons. One is that people don't like to be changed. They are much more receptive to change when they are in control of it. Another is that systemic change requires people in a system to adopt different mental models about the system's activities and structure for the change to be successful (Senge 1990, 2000). This approach overlooks this important principle of systemic change.

A second approach is for a school system to design and implement (process) its own new system (product). One variation of this approach is a process in which all stakeholders (as opposed to a small group of leaders) are involved in, and are given ownership over, the change process and the nature of the new system. Another variation of this "invented here" approach is a process that entails just a few schools, rather than all the schools, in a district changing. A third variation is to create several autonomous schools within a building and initially change only one of those schools-within-a-school, as opposed to changing the whole school building. This approach has only recently begun to be tested but appears to be promising.

Regardless of such variation, some principles appear to have empirical support for the "invented here" approach. The following is a small sample of such principles (see Caine and Caine 1997; Duffy, Rogerson, and Blick 1999; Jenlink et al. 1996):

- Systemic change should not be undertaken unless the district is at a sufficient level of readiness for systemic change. Time can be spent helping to build such readiness.
- The school district's organizational design must be changed from a bureaucratic one to a participative and collaborative one early in the change process. This entails using principles of transformational leadership.
- A facilitator experienced in the process of systemic change should be used.
- Effort should be focused primarily on helping stakeholders (particularly those most responsible for implementing the new system) to evolve their thinking or mental models about education.
- The school district (including the community) must be the unit of change.
- Learning must be at the center of the change effort.
- Change must occur simultaneously in the core work process, internal social structure, and community relationships.
- Efforts should be made early to build political support for systemic change.
- The superintendent and teachers' association president must provide early advocacy for the change effort.
- A leadership team should be created to provide political support for the change effort.
- A school design team should be formed to lead the design work. It should be comprised of seven to nine people from all groups who will be affected by the change (teachers, parents, students).
- No one should be forced to change.

- All who are going to change should be given training to prepare them for the change (students, parents, teachers, and administrators).

Both the product and process knowledge bases in ESD are in the relatively early stages of development.

Charles M. Reigeluth

See aslo Diffusion of Innovations; School Reform

References

Ackoff, Russell L. 1981. *Creating the Corporate Future.* New York: John Wiley and Sons.

Banathy, Bela H. 1991. *Systems Design of Education: A Journey to Create the Future.* Englewood Cliffs, NJ: Educational Technology Publications.

———. 1996. *Designing Social Systems in a Changing World.* New York: Plenum Press.

Bell, Daniel. 1973. *The Coming of Post-Industrial Society: A Venture in Social Forecasting.* New York: Basic Books.

Berends, Mark, Susan Bodilly, and Sheila N. Kirby. 2002. "Looking Back over a Decade of Whole-School Reform: The Experience of New American Schools." *Phi Delta Kappan* 84(2): 168–176.

Bohm, David. 1983. *Wholeness and the Implicate Order.* London: Routledge and Kegan Paul.

Caine, Renate N., and Geoffrey Caine. 1997. *Education on the Edge of Possibility.* Alexandria, VA: Association for Supervision and Curriculum Development.

Capra, Fritjof. 1988. *The Turning Point: Science, Society, and the Rising Culture.* New York: Simon and Schuster.

Checkland, Peter. 1981. *Systems Thinking, Systems Practice.* New York: Wiley and Sons.

Churchman, West. 1979. *The Systems Approach and Its Enemies.* New York: Basic Books.

Cross, Nigel. 1984. *Developments in Design Methodology.* New York: Wiley and Sons.

Davis, Paul. 1989. *The Cosmic Blueprint.* New York: Touchstone.

Duffy, Francis M., Lynda G. Rogerson, and Charles Blick. 2000. *Redesigning America's Schools: A Systems Approach to Improvement.* Norwood, MA: Christopher-Gordon Publishers.

Gustafson, Kent L., and Robert M. Branch. 1997. *Survey of Instructional Development Models.* 3rd ed. Syracuse, NY: ERIC Clearinghouse on Information and Technology.

Hutchins, C. Larry. 1996. *Systemic Thinking: Solving Complex Problems.* Aurora, CO: Professional Development Systems.

Jenlink, Patrick M., Charles M. Reigeluth, Alison A. Carr, and Laurie M. Nelson. 1996. "Facilitating Systemic Change in Public Schools." *Tech Trends* 41(1): 21–30.

———. 1998. "Guidelines for Facilitating Systemic Change in School Districts." *Systems Research and Behavioral Science* 15(3): 217–233.

Michaels, M. 1994. *Seven Fundamentals of Complexity.* Savoy, IL: People Technologies and the Chaos Network.

Nadler, Gerald. 1981. *The Planning and Design Approach.* New York: Wiley and Sons.

Reich, Robert B. 1991. *The Work of Nations.* New York: Alfred A. Knopf.

Reigeluth, Charles M., ed. 1983. *Instructional-Design Theories and Models: An Overview of Their Current Status.* Hillsdale, NJ: Lawrence Erlbaum Associates.

———. 1999. *Instructional-Design Theories and Models, Volume 2: A New Paradigm of Instructional Theory.* Mahwah, NJ: Lawrence Erlbaum Associates.

Reigeluth, Charles M., and Robert J. Garfinkle. 1994. "Envisioning a New System of Education." In Charles M. Reigeluth and Robert J. Garfinkle (eds.), *Systemic Change in Education* (pp. 59–70). Englewood Cliffs, NJ: Educational Technology Publications.

Reigeluth, Charles M., and Roberto Joseph. 2002. "Beyond Technology Integration: The Case for Technology Transformation." *Educational Technology* 42(4): 9–13.

Senge, Peter M. 1990. *The Fifth Discipline.* New York: Doubleday.

———. 2000. *Schools That Learn.* New York: Doubleday.

Toffler, Alvin. 1980. *The Third Wave.* New York: Bantam Books.

Warfield, John. 1990. *A Science of General Design.* Salinas, CA: Intersystems Publishers.

Wheatley, Margaret. 1992. *Leadership and the New Science.* San Francisco: Berrett-Koehler.

Elaboration Theory of Instruction

The elaboration theory of instruction offers a holistic approach to sequencing instruction that helps make learning more meaningful and is more motivating for learners (Reigeluth 1999b; abstracted here with permission of the publisher). It also can allow learners to have more control over some scope and sequence decisions during the learning process. This stands in sharp contrast to the parts-to-whole sequencing, superficial coverage of content, and teacher control over scope and sequence decisions that have been typical of education and training since 1900. (The term "content" is used here to refer to everything that comes under "what to teach." It therefore includes whatever tasks you might teach, as well as whatever knowledge; the term "content analysis" includes "task analysis.")

The elaboration theory recognizes that different kinds of sequences are needed for different instructional situations. Thus it synthesizes recent work on scope and sequence into a single coherent framework, extending that work where holes were found. It currently deals only with the cognitive and psychomotor domains and not the affective domain. (However, there are strong indications that it can be, and indeed is already intuitively being, applied in the affective domain; see, e.g., Greenberg and Kusché 1993; Goleman 1995.)

To understand the elaboration theory, it is helpful to begin with some general issues about the form of the elaboration theory and the nature of instructional sequencing. Then the kinds of situations that call for different kinds of elaboration sequences are discussed. Finally, the three major kinds of sequences offered by the elaboration theory are described.

General Issues about Elaboration Sequencing

One general issue is that elaboration theory is a design theory rather than a descriptive theory (Simon 1969); this means that it is oriented toward achieving goals and making decisions rather than making descriptions and conclusions (Cronbach and Suppes 1969). Its purpose is to offer guidance on the best means for accomplishing a given goal, where "best" is determined by a set of criteria appropriate to the situation at hand. Therefore, the major parts of elaboration theory are (1) methods and (2) the situations under which each method is likely to be best.

Also, elaboration theory is an instructional theory, which means that its purpose is to offer guidance on what methods of instruction are likely to be best for different situations. The elaboration theory deals only with macro-level (broad) methods: guidance for making scope and sequence decisions—decisions about what to teach and what order to teach it.

Sequencing is Based on Relationships

The second general issue is that each method of sequencing is based upon a single type of relationship among parts of the content. For instance, a historical sequence is based upon the chronological relationship—the actual sequence of events. A procedural sequence, the most common kind of sequencing in training, is based upon the relationship of "order of performance" of the steps in the procedure. A hierarchical sequence is based upon the relationship of learning prerequisites among the various skills and subskills that comprise a task. And the "simplifying conditions" sequence (described later) is based upon the relationship of the degree of complexity of different versions of a complex task.

Topical and Spiral Sequencing

A third general issue is that two basic patterns of sequencing can be used that are fundamentally different: topical and spiral (see Figure 1). In topical sequencing, a topic (or task) is taught to whatever depth of understanding (or competence) is desired before moving to the next one. In spiral sequencing (Bruner 1960), the learners master a topic (or task) gradually in several passes. They learn the basics of one topic (or task), then another, and another, and so on, before they return to learn more about each topic. This pattern continues until the desired depth and breadth are reached for all of them.

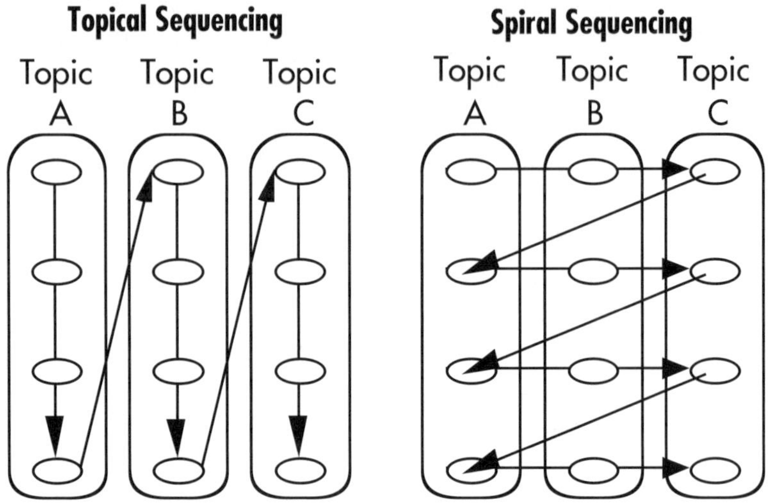

FIGURE 1: Topical and Spiral Sequencing

SOURCE: Created by the author.

Rather than thinking of spiral and topical sequencing as two separate categories, it is useful to think of them as the two endpoints along a continuum. The instructional designer's (or the learner's) decision, then, is where on the continuum to be for any given training program or curriculum and for any given group or individual learner—or when to be at any given point on the continuum.

Different Sequences for Domain Expertise and Task Expertise

The elaboration theory is founded on the notion that different sequences are best for different situations, because different sequencing methods are based on different kinds of relationships within the content, and different relationships are important for different kinds of expertise. So the kind of sequence that will most facilitate learning depends on the kind of expertise one wants to develop.

Elaboration theory distinguishes between task expertise and subject-domain expertise (Reigeluth 1999a). With task expertise the learner becomes an expert in a task, such as managing a project, selling a product, or writing an annual plan. With domain expertise the learner becomes an expert in a subject area not tied to any specific task, such as economics, electronics, or physics (but often relevant to many tasks). This is different from the distinction between procedural and declarative knowledge (J. R. Anderson 1983).

Task Expertise

Tasks range from simple to complex. The elaboration theory is intended only for more complex tasks. It is based on the observation that complex cognitive and psychomotor tasks are done differently under different conditions, that each set of conditions defines a different version of the task, and that some of those versions are much more complex than others. Thus, the elaboration theory offers the simplifying conditions method (SCM) to design a holistic, simple-to-complex sequence by starting with the simplest real-world version of the task and gradually progressing to ever-more complex versions as each is mastered. Problems or projects that learners tackle should be ones that are within the so-called zone of proximal development (Vygotsky 1978)—close enough to the learner's present competence for the learner to be able to deal with successfully—and the problems or projects should gradually increase in complexity.

Domain Expertise

Domain expertise ranges from simple to complex, but also from general to detailed. The general-to-detailed nature of domain expertise allows the design of a holistic sequence that goes from simple to complex. (The elaboration theory's sequencing guidance for domain expertise was derived primarily from Bruner's 1960 spiral curriculum and Ausubel's 1968 advance organizers and progressive differentiation. But it differs in several important ways from each, and it also provides greater guidance on how to design such a sequence.) An elaboration sequence starts with the broadest, most inclusive, most general ideas (which are also the simplest and generally among the first to have been discovered). Examples include the law of supply and demand in economics and Ohm's law in electricity. The sequence gradually progresses to more complex, precise ideas. Examples include ideas related to maximizing profits on the supply side (marginal revenues and marginal costs) and to consumer preferences on the demand side. This makes an elaboration sequence ideal for discovery learning, inquiry learning, and other approaches to the construction of knowledge.

The elaboration theory recognizes two major kinds of domain expertise: conceptual (understanding what) and theoretical (understanding why). In their simplest forms, these are concepts and principles, respectively. In their more complex forms, they are *conceptual knowledge structures* (or concept maps) for "understanding what," and both causal models and *theoretical knowledge structures* for "understanding why."

The conceptual elaboration sequence is briefly described next, followed by the theoretical elaboration sequence, and finally the SCM sequence.

The Conceptual Elaboration Sequence

The conceptual elaboration sequence (Reigeluth and Darwazeh 1982) is based on several observations. The first is that concepts are groupings or classes of objects, events, or ideas that share certain characteristics. For example, "tree" is a concept that includes all individual plants that have certain characteristics, most notably a woody stem. The second observation is that concepts can be broken down into either parts or kind, which are narrower, less inclusive concepts. For example, parts of trees include the trunk, roots, branches, and leaves. Kinds of trees include deciduous and evergreen. And each of those parts and kinds can be further broken down into parts and kinds. The third observation is that people tend to store a new concept under a broader, more inclusive concept in their heads (cognitive structures). The broader concept provides "cognitive scaffolding" (Ausubel 1968). The process of learning that proceeds from broader, more inclusive, and general concepts to narrower, more detailed concepts was called "progressive differentiation" (Ausubel 1968) because it entails a process of making progressively finer distinctions.

The kind of relationship upon which the conceptual elaboration sequence is based is that of parts or kinds of concepts (called the relationship). Such relationships include superordinate, coordinate, and subordinate relationships. In Figure 2, classical music is subordinate to music, is coordinate to medieval music, and is superordinate to instrumental classical music. As you go farther down in the conceptual structure to kinds of kinds of kinds (or parts of parts of parts), the concepts become ever narrower and more detailed. David Ausubel (1968) proposed that new concepts are organized in our heads under more inclusive concepts. Thus if one learns a broader, more inclusive concept before its subordinate concepts, the cognitive structure is more likely to be a sound one that will not have to be reorganized to accommodate new learning.

The conceptual elaboration sequence is one that starts by teaching (or discovering) the broadest, most inclusive, and general concepts that the learner has not yet learned and proceeds to ever more narrow, less inclusive, and more detailed concepts until the desired level of detail has been reached. This kind of sequence might be used by a high school student interested in learning about the kinds and parts of animals and plants or by an employee interested in learning about the kinds and parts of equipment that the company sells.

How do you identify all these concepts and their inclusivity relationships? This is the purpose of a conceptual analysis. The result of such an analysis is a conceptual knowledge structure (see Figure 2), sometimes called a taxonomy. The term "hierarchy" is sometimes used, but that term usually refers to a learning hierarchy (Gagné 1968).

The conceptual elaboration sequence may be designed in either a topical or spiral manner. For a topical sequence, one could go all the way down one leg of the conceptual structure and gradually move on to other topics, one leg at a time. For a spiral sequence, one could go completely across the top row, then across the next row down, and so forth.

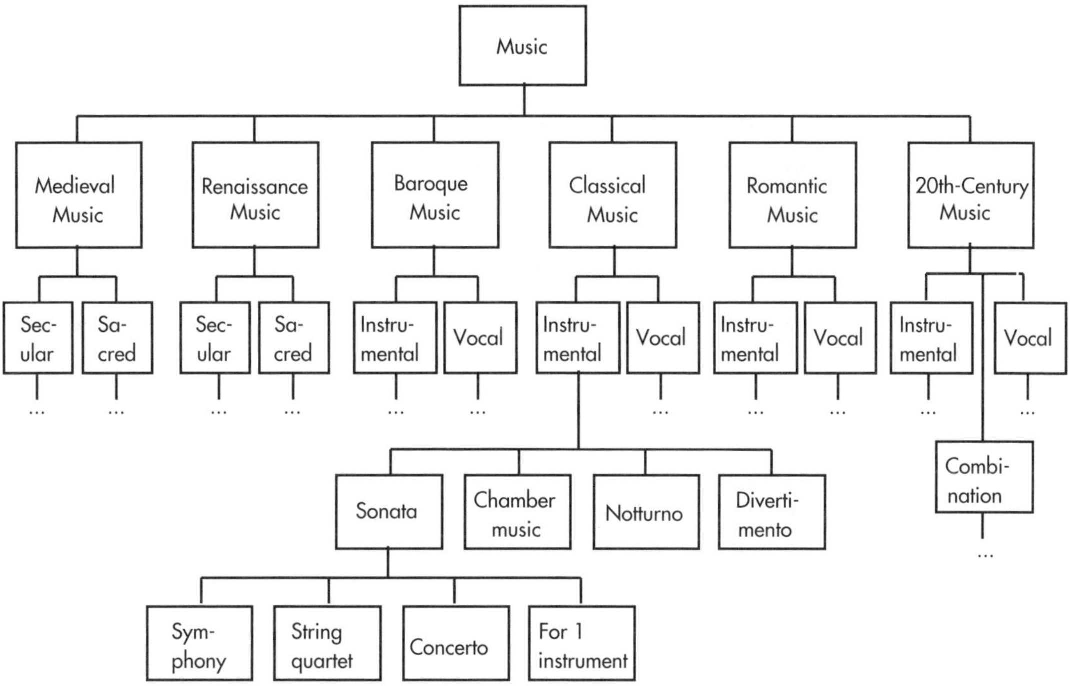

FIGURE 2: An Example of a Conceptual Structure

SOURCE: Reprinted from Charles M. Reigeluth, ed. (1999), *Instructional Design Theories and Models, Volume 2: A New Paradigm of Instructional Theory* (Mahwah, NJ: Lawrence Erlbaum Associates).

The Theoretical Elaboration Sequence

The theoretical elaboration sequence is the second of the two sequencing methods currently offered by the elaboration theory for building domain expertise. It is intended for courses that focus on interrelated sets of principles, which are usually elaborations of each other, such as a high school biology course that focuses on principles of genetics, life cycles, and bodily functions, or a corporate training program on how and why a piece of equipment works.

This sequencing method is based on several observations. The first is that principles are either causal relationships or natural-process relationships (both of which concern changes in concepts). For example, the law of supply and demand says how changes in the supply of, and demand for, something influence its price, and vice versa (how changes in its price influence its supply and demand).

The second observation is that principles, like concepts, exist on a continuum from broader, more general, and more inclusive ones to narrower, more specific, and less inclusive ones. For example, a fairly general principle is: Temperature change in an environment causes behavioral changes in certain organisms within that environment. And two subordinate principles are: High temperatures in a desert environment cause certain organisms to be nocturnal; and high temperatures in a desert environment cause certain organisms to undergo a period of estivation (a summertime equivalent of hibernation). This last principle could be further elaborated by identifying specific physiological changes that occur in a particular species when it estivates. Figure 3 shows another example. So, unlike concepts, the broader principles are generally simpler and easier to learn than the narrower ones. This quality led principles to be the focus of the spiral curriculum (Bruner 1960).

The third observation is that people tend to store a new principle under a broader, more inclusive one in their cognitive structures as they do for a new concept. Again, Ausubel (1968) discovered that the broader principle provides "cognitive scaffolding" for the narrower, more complex principles and therefore recommended the general-to-detailed sequencing method he called "progressive differentiation."

But there is a fourth observation for principles that does not hold for concepts. Principles can be combined into causal models that reflect the complex, systemic, and often seemingly chaotic nature of most phenomena in the world. A causal model is a set of interrelated cause-effect relationships, in which there are chains of causes and effects, and there are usually multiple causes of the effects and multiple effects of the causes (see Figure 4). These causal relationships are usually probabilistic rather than deterministic, meaning that the cause will increase the chances of the effect occurring rather than making it happen.

Figure 4 shows part of a complex causal model related to the water cycle. Each box shows a change—either an increase (shown by a rising arrow) or a decrease (shown by a declining arrow) in some activity or condition. The arrows between boxes show the direction of causality. So looking at the top of the diagram, one would read, "An increase in surface temperature causes (or more accurately increases the chances of) an increase in evaporation."

The theoretical elaboration sequence starts by teaching the broadest, most inclusive, most general principles that the learner has not yet learned in a theoretical structure (which are also the simplest principles and generally the first to have been discovered); and it gradually progresses to ever more narrow, less inclusive, more detailed, more precise principles (which are also more complex and were generally discovered later). Examples for economics (the law of supply and demand) and electricity (Ohm's law) are relevant. This sequence continues until the desired level of complexity has been reached. The fact that this order reflects the order in which the principles were usually discovered, and could be most easily discovered by learners, makes this sequence ideal for inquiry learning and other discovery methods.

When light rays pass from one medium into another (of different optical density):

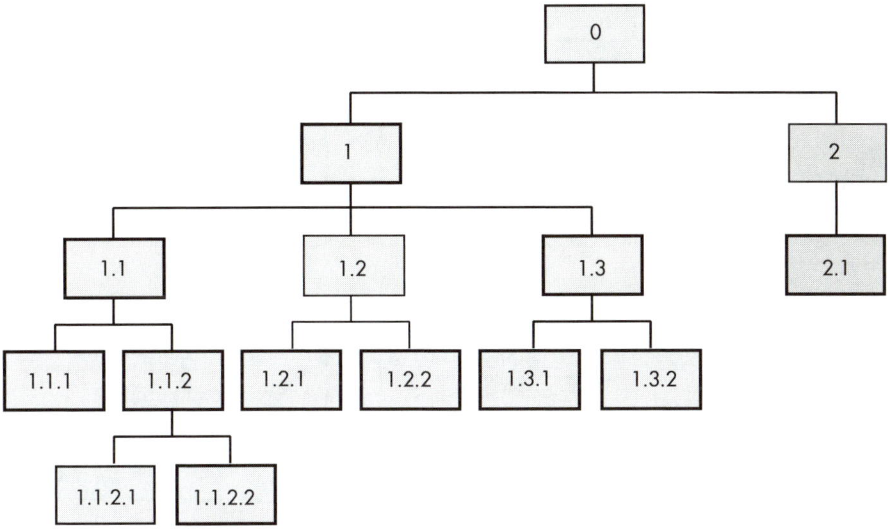

0 they behave unexpectedly,
1 they bend at the surface,
2 a straight object in both media looks bent at the surface,
1.1 the rays bend because they slow down in a denser medium or speed up in a less dense medium (C),
1.2 rays bend and change their distance from each other but remain parallel to each other (A),
1.3 a portion of each ray is reflected off the surface, while the rest is refracted into the new medium (A),
2.1 the apparent position and size of an object usually change (A),
1.1.1 if they pass into a denser medium, the light rays bend toward the normal (B, D),
1.1.2 the greater the difference in optical density between two media, the more the light rays bend (D),
1.2.1 when rays bend toward the normal, they become farther apart (B, D),
1.2.2 the sharper the angle between a light ray and the surface, the more the ray bends (D),
1.3.1 the sharper the angle between a light ray and the surface, the more of each ray that is reflected and the less that is refracted (D),
1.3.2 if the angle is equal to, or sharper than, the critical angle, all of the light ray is reflected (B, E),
1.1.2.1 the index of refraction $(n) = c_i/c_r = (\sin i)/(\sin r)$ (D, E),
1.1.2.2 the relationship between the critical angle and the index of refraction is: $\sin i_c = 1/n$ (D, E).

Codes:
(A) What else happens? (B) When? (B) Why? (C) Which way? (D) How much?

FIGURE 3: An Example of a Theoretical Structure

SOURCE: Reprinted from Charles M. Reigeluth, ed. (1999), *Instructional Design Theories and Models, Volume 2: A New Paradigm of Instructional Theory* (Mahwah, NJ: Lawrence Erlbaum Associates).

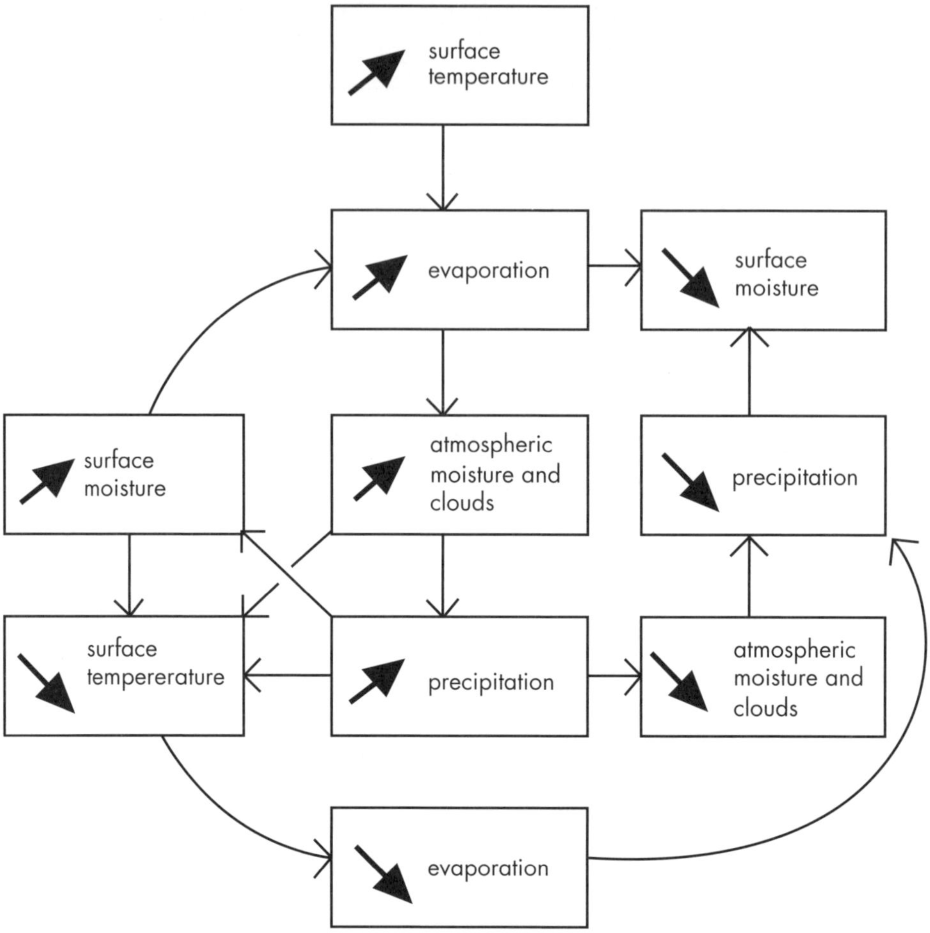

FIGURE 4: A Partial Example of a Causal Model Related to the Water Cycle

SOURCE: Reprinted from Charles M. Reigeluth, ed. (1999), *Instructional Design Theories and Models, Volume 2: A New Paradigm of Instructional Theory* (Mahwah, NJ: Lawrence Erlbaum Associates).

How does a teacher or designer identify all these principles and their inclusivity/complexity relationships? This is the purpose of a theoretical analysis. The result of such an analysis is a theoretical structure (see Figure 3), which is different from a causal model (see Figure 4) in that it shows principles that elaborate on other principles (that provide more complexity or guidance on the same phenomena), whereas a causal model shows principles that combine with other principles (add new phenomena), usually at a similar level of complexity. In Figure 3, principles 1 and 2 elaborate on principle 0 because they each provide more complex information about what happens when light rays pass from one optical medium into another of different optical density.

It should be noted that more detail can be provided by elaborating on either the causal factors or the resultant factors (effects) or both. And elaboration can occur by answering several different kinds of questions, such as:

- What else happens? or What else can cause this?
- When does this cause have this effect?
- Which way (direction) do things change?
- Why do they change?
- How much do they change?

The theoretical elaboration sequence may also be done in either a topical or spiral manner. For a topical sequence, one could go all the way down one leg of the theoretical structure, then gradually broaden out from there. For a spiral sequence, one could go completely across the top row, then across the next row down, and so forth.

The Simplifying Conditions Method

For building task expertise, SCM is a relatively new approach (though practitioners have long used it intuitively) that offers guidance for analyzing, selecting, and sequencing the "what to learn" (content). Briefly, SCM provides practical guidelines to make a very different kind of simple-to-complex sequence from the parts-to-whole (hierarchical) sequence—one that is holistic rather than fragmented. Any complex task has some conditions under which it is much easier to perform than others. For example, driving a car is easier when you have an automatic shift, no traffic, good weather, no need to start on a hill, and no need to parallel park. An SCM sequence begins with mastery of the simplest version of the task that is still fairly representative of the task as a whole; then it teaches ever more complex versions of the task until the desired level of complexity is reached, making sure that the learner is aware of the relationship of each version to the other versions. Each version of the task is a class or group of complete, real-world performances of the task. This process contrasts sharply with the hierarchical approach to sequencing, which teaches all the prerequisites first and does not teach a complete, real-world task until the end of the sequence. Figure 5 shows the differences between the hierarchical approach and the SCM approach. Note that as you conduct a hierarchical task analysis, the subskills become ever more varied (diverse) yet steadily simpler. In contrast, when you conduct a simplifying conditions analysis, the subtasks become ever more varied yet steadily more complex.

For procedural tasks, the focus is on the steps (mental and/or physical) that experts use to decide what to do when. The SCM's selection (scope) and sequencing methodology were derived primarily from the work on "path analysis" of a procedure (Scandura 1973; Merrill 1976, 1980). Every

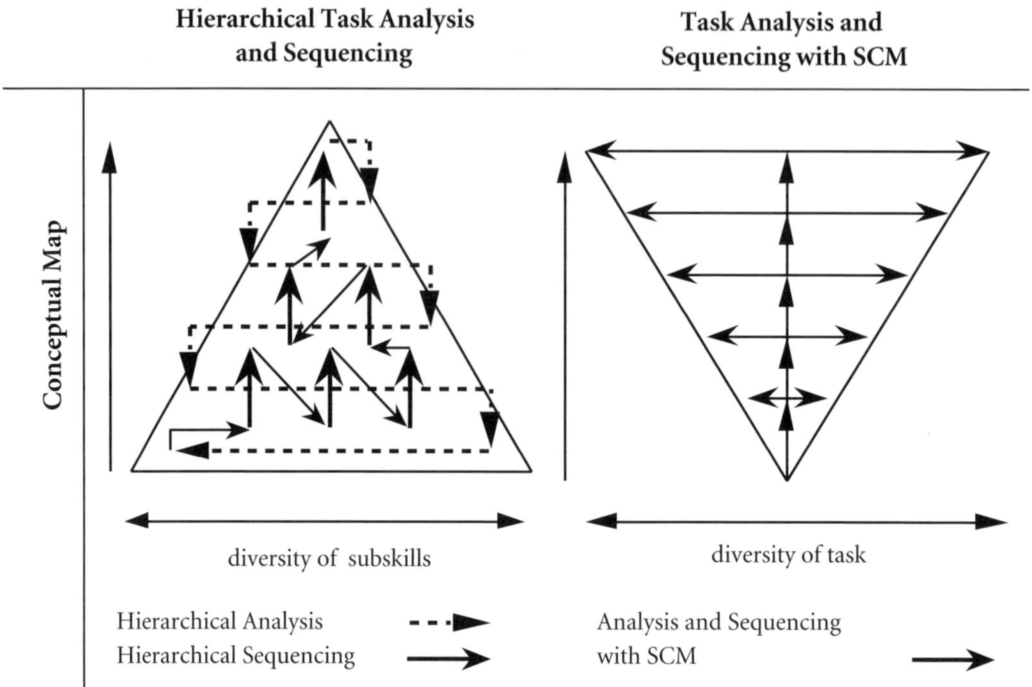

	Hierarchical Task Analysis and Sequencing	Task Analysis and Sequencing with SCM
Underlying Logic	Part to whole/simple to complex (Subskills to main skills)	Simple to complex (simple task to complex task)
For Designer	Task analysis should be done prior to sequencing as separate task.	Task analysis and sequencing can be done simultaneously. The prototype can be developed rapidly.
For Learner	Facilitates the learning of higher-order skills.	From the very first lesson, it provides (1) the flavor of the whole task, (2) a simple but applicable skill, and (3) enhanced motivation.

The hierarchical approach is necessary but not sufficient. It also introduces a very fragmentary approach.

FIGURE 5: Hierarchical Approach and the SCM Approach

SOURCE: Created by the author.

decision step in a complex procedure signals at least two different paths through the flowchart of the procedure (one of which is almost always simpler than the other). It also represents at least two different conditions of performance.

In contrast, for heuristic tasks (Reigeluth 1992; Reigeluth and Kim 1993) the focus is on principles, guidelines, and/or causal models that experts use to decide what to do when (rather than using a set of steps). Such heuristic tasks differ greatly in the nature of an expert's performance, depending on the conditions of performance. Thus, experts do not think in terms of steps when they perform the task. This sequencing methodology was derived by Reigeluth primarily from the procedural SCM sequence.

Both types of SCM sequences are used simultaneously when the task is a combination of both types of knowledge (procedural and heuristic). And the SCM and domain-elaboration sequences can be used simultaneously as well. These are referred to as multiple-strand sequences (Beissner and Reigeluth 1994).

The SCM (for both procedural and heuristic tasks) is composed of two parts: epitomizing and elaborating. Epitomizing involves identifying the simplest version of the task that is still fairly representative of the whole task. Elaborating involves identifying progressively more complex versions of the task.

The principles of epitomizing are based upon the notions of holistic learning and schema-building. Epitomizing utilizes: (1) a whole version of the task rather than a simpler component skill; (2) a simple version of the task; (3) a real-world version of the task (usually); and (4) a fairly representative (typical or common) version of the task. The epitome version of the task is performed by experts only under certain restricted (but usually real-world) conditions, referred to as the simplifying conditions.

The principles of elaborating are similarly based on the notions of holistic learning and assimilation-to-schema. Each subsequent elaboration should be: (1) another whole version of the task; (2) a slightly more complex version of the task; (3) equally authentic (or more so); and (4) equally or slightly less representative (typical or common) of the whole task. The simplifying conditions are removed one by one to define each of the more complex versions of the task.

An SCM sequence is designed by integrating task analysis with design. The analysis/design process centers around the questions, "What is the simplest version of the task that an expert has ever performed?" and "What is the next simplest version?" and so forth. As each version is identified, its place in the sequence is simultaneously determined. (More detailed guidance for analyzing and designing an SCM sequence is provided by Reigeluth 1999a.) Since designing an SCM sequence is more of a heuristic than a procedural process, the guidelines include heuristics as well as steps.

There tend to be more procedural elements at the upper levels of analysis (the major phases of the task). However, there comes a point at which it is no longer productive to break a given step into substeps, for that is not the way an expert thinks. Rather, one must identify the heuristics upon which an expert's performance of the step is based.

The Importance of the Elaboration Theory

The paradigm shift from teacher-centered and content-centered instruction to learner-centered instruction is creating new needs for ways to sequence instruction. In the industrial age paradigm of education and training, the need was to break the content or task down into little pieces and teach those pieces one at a time (Reigeluth 1999b). But most of the new, learner-centered approaches to instruction, including simulations, apprenticeships, goal-based scenarios, problem-based learning, and other kinds of situated learning, require a more holistic approach to sequencing, one that can simplify the content or task, not by breaking it into pieces but by identifying simpler real-world versions of the task or content domain. Elaboration sequences accomplish this and simultaneously make the learning process more meaningful and motivational to learners.

Charles M. Reigeluth

See also Analysis; Bruner, Jerome S.; Gagné, Robert Mills; Instructional Design; Vygotsky, Lev

References

Anderson, John R. 1983. *The Architecture of Cognition.* Cambridge, MA: Harvard University Press.

Anderson, Richard C. 1984. "Some Reflections on the Acquisition of Knowledge." *Educational Researcher* 13(9): 5–10.

Ausubel, David P. 1968. *Educational Psychology: A Cognitive View.* New York: Holt, Rinehart, and Winston.

Beissner, Katherine L., and Charles M. Reigeluth. 1994. "A Case Study on Course Sequencing with Multiple Strands Using the Elaboration Theory." *Performance Improvement Quarterly* 7(2): 38–61.

Bruner, Jerome S. 1960. *The Process of Education.* New York: Random House.

Cronbach, Lee J., and Patrick Suppes, eds. 1969. *Research for Tomorrow's Schools: Disciplined Inquiry for Education.* Toronto: Macmillan.

Gagné, Robert M. 1968. "Learning Hierarchies." *Educational Psychologist* 6(1): 1–9.

Goleman, Daniel. 1995. *Emotional Intelligence: Why It Can Matter More than IQ.* New York: Bantam Books.

Greenberg, Mark T., and C. A. Kusché. 1993. *Promoting Social and Emotional Development in Deaf Children: The PATHS Project.* Seattle: University of Washington Press.

Lee, Ji Yeon, and Charles M. Reigeluth. Submitted. "Formative Research on the Heuristic Task Analysis Process." *Educational Technology Research and Development.*

Merrill, Paul F. 1976. "Task Analysis: An Information Processing Approach." *NSPI Journal* 15(2): 7–11.

———. 1980. "Analysis of a Procedural Task." *NSPI Journal* 19(2): 11–15.

Reigeluth, Charles M. 1987. "Lesson Blueprints Based on the Elaboration Theory of Instruction." In Reigeluth (ed.), *Instructional Theories in Action: Lessons Illustrating Selected Theories and Models* (pp. 245–288). Hillsdale, NJ: Erlbaum Associates.

———. 1992. "Elaborating the Elaboration Theory." *Educational Technology Research and Development* 40(3): 80–86.

———. 1999a. "The Elaboration Theory: Guidance for Scope and Sequence Decisions." In Reigeluth (ed.), *Instructional-Design Theories and Models, Volume 2: A New Paradigm of Instructional Theory* (pp. 425–453). Mahwah, NJ: Lawrence Erlbaum Associates.

———. 1999b. "What Is Instructional-Design Theory and How Is It Changing?" In Reigeluth (ed.), *Instructional-Design Theories and Models, Volume 2: A New Paradigm of Instructional Theory* (pp. 5–29). Mahwah, NJ: Lawrence Erlbaum Associates.

———. In final preparation. *Scope and Sequence Decisions for Quality Instruction.*

Reigeluth, Charles M., and Afnan N. Darwazeh. 1982. "The Elaboration Theory's Procedure for Designing Instruction: A Conceptual Approach." *Journal of Instructional Development* 5(3): 22–32.

Reigeluth, Charles M., and Kim, Young Hwan. April 1993. "Recent Advances in Task Analysis and Sequencing." Paper presented at the NSPI national conference, Chicago, IL.

Rummelhart, D. E., and Anthony Ortony. 1977. "The Representation of Knowledge in Memory." In Richard C. Anderson, Rand J. Spiro, and William E. Montague (eds.), *Schooling and the Acquisition of Knowledge* (pp. 99–135). Hillsdale, NJ: Lawrence Erlbaum Associates.

Scandura, Joseph M. 1973. "Structural Learning and the Design of Educational Materials." *Educational Technology* 13 (August): 7–13.

Simon, Herbert A. 1969. *Sciences of the Artificial.* Cambridge, MA: MIT Press.

Vygotsky, L. S. 1978. *Mind in Society: The Development of Higher Psychological Processes.* Ed. M. Cole, V. John-Steiner, S. Scribner, and E. Souberman. Cambridge, MA: Harvard University Press.

Electronic Books (e-books)

An electronic book is the result of combining a digital text with an electronic reading device so that the text can be read in the same manner as a paper-based book. The advent of e-books signals an advance in on-screen reading. Computers and other electronic devices have always supported the reading of text files. However, the text that makes up an e-book is coded to provide many of the features that book lovers and scholars value in printed works. Users "turn" electronic pages by tapping the screen or by pressing a button on a computer or on a hand-held device. Users can also highlight important passages and make annotations or sketches as they read. Other features take advantage of the technology inherent in the computers and electronic readers on which e-books are read. Users can acquire updated versions of texts in minutes, use hyperlinks to move about in the textual world, view multimedia, access definitions and related texts, adjust text size and screen backlighting, and even hear the e-book read by a digitized voice. Furthermore, dozens, perhaps hundreds, of e-books can be stored on a computer or other device, making it possible for an e-book user to have a portable and current library of information constantly available.

The convergence of printed text and electronic devices has potential in education. This potential is particularly apparent at the postsecondary level, where self-directed learning and the need for current content thrives. Currently textbook publishers like Houghton Mifflin are moving to provide up-to-the-minute, tailor-made versions of textbooks that can be downloaded to electronic reading appliances. Students attach an electronic reader to a phone line or a computer through a USB connection, visit the appropriate website, purchase the e-textbook, and receive the book in a matter of minutes. Students read the text as required and can use the highlighting features or add notes. Annotations can be shared with study groups or uploaded for review by the instructor. Instructors can also arrange for their annotations to be available to guide student reading or to relate textbook materials to class lectures. In addition, instructors can customize textbooks by selecting specific chapters rather than requiring students to purchase an entire book.

In K-12 settings, e-books have had little exposure thus far. Beyond the issue of the expense and delicate nature of electronic reading devices are issues surrounding the instructional purposes of K-12 settings. Though e-books may be appropriate for the self-directed, content-rich mode of instruction typical in postsecondary settings, it is unclear whether e-books will be advantageous for younger students or for students who need to develop focused, sustained attention and critical thinking skills. Electronic-based texts may not support or encourage sustained reading, a key component in developing as a sophisticated, critical reader of alphabetic texts (Crawford and Gorman 1995). On the other hand, e-books invite active reading and involvement with texts, strategies that proponents of alphabetic literacy have long held in high regard (Schilit 1999). This debate is not likely to disappear, but neither are e-books. Ultimately, the future of e-books in K-12 settings will depend on how understandings of literacy change and how e-books themselves evolve.

Barbara Pace

See also E-texts and Readers; Handheld Technologies; Virtual Library

References

Crawford, Walt, and Michael Gorman. 1995. *Future Libraries: Dreams, Madness, and Reality.* Chicago: American Library Association.

Murray, Janet H. 1997. *Hamlet on the Holodeck: The Future of Narrative in Cyberspace.* New York: The Free Press.

Schilit, Bill. 1999. "Why e-Read?: Finding Opportunity in the Merger of Paper and Computers." *Future of Print Media Journal.* Available online at www.futureprint.kent.edu/articles/schilit01.htm (April 4).

Tyner, Kathleen. 1998. *Literacy in a Digital World: Teaching and Learning in the Age of Information.* Mahwah, NJ: Lawrence Erlbaum Associates.

Electronic Emissary

The Electronic Emissary (emissary.wm.edu) is a web-based service and resource center that helps K-12 teachers and students with Internet access locate expert mentors in various disciplines for the purpose of setting up curriculum-based, extended electronic exchanges among teachers, students, and experts. In this way, the face-to-face teacher-student interaction in the classroom is supplemented and extended by e-mail, chat, and teleconferencing exchanges.

Resources available via telecommunication technologies are providing creative K-12 teachers with new ways to engage their students in authentic learning experiences—those that reflect how knowledge is built and used in the world outside school. Today a teacher no longer needs to be the sole expert in the classroom. It is possible, for example, for students to learn about current weather phenomena from meteorologists, or to discuss the paleontological implications of a recent T-rex skeleton discovery with evolutionary scientists, using simple telecomputing tools such as e-mail and chat. Volunteer subject-matter experts, such as the meteorologists and paleontologists mentioned above, can work virtually with students over an extended period of time, developing and sustaining mentor-protégé relationships that contribute to the richness and relevance of curriculum-based learning in K-12 classrooms. This practice has come to be known as telementoring.

The Electronic Emissary is both a telementoring service and a research effort—its staff both facilitates and investigates the nature of primarily e-mail communication between adult subject-matter-expert volunteers and elementary, middle-level, and secondary students and their teachers. It has been online since February 1993 and on the World Wide Web since December 1995. It serves students and teachers globally, but the majority of its participants are in North America. Emissary-related research has focused upon the nature of telementoring interactions in which K-12 students are active inquirers, the motivations and perceptions of their volunteer subject-matter mentors, effective telementoring facilitation techniques, and what teachers learn as they help students participate in curriculum-oriented telementoring projects.

Though simple and appealing in concept, successfully planning, implementing, and completing telementoring projects is challenging. Online communication lacks the full spectrum of visual and audible information that we depend upon, often unconsciously, in face-to-face exchanges. Therefore, it requires somewhat different interaction strategies if it is to be used to create maximal educational benefit by and for students and teachers. These techniques can be modeled and made explicit by someone closely following online conversations in the role of facilitator, helping participants to construct the online teaching/learning experience in mutually beneficial ways. The Electronic Emissary's years of customized project assistance have shown that the people best prepared have experience in online communication and education. The most important and valuable part of the Electronic Emissary's services is its individualized online facilitation for each and every telementoring project, providing just-in-time teaching and learning assistance to teachers, students, and subject-matter experts.

Sample Telementoring Projects

Table 1 provides some examples of curriculum-related work supported by Electronic Emissary–facilitated telementoring.

Members of Emissary-supported telementoring teams are engaged in in-depth, dynamic exchange. Project evaluation results provided by team members have emphasized the importance of the relationships that have developed among participants. Subject matter "came alive" for students who could interact with someone for whom curriculum content is part of everyday life—and a passionate interest. Many participating teachers develop close, apprentice-like relationships with the experts, requesting and receiving assistance with content-related concepts, resources, and activity design. Subject-matter experts often delight in opportunities to revisit and delve deeper into their disciplinary specializations by interacting with interested others. Online facilitators express fascination with the often challenging, personal, and in-depth communication created by people who know each other only through pixels on a screen.

TABLE 1: Telementoring Examples from the Electronic Emissary

Examples from Elementary Schools

- A ten-year-old student in Connecticut corresponded frequently with a professor in his seventies in Arizona. They continued their study of Arthurian legends that began in the spring semester of 1995 for more than three years. The student, the professor, their online facilitator, and the Emissary's director coauthored an article describing their online educational experiences (Harris, O'Bryan & Rotenberg, 1996).
- Nineteen fourth- and fifth-grade students in McAllen, Texas, compared the experiences of their families on the Texas "La Frontera" to colonial life in

the original thirteen U.S. colonies with the help of the director of a historic preservation center and museum in Fredericksburg, Virginia.

- Second-grade students in Reading, Massachussetts, who were studying about magnetism posed questions that were answered by a physicist from Arizona State University.
- A marine biologist from California State University–Monterey Bay helped thirty-two third-grade students learn about coral reefs, in particular, damage to reefs from pollution and the diversity of life on a coral reef. The students originally planned to create a saltwater aquarium in their classroom but decided to create a freshwater aquarium with local species instead. They gathered plant and fish specimens from a local stream, monitoring their activity in the classroom aquarium. The mentor guided this process, providing scientific and technical information, answering questions posed by students, and directing them to web-based and paper resources related to their marine biology interests. He also assisted the teacher when she requested information specific to the marine unit that she was building with her students.

Examples from Middle Schools

- Ninth-grade students from San Angelo, Texas, corresponded with an anthropologist from Los Angeles about civil rights. The topic was explored with reference to the first Rodney King trial (taking place at the time of the exchange) and historically by examining the struggle for African-American rights during the late 1950s and early 1960s, emphasizing the contributions of Dr. Martin Luther King Jr.
- Sixth-grade students in Houston, Texas, who were engaged in multidisciplinary study of the Middle Ages posed questions to a medieval history professor who worked at the University of Illinois, addressing her as "Learned Sage." She, in turn, answered their questions, calling them "Seekers of Knowledge."
- Ninth-grade students in Hart, Texas, corresponded with an engineering professor from Boston University about waves and wave phenomena, including radar, sonar, light, sound, radio, seismic waves, ultrasound, and water. The focus of the communication was discussion of applied physics experiments and activities that the students conducted about different types of waves and their interactions.

Examples from High Schools

- High school students in Delaware who were studying Nathaniel Hawthorne's *The Scarlet Letter* communicated with the character Arthur Dimsdale, who was actually an American literature professor at the U.S. Naval Academy. During the following semester, the students communicated with the professor himself about Mark Twain's *Huck Finn*, culminating their exchange by creating a newspaper that they called *The Mississippi Times*, an idea first suggested by the expert. The teacher and the professor shared instructional ideas, resources, and perspectives about Mark Twain's works and views.

- An advanced placement Spanish literature class in Ross, California, communicated with a professor of Spanish at Ball State University. All communication was conducted in Spanish. Topics addressed included the *nivolas* of Miguel de Unamuno and how those works fit into the cultural and historical contexts of Spain.
- Sixteen- to eighteen-year-old students from Salmon Arm, British Columbia, who were curious about virtual reality technologies corresponded with a computer scientist working for Boeing and NASA, later commenting upon his skill in using humor and professional anecdotes to help them to understand technical information.

SOURCE: Created by the author.

Why Telementoring?

We have learned that students and teachers exploring real-world, multifaceted, curriculum-based topics need to actively build deep and sophisticated understanding. One of the most effective ways to do this is by engaging in ongoing dialogue with knowledgeable others, as the students form, refine, and expand their comprehension. Classroom teachers typically serve as the subject-matter experts with whom students interact in such complex areas of inquiry. Yet when the issues being explored are multidisciplinary, technically and conceptually sophisticated, or dependent upon current and highly specialized research and theory, additional expertise must be made directly available to students and teachers longitudinally and on an as-needed basis. This is what telementoring offers to learners and educators today—and what the Electronic Emissary Project brings to students and teachers worldwide.

Judi Harris

See also Cognitive Apprenticeship; Collaborative Technologies; Computer-Mediated Communication; Computer-Supported Collaborative Learning; Telementoring

References

Electronic Emissary Project (emissary.wm.edu).

Harris, Judith. 1998. *Virtual Architecture: Designing and Directing Curriculum-Based Telecomputing.* Eugene, OR: International Society for Technology in Education.

Harris, Judith B., and Candace Figg. 2000. "Participating from the Sidelines, Online: Facilitating Telementoring Projects." *Journal of Computer Documentation* 24(4): 227–236.

Harris, Judith B., and Greg Jones. 1999. "A Descriptive Study of Telementoring among Students, Subject Matter Experts, and Teachers: Message Flow and Function Patterns." *Journal of Research on Computing in Education* 32(1): 36–53.

Harris, Judith, Ellen O'Bryan, and Lena Rotenberg. 1996. "It's a Simple Idea, but It's Not Easy to Do: Practical Lessons in Telementoring." *Learning and Leading with Technology* 24(2): 53–57.

Electronic Performance Support System (EPSS)

During the past ten years there has been a paradigm shift revolutionizing the training industry. Both the public sector and the private sector have been changing their focus from developing high-quality learning solutions to supporting human performance with alternate approaches. One of the many solutions available to improve human performance is Electronic Performance Support Systems.

EPSSs include tools with an electronic infrastructure that can support a user and allow him to complete a job, function, or task in the time of need. The primary goal of an EPSS is to enhance human performance and productivity regardless of a person's preexisting knowledge or skills (Stevens and Stevens 1995). This goal is accomplished by developing a number of different components or structures that make up an EPSS (see Figure 1). The underlying foundations used to develop these structures are based on two types of electronic performance support: extrinsic and intrinsic.

Extrinsic Support

An extrinsic EPSS is performance support that is integrated with a system but is not inherent to the system itself (Gery 1995). In other words, a person has to break away from the task at hand in order to access information that supports performance. Extrinsic support can be grouped into the following categories/structures: (1) online learning experience; (2) context-sensitive tools; (3) online reference; and (4) knowledge management.

Online Learning Experience

Traditionally, online learning experiences are in the form of computer-based training (CBT) or self-paced modules of instruction. These courses last anywhere from thirty minutes to two hours. However, when a person needs just-in-time support, a traditional CBT is not practical. Granular CBT has become a popular form of extrinsic support. This type of CBT lasts no more than five to ten minutes, therefore truly falling under the category of just-in-time support. Granular CBT allows a person to become oriented and to develop skills through explanations, demonstrations, tips, or practice activities (Gery 1995). Granular CBTs are exemplified by Lotus Screen Cam technology. This software provides an animated sequence that demonstrates an application's functionality when performing a certain task. People can learn how to perform tasks by viewing demonstrations, which can be reviewed as many times as necessary.

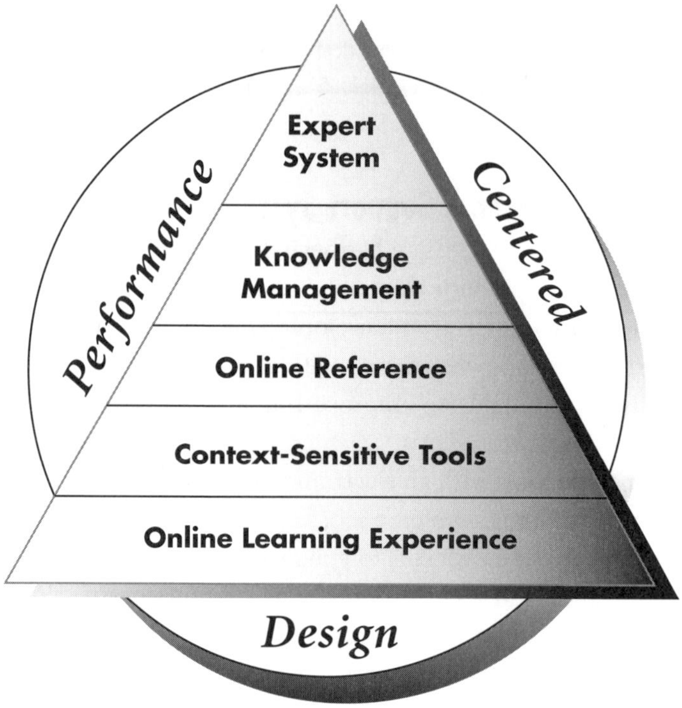

Figure 1: Illustration Depicting the Structures and Designs That Reflect the Different Intrinsic and Extrinsic Types of EPSS

SOURCE: Created by the author.

Context-Sensitive Tools

While learning experiences are structured around demonstrations and practice activities, context-sensitive tools are most commonly structured as online coaches, templates, and task wizards. The common use for this type of extrinsic support is task guidance and task execution, which ultimately lead to consistent and rapid task completion (Gery 1995). Online coaches provide task guidance when using an application. This technology is particularly helpful when someone is unfamiliar with the nuances that are inherent to an application. It allows people to quickly learn by referencing the support.

Microsoft Word's Mr. Paper Clip is an example of this type of extrinsic support. This help provides procedures or troubleshooting guidelines based on a person's answers to a series of questions. After running through a number of options or choices, the user is given a step-by-step procedure plan on how to perform a task. This is particularly useful for someone who is a novice at using a software system.

Templates and wizards allow people to develop documents using a particular style or format, based on a series of questions and answers posed to users. Again the advantage of using this performance support is for rapid development. Microsoft PowerPoint is a perfect example of the template and wizard support. Each time a user launches this application he is given the choice of developing his presentation using the wizard or template option. Users are led through a series of questions, beginning with what type of presentation must be developed through the process of actually developing each slide in the presentation.

Online Reference

Online reference systems (generally known as online help) usually provide step-by-step instructions on how to perform a specific task or function, whether technical or nontechnical. Online help systems can be stand-alone or embedded within a software application. An example of this extrinsic support is the online help embedded in Microsoft Word. When accessing this help, people can drill down to the step-by-step procedures on how to perform a given task. Even though the help is part of the application, it is considered extrinsic because people must break away from their tasks in order to access the information located in the online help.

Knowledge Management

Where online learning experience, context-sensitive tools, and online reference may support specific tasks, knowledge management encompasses a higher level of extrinsic support. Knowledge management supports the creation, archiving, and sharing of "intellectual capital" and best practices within and across communities of people and organizations (Rosenberg 2001). Intellectual capital is the compilation of people's collective knowledge, within a given organization, that may reside in policies, procedures, instructional materials, or within an individual's experiential wisdom. The knowledge management system's main purpose is to consolidate this information into one central location for the purpose of rapid reuse. Generally, public- and private-sector organizations use web-based technology as the infrastructure for knowledge management systems. The structure of a knowledge management system would look like that of a search engine, organized for flexible search, review, and navigation (Gery 1995).

Intrinsic Support

Intrinsic EPSSs are those systems that are tightly integrated within a system or application. In intrinsic support, users do not see the difference between the support and the actual application. The higher the level of intrinsic support, the less aware the user is of its presence to the point that it is invisible (Gery 1995). Ultimately, the actual work environment (within an application) is centered on the tasks a user is responsible for completing. The development of this type of performance-based environment is known as performance-centered design.

Performance-Centered Design

Performance-centered design is an approach to designing software that inherently supports work processes. The needs of the user are placed at the center of the development process (Rosenberg 2001). An expert system can be categorized as an EPSS created using a performance-centered approach. The two main attributes of an expert system include the following: The system should replace one or more experts; and the system should take a person beyond his current knowledge. In an expert system, the user has the opportunity to seek expert advice when executing important decisions. The idea behind having an "expert" built into the system is to allow a person to perform a task with minimal outside guidance.

Usually, a person using an expert system would not have the subject-matter knowledge or skill to perform a given task or function. That is why a person can perform above and beyond his current knowledge base when using this type of system. In this design approach extrinsic support structures may also be embedded, such as coaches, wizards, and online help systems.

A classic example of an expert system developed using a performance-centered approach would be Intuit's TurboTax. This expert system provides advice through a series of questions and answers and through the use of online help and coaches. TurboTax allows people to prepare tax returns using a self-guided approach. The software provides a series of questions formulated by tax experts that drive the creation and completion of a person's tax return. The question-and-answer dialog is used from beginning to end. Anytime in between, the user can choose to use the embedded help system or access an online coach to help understand certain concepts.

Sanjay K. Dua

See also Analysis; Just-in-Time Training; Human-Computer Interaction; Knowledge Management; Performance Support

References

Bahill, A. Terry. 1991. *Verifying and Validating Personal Computer-Based Expert Systems.* Englewood Cliffs, NJ: Prentice-Hall.

Bielawski, Larry, and Robert Lewand. 1990. *Intelligent Systems Design: Integrating Expert Systems, Hypermedia, and Database Technologies.* New York: John Wiley and Sons.

Gery, Gloria. 1991. *Electronic Performance Support Systems.* Boston: Weingarten.

———. 1995. "Attributes and Behaviors of Performance-Centered Systems." *Performance Improvement Quarterly* 8(1): 47–93.

Horton, W. 1990. *Designing and Writing Online Documentation, Help Files to Hypertext.* New York: John Wiley and Sons.

Microsoft Office Applications. Microsoft Corporation. One Microsoft Way, Redmond, CA 98052.

Rosenberg, Marc. 2001. *E-Learning: Strategies for Delivering Knowledge in the Digital Age.* McGraw-Hill.

Stevens, Emily F., and George H. Stevens. 1995. *Designing Electronic Performance Support Tools.* NJ: Educational Technology Publications.

Stewart, Thomas. 1997. *Intellectual Capital—The New Wealth of Organizations.* New York: Currency.

TurboTax. Intuit. Box 50630, Palo Alto, CA 04303.

Electronic Portfolios

An electronic portfolio is a purposeful collection of artifacts in a variety of media formats (audio, video, graphics, text, etc.). Artists have maintained portfolios for years, often using their collections for seeking further work, or simply for demonstrating their art. An artist's portfolio usually includes only her best work. Financial portfolios contain a comprehensive record of fiscal transactions and investment holdings that represent a person's monetary worth. By contrast, an educational portfolio contains work that a learner has selected and collected to show growth and change over time. A critical component of an educational portfolio is the learner's reflection on the individual pieces of work (often called "artifacts"), as well as an overall reflection on the story that the portfolio tells. The traditional formats for portfolios in education are paper-based documents stored in manila folders, three-ring notebooks, or larger containers. Most often, the artifacts are in the form of text and images on paper, but the use of electronic media is quickly emerging.

A commonly accepted definition of a portfolio is provided by educators in the Pacific Northwest who form the Northwest Evaluation Association (Paulson, Paulson, and Meyer 1990): "A portfolio is a purposeful collection of student work that exhibits the student's efforts, progress and achievements in one or more areas. The collection must include student participation in selecting contents; the criteria for selection; the criteria for judging merit; and evidence of student self-reflection." The learner's reflections provide the rationale that specific artifacts are evidence of achieving the stated standards or goals. Portfolios that use a paper-based organizer, a database, or hyperlinks to clearly show the relationship between standards or goals, artifacts and reflections, may also be referred to as a standards-based portfolio. Standards-based portfolios are frequently used in education and can take a variety of forms, including traditional paper-based portfolios, electronic portfolios, and digital portfolios. The terms "electronic portfolio" and "digital portfolio" often are used interchangeably; however, there is a distinction. An electronic portfolio contains artifacts that may be in analog (i.e., video- or audiotape) or digital (i.e., computer-based) form. In a digital portfolio, *all* artifacts have been transformed into computer-based form.

An electronic portfolio is not a haphazard collection of artifacts (i.e., a digital scrapbook or a multimedia presentation) but rather a reflective tool that demonstrates growth over time (Barrett 2000). An electronic portfolio is different from other collections such as digital scrapbooks, online resumes, and multimedia presentations because they (1) are organized around a set of standards or learning goals; (2) include learner reflection on her achievement of each standard or goal; (3) include a rationale for why each artifact was selected; and (4) provide an overall reflection of the process of putting together the portfolio.

There are three general purposes for developing portfolios: (1) learning (formative) portfolios, which are usually developed on an ongoing basis supporting professional development; (2) assessment (summative) portfolios, which are usually developed within the context of a formal evaluation process; and (3) employment (marketing) portfolios, which are used for seeking employment (Hartnell-Young and Morriss 1999; Wolf 1999). The benefits of developing electronic portfolios for either students or teachers include: minimal storage space, easily created backup files, portability, long shelf life, learner-centered activities, improved technology skills, multiple ways to demonstrate achievement, and accessibility (especially web portfolios) (Kankaanranta, Barrett, and Hartnell-Young 2000).

Development Process

Creating an electronic portfolio can seem daunting but is less arduous if viewed as a series of stages. A useful framework for electronic portfolio development is anchored in two bodies of literature: multimedia development, and portfolio development in K-12 education (Barrett 2000). Teachers and students gain a powerful tool for demonstrating growth over time if they understand how these processes fit together and how standards or goals contribute to electronic portfolio development. The multimedia development process typically requires the following stages (Ivers and Barron 1998):

- *Assess/decide:* The focus is on needs assessment of the audience, the presentation goals, and the appropriate tools for the final presentation.
- *Design/plan:* The focus is on organizing or designing the presentation. Tasks may include determining audience-appropriate content, software, and storage medium, developing a presentation sequence, and constructing flowcharts or storyboards.
- *Develop:* Materials are gathered and organized for the best presentation using appropriate software.

- *Implement:* The developer makes the presentation to the intended audience.
- *Evaluate:* The focus is on evaluating the presentation's effectiveness in light of its purpose and the assessment context.

Following are stages of portfolio development (Danielson and Abrutyn 1997):

- *Collection:* Teachers and students save artifacts that represent the successes and growth opportunities from their day-to-day teaching and learning.
- *Selection:* Teachers and students review and evaluate the artifacts they have saved and identify those that demonstrate achievement of specific standards, outcomes, or goals.
- *Reflection:* Teachers and students become reflective practitioners by evaluating their growth over time, their achievement of the standards, and gaps in their development.
- *Projection (or Direction):* Teachers and students compare their reflections to the standards, outcomes, or goals and performance indicators and set learning goals for the future. This is the stage that turns portfolio development into professional development and supports lifelong learning.
- *Presentation:* Teachers and students share their portfolios with peers. This stage can encourage collaboration and commitment to professional development and lifelong learning.

Five Stages of Electronic Portfolio Development

Combining the multimedia development and portfolio development processes, five stages of electronic portfolio development emerge.

Defining the Portfolio Context and Goals. The primary task in this stage is to identify the assessment context. This includes identifying the goals or standards to be addressed in the portfolio (these often come from state or national standards), the audience for the portfolio, and the resources available to complete the portfolio. This important step helps frame the rest of the portfolio development process.

The Working Portfolio. This critical stage occupies the longest span of time and involves determining the types of artifacts that best demonstrate the required standards, outcomes, or goals and gathering appropriate evidence. Portfolio artifacts should be selected from different points in time to demonstrate the growth and learning that has taken place, and a short reflective statement should be placed with each artifact at the time it is saved. These statements are revisited and expanded in subsequent stages.

It is also necessary at this stage to determine the software development tools most appropriate for the portfolio context, the content, and the resources available. The software used to create the electronic portfolio will likely control, restrict, or enhance the portfolio development

The Reflective Portfolio. This stage usually precedes evaluation reviews (for summative portfolios) or employment applications (for marketing portfolios). In this stage, portfolio reflections are derived from statements in the working portfolio, highlight significant points in the learning process, and ensure that formative self-assessment is occurring. Reflection on one's work is requisite if the portfolio owner is to learn from the process. Three questions should guide this reflective process (Campbell et al. 2000, 22): "What?" "So what?" and "Now what?" To use these questions, the learner would first summarize the artifact that documents the experience, in order to answer the question "What?" Second, the learner would reflect on what she learned and how this leads to meeting the standard, which answers the question "So what?" And third, the learner would address implications for future learning needed and set forth refinements or adaptations, in order to answer "Now what?" It is this process of setting future learning goals that turns electronic portfolio development into a powerful tool for professional development. A portfolio system encourages students to take an active role in their learning.

The Connected Portfolio. This stage is somewhat unique to the electronic portfolio because it assumes students have the ability to create hypertext links between documents, goal statements, work samples, assessment rubrics, and reflections. This stage may also include the use of appropriate multimedia artifacts and usually requires the creation of a table of contents to structure the portfolio. When using the portfolio for assessment, the transformation from "artifacts" to "evidence" is not always clear. The connected portfolio makes this thinking process more explicit. The ability to create links from multiple perspectives (and multiple goals) also overcomes the linearity of two-dimensional paper portfolios and permits a single artifact to demonstrate multiple standards (i.e., national technology standards and state teaching standards).

The Presentation Portfolio. At this stage, the portfolio is presented before an audience and is stored in an appropriate medium for future use and reflection. This individual strategy depends on the context, the type of portfolio, and the level of feedback and collaboration available. In an environment of continuous improvement, a portfolio should be viewed as an ongoing learning tool, and its effectiveness should be reviewed on a regular basis to be sure that it is meeting the goals set.

There are many technology-based tools that can be used to develop electronic portfolios throughout the stages that have been outlined in this article. The value of creating an electronic portfolio should exceed the efforts expended. Above all else, the electronic portfolio should showcase learner achievements and the growing capabilities of using technology to support lifelong learning.

Electronic Portfolios as Alternative Assessment

Portfolios are often used as a form of alternative assessment. The terms "alternative assessment," "authentic assessment," and "performance-based assessment" may be used synonymously "to mean variants of performance assessments that require students to generate rather than choose a response" (Herman, Aschbacher, and Winters 1992, 2). There are two central features to alternative assessments. "First, all are viewed as *alternatives* to traditional multiple-choice, standardized achievement tests; second, all refer to direct examination of student performance on significant tasks that are relevant to life outside of school" (Worthen 1993, 445). The characteristics of this type of assessment are: meaningful performance tasks; clear standards and criteria for excellence; emphasis on metacognition and self-evaluation; student-produced products and performances; and positive interaction between assessor and assessee (Burke 1999). Kay Burke (1999) and Robin Fogarty (1998) advocate a balanced approach to assessment, with a focus on three components:

1. Traditional assessment, which focuses on grades, rankings, knowledge, curriculum, and skills. Traditional assessment can be implemented through classroom assessments (tests, quizzes, homework assignments), as well as standardized tests (either norm-referenced or criterion-referenced).
2. Performance assessment, which focuses on observable results, standards, application, and transfer. Performance assessment is often implemented through standards-related tasks, criteria, and scoring rubrics.
3. Portfolio assessment, which focuses on growth and development over time. Portfolio assessment is often implemented through selection, reflection, and inspection of class work, goal-setting, and self-evaluation.

Nevertheless, there are importance differences between performance assessments and portfolios. "Performance assessment focuses on the direct observation of a student's performance" (Fogarty 1998, 10). Students create projects or perform tasks based on predetermined standards, criteria, and indicators, which are evaluated by scoring rubrics. A portfolio is a container that holds examples of student or teacher work (the "artifacts") and reflections on that work that transform the artifacts into "evidence" of achievement. Although many of these artifacts could be the results of performance assessments, it is the reflective components in the portfolio that provide the rationale for converting artifacts into evidence of learning.

Electronic Portfolios or Online Assessment Management Systems?

Many teacher education programs are adopting electronic portfolios to meet the National Council for Accreditation of Teacher Education 2000 Standard no. 2, Assessment System, and the implementation often resembles more of a grading or record-keeping system than the traditional paper-based portfolio (Barrett 2003). A portfolio that closely emulates a paper version and just happens to be stored in an electronic container is a very different document from the current implementation of these online database-structured systems, whether developed in-house or purchased commercially.

Electronic portfolios tend to be more learner-centered, with appearance, format, and structure controlled by the portfolio developer, requiring and demonstrating higher technology skills. Electronic portfolios may be constructed with a variety of software and may be published in a variety of formats, including CD-ROM, Internet, videotape, or DVD.

Online assessment management systems tend to be more institution-centered, with format and structure controlled by the underlying database, requiring and demonstrating lower technology skills. These systems are usually accessed using a web browser and published on a secure web server. More research is needed on examples of implementations that clearly differentiate between learner-owned electronic portfolios and the assessment systems used by institutions to record evidence of progress toward meeting standards, outcomes, or goals.

Helen Barrett

See also Alternative Assessment; Evaluation; Weblog

References

Barrett, Helen. 2000. "Create Your Own Electronic Portfolio." *Learning and Leading with Technology* 27(7): 14–21.

———. 2003. "e-Portfolios: Issues in Assessment, Accountability, and Preservice Teacher Preparation." Paper presented at the American Educational Research Association Conference, April 22. Available online at http://electronicportfolios.org/portfolios.html.

Burke, Kay. 1997. *Designing Professional Portfolios for Change.* Palatine, IL: IRI/SkyLight Training and Publishing.

———. 1999. *How to Assess Authentic Learning.* Arlington Heights, IL: Skylight Professional Development.

Campbell et al. 2000. *Portfolio and Performance Assessment in Teacher Education.* Boston: Allyn and Bacon.

Danielson, Charlotte, and Leslye Abrutyn. 1997. *An Introduction to Using Portfolios in the Classroom.* Alexandria, VA: Association for Supervision and Curriculum Development.

Fogarty, Robin. 1998. *Balanced Assessment.* Arlington Heights, IL: Skylight Professional Development.

Hartnell-Young, Elizabeth, and Maureen Morriss. 1999. *Digital Professional Portfolios for Change.* Arlington Heights, IL: Skylight Professional Development.

Herman, Joan L., Patricia R. Aschbacher, and L. Winters. 1992. *A Practical Guide to Alternative Assessment.* Alexandria, VA: Association for Supervision and Curriculum Development.

Ivers, Karen, and Ann E. Barron. 1998. *Multimedia Projects in Education.* Englewood, CO: Libraries Unlimited.

Kankaanranta, Marja, Helen Barrett, and Elizabeth Hartnell-Young. 2000. "Exploring the Use of Electronic Portfolios in International Contexts." Paper submitted to ED MEDIA 2001 World Conference on Educational Multimedia, Hypermedia, and Telecommunications, Tampere, Finland, June 25–30.

Linn, Robert L., and Norman E. Gronlund. 2000. *Measurement and Assessment in Teaching.* Upper Saddle River, NJ: Merrill.

Paulson, F. L., P. R. Paulson, and C. A. Meyer. 1991. "What Makes a Portfolio a Portfolio?" *Educational Leadership* 58(5): 60–63.

Wolf, Kenneth. 1999. *Leading the Professional Portfolio Process for Change.* Arlington Heights, IL: Skylight Professional Development.

Worthen, Blaine R. 1993. "Critical Issues That Will Determine the Future of Alternative Assessment." *Phi Delta Kappan* (February): 444–456.

Entertainment-Education (e-e)

The term "entertainment-education" describes the use of traditional entertainment media formats to deliver educational, prosocial message content. Terms such as "enter-educate," "infotainment," and "pro-development entertainment" are also used to describe this practice, with entertainment-education becoming most acceptable in recent years. This communication strategy employs a range of media and technology formats, such as radio and TV soap operas, street theater, songs, and films, among others. The strategy has been used by various institutions that have a message to convey to a specific target audience. The field of health communication, for instance, has used e-e extensively to raise awareness about problems such as population control and diseases such as HIV/AIDS. E-e is the process of "purposely designing and implementing a media message both to entertain and educate, in order to increase audience members' knowledge about an educational issue, create favorable attitudes and change overt behavior" (Singhal and Rogers 1999, 9).

Although there is still some debate on the definition of e-e, the above is currently acknowledged as an acceptable guideline. However, the boundaries continue to be redrawn as the strategy continues to develop and newer formats appear in the communication spectrum. An example of this is the growing popularity of the talk-show format, which has elicited a lot of interest among e-e strategists due to its appeal among audiences and its educational potential. Many projects fall outside the boundaries of e-e; although they may contain elements of an e-e strategy, they do so intuitively and were not purposely designed to educate. Meanwhile, some professionals in the field consider only those projects that successfully increase knowledge, change attitudes, and influence behavior as e-e.

Furthermore, the definition provided above is based on the three-stage model of behavior change, whereas other models add additional stages. Martine Bouman (1999, 25) modifies the above definition and says that e-e is "the process of purposively designing and implementing a mediating communication form with the potential of entertaining and educating people, in order to enhance and facilitate different stages of pro-social (behavior) change." In short, e-e is considered to be any project that contains educational messages embedded into an entertainment format, particularly projects driven by theories of social marketing, persuasion, and social learning. For instance, social marketing is based on the idea that social change can be communicated to audiences in the same way that commercial products are advertised. This involves an assessment of the target audience's characteristics in order to find out how best to make the message appealing.

The idea of combining entertainment and education goes as far back in history as the oldest oral cultures. Mythologies, folktales, parables, tragedies, comedy, and games have all influenced thought, belief, and action, providing entertainment and simultaneously articulating, reinforcing, and shaping educational goals and value systems. However, e-e as a deliberate strategy grounded in communication and social psychological theories has moved beyond the oral traditions of earlier times to include the audiovisual mass media of today.

The e-e approach draws heavily on Albert Bandura's social cognitive theory, which suggests people learn by observing the behavior of models. E-e programs thus capitalize on parasocial interaction, the quasi-interpersonal relationship between an audience member and a media character. This is evident among avid followers of television soap operas who cultivate a personal relationship with the characters, sharing all manner of emotions with them as the plot unfolds and engaging in discussion about episodes as if they were a part of their everyday lives. The idea is that by rewarding positive and by punishing negative behaviors audience members will identify with and model their behavior on the characters they observe.

There has been substantial growth in the application of this communication strategy globally. This is in part due to changes in the predominant development paradigms—from those that focus on technology and massive infusions of financial resources to those that focus more on education and communication to change behavior and aim at self-sustaining development. The e-e strategy attempts to create a balance between entertainment-degradation and boredom-education. The former refers to a tendency to degrade a message to increase its attractiveness (e.g., the increasing use of sex and violence in entertainment television). But boredom-education programs are those that frequently overemphasize educational content without engaging audiences. With increased availability of different media content, audiences' choice of what to watch or listen to plays a big role.

The Archers is one of the earliest examples of e-e. Conceptualized and produced by the British Broadcasting Corporation in 1945, this radio serial was designed to promote the diffusion of agricultural innovations to British farmers and to help urban listeners understand rural problems.

Other innovative endeavors include the work of Miguel Sabido, the Mexican TV writer-producer-director who helped lay the theoretical foundations of the strategy. Sabido's telenovelas were inspired by the 1971 Peruvian soap opera *Simplemente Maria,* which led to increased enrollment of young women in literacy and sewing classes. Sabido's soaps were exceptional in that their message construction was guided by theories of human communication.

Examples of e-e from the United States include *Sesame Street,* a TV series produced by the Children's Television Workshop that helps prepare preschoolers for classroom learning. In the 1980s the TV soap opera *Hum Log* (We People) in India addressed issues of family planning, gender, and intergenerational values. In the 1990s in Tanzania, the radio program *Twende na Wakati* (Let's Go with the Times) focused on family planning and HIV prevention. In recent years the Johns Hopkins University Center for Communication Programs has made use of rock music to promote sexual responsibility for youth in Latin America and the Philippines.

One analysis (Bosch and Ogada 2000) found that e-e projects are concentrated primarily in Africa, followed by North America and Asia. In Africa and Asia most projects are centered on reproductive health and family planning, followed closely by HIV/AIDS. Most projects in North America deal with the environment, followed by HIV/AIDS. Globally projects utilize mass-media channels, frequently adopting an integrated multichannel approach. Radio and television are most used, and the soap opera genre is predominant. Research has further shown that mass-media channels are effective for disseminating knowledge, whereas interpersonal channels are effective in changing attitudes and behavior. Folk media, mass media, and the new media are all being utilized, either exclusively or in varying combinations.

Soul City, a multimedia initiative based in South Africa that has been successful in changing attitudes and behavior through entertainment, is an example of a multiyear, strategic communication intervention aimed at supporting social development in a new political climate. The multimedia approach that makes use of print and electronic media, such that they are mutually reinforcing, is more effective in its reach as well as its shelf life.

The themes and issues addressed by e-e interventions support the action agenda of the international development community. With the growing importance of research methods and processes that actively engage the target populations to be the principle agents of their own development, e-e programs are allowing the voices of primary stakeholders to be heard, an increasingly vital element of sustainable human development. E-e programs are beginning to contribute to national, regional, and global public agendas, demonstrating their utility and potential in nurturing the participation necessary for sustainable development. With the increasing application of new technologies in the field of education, the e-e strategy has an important role to play. The novelty of the technology alone cannot enhance the effectiveness of educational content. Attention has to be paid to aesthetics and more entertaining formats that will attract and hold the attention of a target audience that is already inundated with an abundance of more entertaining but less useful information.

Tanja Bosch
Joshua Ogada

See also Communication Theory; Diffusion of Innovations; Instructional Communications; Schramm, Wilbur; Television and Learning

References

Bosch, Tanja E., and Joshua O. Ogada. 2000. "Entertainment-Education around the World (1989–2000)." A preliminary report to the Third International Conference on Entertainment-Education and Social Change, Arnhem and Amsterdam, the Netherlands, September 17–22.

Bouman, Martine. 1999. "The Turtle and the Peacock: The Entertainment Education Strategy on Television." Thesis, Wageningen Agricultural University, Holland.

Singhal, Arvind, and Everett M. Rogers. 1999. *Entertainment Education: A Communication Strategy for Social Change.* Mahwah, NJ: Lawrence Erlbaum Associates.

E-texts and Readers

The idea of a handheld reading device has been the dream of engineers and science-fiction writers for many years. In 1988 Apple Computer invited students from twelve universities to participate in a competition to imagine how computing technology would be used in education in the year 2000. The judges were Steve Wozniak, Alvin Toffler, Alan Kay, Diane Ravitch, and Ray Bradbury. The winning entry, from the University of Illinois, described a notebook-sized tablet that contained all of the students' texts as well as capabilities for multimedia and communications functions. A number of companies have attempted to achieve this vision. Although most of those ventures disappeared or were absorbed by larger companies, the development of a reading device has moved steadily forward.

The evolution of the electronic book and digital reading devices is a journey of courageous, bright, and visionary individuals in startup companies and corporate and university think tanks. Unfortunately, no large company—publisher or manufacturer—has been successful in bringing a reading device and the content that would support it to the market in a way that would allow it to proliferate. Recent corporate attempts to market e-books have been stymied by digital rights management, slow publisher commitment of content, and the high cost of the devices. The devices marketed so far have been closed systems focused on commercial content, not open systems that would support education. In spite of that, schools, administrators, and libraries have shown great interest in electronic texts and readers.

Variations of a text-reading device have been seen in movies such as *2001: A Space Odyssey* and on television in *Star Trek*. Alan Kay, who led the Learning Research Group at the Xerox Palo Alto Research Center in the 1970s, defined a DynaBook (a dynamic book), during the late 1960s. Alan Kay and his device are often mentioned as inspiring the recent wave of e-book readers.

The first generation of widely used e-text reading devices was the personal computer. At first, educators used special e-text software or plain-text programs to give students the experience of reading hypertext or plain text on the screen. To support the interest in desktop e-book reading, several extraordinary collections of books in the public domain developed. Project Gutenberg and Bartleby are probably two of the most well known.

Project Gutenberg was the first free source of electronic books distributed through the Internet. Though most books in the collection are in the public domain, works still in copyright are also distributed by Project Gutenberg under special agreement with the authors and publishers. Project founder and executive coordinator Michael Hart, with the help of many dedicated volunteers, continuously turned paper texts into digital files that are available through websites around the world. The first document Hart typed to distribute was the Declaration of Independence. Project Gutenberg began by producing one book a month and in 2002 had 2,500-plus e-books that readers downloaded at a rate of more than 1 million per month.

Bartleby is a small electronic text collection created for education that specializes in classic and modern reference works and poetry. It originated in 1993 as the personal project of Steven van Leeuwen while he was at Columbia University and was incorporated as a commercial business in 1999.

The Bartleby collection is small, consisting of about 300 titles. The library specializes in reference works, including poetry, quotations, and classic works. Bartleby's sources are mostly from the early twentieth century. Except for the modern references, the average date of publication is reported to be in the 1930s.

The Electronic Text Center of the University of Virginia was the first academic site to allow the downloading of classic books that were in the public domain beginning in August 2001. The collection now includes books formatted to read on a desktop or laptop computer as well as on popular handheld devices with the Palm Reader or Microsoft Reader software. The mission of the center, under the guidance of associate librarian Kendon Stubbs, was to create a community of electronic text users and creators through standards and accessibility and by teaching others to create their own electronic texts.

Making books available to students, faculty, and the public has always been a great priority to the University of Virginia. Thomas Jefferson, architect of the Academical Village at the University of Virginia, placed a library in the Rotunda in the middle of the campus, underscoring the importance of information to academic life. The center continued this tradition by shifting the format of as many of its holdings as possible from print to networked technologies, thereby continuing to make information central to everyday life as well as accessible to the world.

More recently, software such as Adobe Acrobat has gained wide use for reading documents with text, images, and multimedia display on computers. The Adobe product was also developed for reading on a variety of personal digital assistants (PDAs). In the PDA handheld market, the first e-text readers were small-screen, low-resolution devices such as Newton and the Palm. The most widely distributed handheld PDAs today are PalmOS and PocketPC devices, both offering a wide range of e-text content.

E-books were provided to Palm users at first through a company called Peanut Press. As interest in e-book reading grew, this company was acquired by Palm and folded into the Palm Digital Media group. Today Palm offers a broad range of commercial publications for handhelds. Many college bookstores offer required texts in PDA formats. Today PDA readers have come full circle, and Palm offers free reading software for the desktop computer as well as the PDA. The same files that are purchased for the Palm can also be read on a computer as well as a PDA. Although there is not an education category in the Palm e-book offerings, there are many books that could apply to a broad range of courses.

Not to be left out of a market, Microsoft's handheld operating system, PocketPC, also developed software to serve the growing e-book reading market. Microsoft's Reader comes with its devices, and large publishers are now delivering commercial content in the Microsoft format. The Microsoft Reader had a distinct advantage over the Palm Reader software as it employed subpixel rendering and created a much smoother text image on the small screens.

Franklin Electronic Publishers was an early distributor of e-text content on external cards that could be placed in handheld devices. Largely reference works, this content appealed to users in education as well as in the medical professions. In the late 1990s Franklin distributed the first dedicated e-text reader and later developed its own, the eBookman. In addition to playing music and audio books, eBookman included organizer software such as notes and calendars because of the popularity of PDAs.

Dedicated handheld readers were the next generation of e-text devices. Unlike PDAs, these e-book readers did not include any software other than software to support reading features such as a dictionary, e-book rendering, bookmarking, highlighting, and hyperlinking. The first and most popular device was the Rocket eBook followed by the larger and more expensive Softbook.

Launched in 1998 by NuvoMedia, the Rocket eBook could be held comfortably in one hand. It was about the size of a paperback and sold for under $300. The Rocket eBook could receive digital content from a desktop computer or through the computer from the Internet. The device attracted an enthusiastic and dedicated following for commercial e-books, and the e-books contributed to the freely distributed Rocket Library. The device's design, affordability, and ability to load personal content made the Rocket eBook extremely attractive to e-book enthusiasts. Barnes and Noble was an early partner, selling the devices in stores and electronic books through its online enterprise.

Many tests of the educational application of the Rocket eBook took place. In 1999 Resurrection Catholic School in Dayton, Ohio, participated in the first e-book pilot program in a K-12 environment. Electronic readers were used for every subject in a fourth-grade classroom for a minimum of one grading period. Content came from the web, books were in the public domain, and materials were created by the teacher.

The Softbook Reader was the first handheld e-book reading device to operate without being attached to a computer. The Softbook was a large, backlit, touch-screen tablet with an attached brown leather cover and pushbuttons for navigating the software to read the specially prepared digital texts. Reading materials loaded to the Softbook could be annotated, underlined, bookmarked, and searched on-screen. The device included a stylus, an inkless pen-type implement common to handheld devices. Reading material was loaded to the device by connecting its internal modem by telephone line through the Internet to the Softbook server. The Softbook Reader was designed to make the experience of reading digital text as close to reading a print book as possible. Softbook was one of twenty-one technology and curriculum companies that the Texas Education Agency selected to conduct innovative pilot programs using existing and cutting-edge technology in an effort to boost student learning in thirteen school districts.

Primary Research found in 2002 that 13 percent of the libraries that it surveyed lent some type of e-text reading device (excluding CD-ROM/DVD devices). While public libraries were represented in that percentage, the greatest evidence of interest in reading devices was from libraries at medium-sized colleges.

A primary motivation for developing an electronic reading device for education in the 1990s focused on the size and weight of children's backpacks because of the books they were required to carry. Richard Katzmann's goReader was one reading device made specifically for the education market in the late 1990s to replace heavy backpacks. In spite of some promising university and K-12 testing, the goReader quietly folded in 2001.

E-text development will likely continue to move forward, but the idea of a dedicated reader may take a backseat to recent developments in the tablet format. Now that handheld computers can be small, comfortably held, and contain all of the features of a larger computer, electronic readers may simply become another feature of tablet computing. Tablet computers can take advantage of all of the web-based e-texts available, run the software of the e-book readers, participate in e-book collections, and download commercial e-books from online booksellers. As the form of the hardware evolves, the function of reading texts electronically becomes an increasingly accepted way of reading.

Susan Amirian

See also Electronic Books; Handheld Technologies; Virtual Library

References

Adobe Acrobat 5 (www.adobe.com/products/acrobat).

Bartleby (www.bartleby.com).

Bartlett, Mel, et al. 1988. "Academic Computing in the Year 2000." *Academic Computing* (May/June): 7012. Also available online at www3.shore.net/~kht/text/cacm/cacm.htm.

Franklin Electronic Publishers (www.franklin.com).

Gemstar eBook (www.gemstar-ebook.com).

Katzmann, Richard. 2001. "The goReader: A Digital-Age Alternative to the 30-Pound Bookbag." Available online at www.futureprint.kent.edu/acrobat/katzmann01.pdf.

Linzmayer, Owen. 1999. *Apple Confidential: The Real Story of Apple Computer, Inc.* San Francisco: No Starch Press.

Microsoft Reader (www.microsoft.com/reader/default.asp).

NuvoMedia, Inc. 2000. About the Rocket eBook. Available online at www.rocket-ebook.com/Products/index.html.

Open Ebook Forum, International Trade and Standards Organization (www.openebook.org).
Palm Digital Media (www.palmdigitalmedia.com).
Primary Research Group 2002. *Library Use of E-Books.* New York: Primary Research Group.
Project Gutenberg (promo.net/pg/index.html).
Rheingold, Howard. 2000. *Tools for Thought.* Cambridge MA: MIT Press. Available online at www.rheingold.com/texts/tft.
University of Virginia Library Electronic Text Center (etext.lib.virginia.edu).

Evaluation

Evaluation is the systematic process of determining the merit or worth of an object or thing; evaluations are the product of that process. Without evaluation we would not know which educational innovations (such as programs, curricula, and activities) to continue, stop, or modify. Evaluation is a systematic process.

In the broadest sense, evaluation includes all efforts to place value on events, things, processes, or people. In the field of educational technology, evaluation encompasses all efforts to place value on events, things, processes, or people with regard to learning and/or instruction. Evaluation can be conducted before, during, and after the development and/or implementation of an educational intervention. The appropriate time to conduct the evaluation depends upon the purpose of the evaluation.

In general, the purposes for conducting an evaluation are to inform, improve, and/or prove. Evaluation that is conducted to identify needs and to understand the nature of the needs is known as needs assessment or needs analysis. Evaluation that is undertaken to furnish information that will guide improvement to the educational program or activity is known as formative evaluation. For example, a formative evaluation might be intended to (1) help management improve community acceptance of a new distance learning program; (2) improve student learning through the use of a new software package; or (3) help a state improve the cost-effectiveness of a technology integration professional development program for teachers. Formative evaluation should be conducted in the early stages of developing an educational product or program or in the early stages of implementation when there is still sufficient time to make use of the evaluation results.

In contrast to formative evaluation, evaluation undertaken to render a final judgment is known as summative evaluation. For example, a summative evaluation might intend to (1) judge the effectiveness of learning a topic in a group classroom setting compared to an individualized computer-based learning program; or (2) show the extent to which accidents were reduced as a result of a new safety training program. An evaluation conducted to show outcomes of program A will most likely be quite different from an evaluation conducted to ascertain the outcomes of program B, although both evaluations are summative. Summative evaluation is conducted at the end of the educational program, module, or other activity.

Although it is first necessary for evaluators to determine the purpose of the evaluation, this alone will not provide the needed focus to conduct a meaningful evaluation. The evaluator must spend time understanding the definition, goals, objectives, and components of the evaluation, which might include policies, programs, curricula, courses, instructional software, instructional materials, delivery systems, learners, and so on. The focus of the evaluation and the evaluation questions should be reflective of the goals, objectives, definition, and components of the evaluation.

The process of conducting an evaluation requires that there be a methodology and that measurements be developed as well as standards against which measurements are judged. Although the terms "measurement" and "evaluation" are often used interchangeably in education, there is an important distinction. Measurement is concerned with the establishment of the degree or amount to which a characteristic, trait or feature exists. Evaluation goes beyond measurement by combining quantitative and/or qualitative ratings to establish the desirability and/or importance of the phenomena under evaluation. Measurement in any field involves three common steps: (1) identifying and defining the quality or attribute that is to be measured; (2) determining a set of operations by which the attribute may be made manifest and perceivable; and (3) establishing a set of procedures or definitions for translating observations into quantitative or qualitative statements of degree or amount.

Quantitative data are those observations that lend themselves to numerical ratings, such as answers to multiple-choice questions on an exam, number of dropouts in a class, and counts of aye votes for a technology integration project. Qualitative data are usually narrative in nature, although they can also be summarized in numerical form. Because qualitative data are generally collected to provide in-depth understanding of a phenomenon, it often is not desirable to represent qualitative data with numerical ratings. Numerical ratings are often replaced with or accompanied by narrative information that describes the evaluation in detail. Quantitative and qualitative data are used in both formative and summative evaluation situations. Generally speaking, quantitative data are used to gather information from larger groups and qualitative data are used to gather more in-depth information from smaller groups. For example, the objective of conducting a one-on-one formative evaluation of a piece of educational software with a subject-matter expert, stakeholder, or user is to gather unlimited, detailed, and exhaustive information. In that instance, a qualitative approach would be more appropriate for providing the kind of feedback needed. In contrast, if the objective of conducting a summative evaluation of a piece of educational software used by some 5,000 students in thirty-two school districts across the nation is to gather the same information on a limited number of predetermined criteria from a representative, valid sample, then a quantitative approach would be appropriate.

Evaluation includes, but goes beyond, measurement. Evaluation consists of appraisal, or valuing something. In order for an appraisal or valuation to have meaning, the appraisal must go beyond determining or reporting a numerical rating and place a value on that rating. That process requires reasoning and logic to determine the appropriate valuation system. Establishing the criteria for appraisal or valuation is the first step required if measurement is to lead to evaluation.

In the field of evaluation, criteria are often classified using the following categories: inputs, activities, outputs, and outcomes. Program inputs typically include resources and constraints applicable to the program. Activities include the products and/or services that the program is expected to provide. Outputs are the products and/or services that were actually delivered to program participants. And program outcomes are the results of a program's efforts. Commonly, formative evaluation includes evaluating inputs, activities, and outputs, whereas summative evaluation includes evaluating outputs and outcomes, although it is possible for summative evaluation to include inputs and activities as well.

Once the criteria of merit have been identified, it is necessary to distinguish between good and bad, high and low, acceptable and unacceptable, and so on. For example, is a 5 percent improvement in test scores over a five-year period good? If school A receives a 4 percent increase in funding and increases its test scores by 2 percent, is this better than school B, which receives a 3 percent increase in funding and increases its test scores by 1 percent? Evaluating the degree or amount to which an event, program, product, process, or person is good or better than requires establishing a usable standard as a basis for comparison and evaluation. This process of establishing standards against which to judge the evaluation is the job of the evaluators and stakeholders. Stakeholders are constituent groups that have a vested interest in the program or activity. A measurement instrument does not place value on what is good or good enough, and unfortunately many numerical ratings garnered from measurement instruments are often misused and misunderstood. An example of this is the use of standardized test scores as a predictor of success in college. Evaluation is the process of placing value on measurement results.

Once standards are established, the next step in the evaluation process is to establish systematic methods for conducting the evaluation. This includes determining the evaluation design and creating a plan for managing the evaluation project. When designing evaluation, the evaluator must consider the following, often competing, issues: (1) the purpose of the evaluation (formative or summative); (2) what will be accepted as evidence; (3) the scale or size of the program; (4) the purpose of the program; (5) the context or culture; and (6) available time, money, and expertise. The following designs are among the most common: cyclic and comparative, posttest-only, and pretest-posttest.

Cyclic designs include subject-matter-expert review, stakeholder review, small- and large-scale pilots, and field trials with typical users. Data collected through cyclic evaluations can be used for formative and summative evaluation. Another type of cyclic design, more commonly used for summative evaluation, is the time series design. In a time series design outcomes are measured at several predetermined intervals. The predetermined intervals are specific points in time (e.g., before, immediately after, several months after, and several years after a program). Which intervals are appropriate for an evaluation depends upon the program purposes and the evaluation questions. Cyclic designs can require extensive time, money, and expertise but do not necessarily require that the number of program participants be large.

Comparative designs, commonly used for summative evaluation, are used when evidence is needed to demonstrate that observed outcomes are a result of the program. Comparative designs compare criteria from the treatment with the same criteria from a valid control group to discern whether the outcomes are different between the two groups. The time, money, and expertise required to conduct comparative evaluations can vary greatly. Larger numbers of participants are required for comparative designs.

Posttest-only designs, as the name implies, rely on postprogram measures as evidence. The disadvantage to this design is that it is almost impossible to attribute observations to the program. It is entirely possible that there was no change in the evaluation. When a posttest-only design is used, the use of a comparison group can be used to determine if treatment and control groups differ on the posttest. However, the posttest-only comparison design assumes that the treatment and control groups did not differ before the program, which could well be an inaccurate assumption.

Pretest/posttest designs provide stronger evidence by utilizing pretreatment and posttreatment measures to detect change. Although the observed change is often sufficient evidence of program effects, it is possible that observed effects are due to other factors, such as maturation, other conditions, and another program happening at the same time. If evidence based on a pretest/posttest design is not conclusive enough, then it might be necessary to use a pretest/posttest comparison group design. The pretest/posttest comparison group design collects information for both groups both before and after the program. With this design, both the treatment and the control group will experience change due to maturation, other conditions, and so on. Posttest-only designs require less time and money than pretest/posttest designs. The addition of a control group will require that the available participant group be larger. As a rule of thumb, as the need for rigor increases and the evaluation design becomes more sophisticated, the resources required also increase.

Systematic evaluation methods also include establishing a plan for managing the evaluation project. The evaluator should identify personnel required, project phases, inventory timelines, available resources, and constraints. Given this information, the evaluator should construct a detailed plan for managing the evaluation project and then implement the evaluation according to the plan.

Once data are collected, the next step in the formal evaluation process is to synthesize the evaluation results. The synthesis of evaluation results includes analyzing the various types of data that were collected and combining the results in a manner that provides a coherent representation of the evaluation. The purpose of analyzing data is to depict needs, current performance in contrast to desired performance, the effectiveness of a method, what was accomplished, and so on for a given program. Data analysis is the quantification or qualification of outcomes. Because programs are usually comprised of several inputs, activities, outputs, and outcomes, data analysis usually consists of measuring several variables. It is usually best to start by analyzing one variable at a time as well as the progress, then analyzing the patterns of single variables over time and/or relationships among variables. The variables, patterns, and relationships studied should answer the individual evaluation questions as well as form a coherent whole in a manner that can be used to compare the evaluation results to previously established standards.

The final step in the evaluation process is to draw conclusions, render judgments, and/or issue recommendations. With previously established standards as well as complete evaluation data in hand, the evaluator is in a position to complete the appraisal. In the case of formative evaluation, this step often includes determining areas where the program is performing to standard and areas where it is deficient; determining how the program can be improved; and issuing recommendations for improving the program. In the case of summative evaluation, this step often includes determining final outcomes, comparing final outcomes to standards, and rendering final judgments based on outcomes and standards.

Melissa J. Dark

See also ADDIE Model; Alternative Assessment; Analysis; Electronic Portfolios; Instructional Design; Rapid Prototyping

References

Dick, Walter, and Lou Carey. 1996. *The Systematic Design of Instruction,* 4th ed. New York: HarperCollins.

Joint Committee on Standards for Educational Evaluation. 1994. *The Program Evaluation Standards,* 2nd ed. Thousand Oaks, CA: Sage.

Oosterhof, Albert. 1990. *Classroom Applications of Educational Measurement.* Columbus, OH: Merrill.

Preskill, Hallie, and Rosalie T. Torres. 1999. *Evaluative Inquiry for Learning in Organizations.* Thousand Oaks, CA: Sage.

Rossi, Peter, Howard Freeman, and Mark Lipsay. 1999. *Evaluation: A Systematic Approach,* 6th ed. London: Sage.

Shadish, William R., Jr., Thomas D. Cook, and L. C. Leviton. 1995. *Foundations of Program Evaluation: Theories of Practice.* Thousand Oaks, CA: Sage.

Smith, Patricia, and Tilman Ragan. 1999. *Instructional Design,* 2nd ed. Columbus, OH: Merrill.

F

Florida Virtual School (FLVS)

The Florida Virtual School is a pioneer in public school e-learning. Funded entirely by the state of Florida, its courses are available at no charge to Florida residents. Since it began serving high school students on a statewide basis in the spring of 1997 with approximately 200 student enrollments in six course offerings, FLVS has grown dramatically. At the close of the 2001–2002 school year, FLVS had 8,200 student enrollments in sixty courses. (Here, "enrollment" is defined as the number of course seats that students have taken or are currently taking.) The enrollment number is consistently larger than the student count number. This is because many students sign up for more than one course. It is not uncommon for one student to be enrolled in three courses; thus the student count would be one, but the enrollment count would be three.

Offering core curriculum, advanced placement (AP), and a variety of elective courses, FLVS provides students an alternative avenue for earning high school credits. In 2003 FLVS will offer middle school courses and a graduate equivalency diploma (known as GED) prep course. The FLVS's environment offers a choice for all learners. Home-schoolers, athletes, performers, and any students with scheduling conflicts or medical problems benefit from the "any time, any place, any path, any pace" environment. Time is flexible; therefore, students have choice in scheduling time for the online classroom. According to SCANS (1992), "In our current system, time is the constant and achievement the variable. We have it backwards. Achievement should be the constant and time the variable." Students and parents say that flexibility in time is key to student success.

Often, current educational systems ignore the fact that curriculum and teaching style play a major role in whether or not students are successful (Pollak 1999). What is missing in some classrooms is choice. An effective online classroom is not an "incidental pedagogical choice but a choice that shapes how and what students learn and, crucially, how they learn to learn" (Katz 1993). FLVS instructors create effective classrooms, engaging students in relevant applications that offer choices in how to process concepts and for demonstrating proficiency.

Each course begins with a real-world motif or metaphoric construct that allows students to have a specific role as they make connections with course concepts. For example, in the course on American government, through a virtual tour of Washington, D.C., students connect to concepts as they journey to different monuments and buildings. The goal is for students to become politically active and discover how to access the governmental system. Projects include writing letters to senators, researching interest groups, and creating a persuasive product to support a political issue of their choice.

As distance learning experts have emphasized, teacher-student interaction is the most important ingredient in student success. Students may have an interactive role in the courses. Through threaded discussions, e-mail, synchronous chats, phone calls to individual students, and conference calls, students and teacher stay connected.

To involve the students in the learning process, students are given choices on what resources to consult and how to demonstrate learning. For example, in AP Literature and Composition the dinner party motif casts students as guests conversing on literary masterpieces. In this role students enjoy appetizers (short stories and poems), examine entrees (often novels or plays they choose), engage in table talk, and have a choice of whether or not to take dessert (creative projects that are optional). The instructor models for students how to design a dinner party, and then toward the end of the course students become creators and make choices in designing their themed dinner party.

Learning in a context such as that outlined above is easily transferred to long-term memory. By only asking students to "give us right answers without asking how they arrive at them, make correct choices on multiple-choice questions, or distinguish true from false, . . . they will never see the value of working hard" (Glasser 1992, 208). FLVS provides a caring learning environment that gives students choices in how they learn. By establishing high expectations and curriculum choices (projects) for demonstrating proficiency, teachers engage students in learning.

Expectations are focused on academic as well as workplace and lifelong skills. For example, students complete their registration process online and submit the information. This requires students to take the initiative of going to the website and following instructions independently to complete the process.

Throughout all the course work students decide a pace for assignment completion. Again, students are encouraged to take responsibility for learning as they look at their schedule and course requirements in choosing a plan of action. As in the workplace, students are expected to make choices and perform specific roles in a timely manner with minimal supervision.

FLVS courses integrate higher-order thinking and successful workplace skills, especially decisionmaking, throughout the curriculum. According to Florida high school evaluators, "This is in vast contrast to most of the teaching occurring in Florida and the United States today" (Bergquist, Bigbie, and Richardson 1999; National Center for Education Statistics 1999). Teachers create problem-solving activities that allow students to create hypotheses, gather information, and apply information to test hypotheses. For example, in the ninth-grade English course students analyze the effectiveness of advertisements to discover what people buy and what strategies cause them to buy. After researching products online at sites provided by the teacher, students create a chart to determine the target audience, the type of delivery, and the effectiveness of the advertisement. Finally, students choose and create their own product and design an accompanying marketing strategy that they present to their online classmates.

In FLVS courses, to encourage students to take responsibility for their learning, they have the option of submitting assignments for review. Teachers give feedback and invite students to revise and resubmit the assignment for another evaluation. Formative and summative assessments are an integral part of the learning process, but at FLVS students decide when they want to take advantage of formative assessments.

Using technology as a tool to design and deliver content, the FLVS curriculum has changed the way teachers interact with each other and teach students. FLVS educators have reshaped the routine learning modes of the traditional school day to a dynamic, interactive, real-world learning environment that gives choices to students and requires students to take ownership of the learning process as they click through school.

Sharon Johnston

See also Collaborative Technologies; Distance Education

References

Bergquist, Constance, Cindy Bigbie, and Geraldine Richardson. 1999. *Statewide Evaluation of the Florida Area Centers for Educational Enhancement.* Florida Department of Education.

Glasser, William. 1992. *The Quality School,* 2nd ed. New York: HarperCollins.

Katz, Stanley. 1993. "The Humanities and Public Education." In *The Humanities in the Schools, ACLS Occasional Paper* No. 20 (pp. 2–3): New York.

Pollak, M. 1999. "Let's Talk Curriculum: Predictor of Success." *New York Times,* June 30, p. 8.

SCANS (U.S. Department of Labor, Secretary's Commission on Achieving Necessary Skills). 1992. *In Learning a Living: A SCANS Report.* Washington, DC: U.S. Department of Labor.

G

Gagné, Robert Mills (1915–2002)

In addition to instructional hardware, technology is also viewed as "a systematic process of solving problems by scientific means. Hence, educational technology properly refers to a particular approach to achieving the ends of education. Instructional technology refers to the use of such technological processes specifically for teaching and learning" (Ely 1995). Using this definition, Robert Mills Gagné was one of the preeminent scholars in the field of educational technology. His work in learning research focused on the applications of psychology to the design of instruction, especially the role of information processing as it affects learning.

Born in 1915 and raised in North Andover, Massachusetts, Gagné decided while in high school that he wanted to study psychology. He earned his B.A. at Yale in 1937 and his M.S. and Ph.D. from Brown University in 1939 and 1940. Gagné was influenced at Brown by Walter Hunter, who was interested in cognitive processes. In 1941 Gagné enlisted in the military and served in the aviation psychology program until 1945. In the military he continued his interest in perceptual and motor learning and worked at two different Air Force laboratories on simulator design and other training devices. There he developed an early conception of how information processing affected human performance.

Gagné taught at Princeton University between 1958 and 1962. In 1962 he became director of the American Institutes for Research. At Princeton he expanded on the definition of domains of learning and developed the concept of learning hierarchies. Learning hierarchies are important because they establish the importance of learning readiness in the form of prerequisite skills. In 1965 he published *Conditions of Learning,* which established a framework for looking at instructional requirements for different types of learning outcomes. Gagné tried out his ideas in applied settings. He collaborated on the development of a mathematics curriculum at the University of Maryland, which further strengthened his belief in prerequisite skills.

In 1966 he moved to and taught at the University of California–Berkeley, and in 1969 he moved to the Department of Educational Research at Florida State University, after he was recruited by Robert Morgan, to help develop a new curriculum in instructional systems. While at Florida State, Gagné focused on instructional research, continuing to develop his ideas about the internal and external conditions of learning. In 1974 he coauthored *Principles of Instructional Design* with Leslie J. Briggs. This book was one of the first to develop an integrated model of instructional design based on a psychological model of learning. In this book, Gagné and Briggs explicated the idea of "events of instruction," which were correlated with phases of the information-processing model. Gagné was also interested in schema theory. He felt that schema were structures for organizing, storing, and recalling knowledge and that schema played both proactive and reactive roles in learning. Later, in 1990, Gagné and David Merrill explored the relationships of goals and schema in the form of enterprise schema. Enterprise schema probably are as close as either scholar would come to embracing some of the ideas associated with constructivist theory today. One of his last publications, *The Conditions of Learning: Training Applications,* with Karen Medsker, integrated instructional theory with workplace training (Gagné and Medsker 1996). Gagné was a prolific writer, and in 1989 he gathered a sample of what he felt was his most important research into an annotated anthology (Gagné 1989).

Gagné is often categorized as a behavioral psychologist because of some of his early writing and affiliation with other behavioral psychologists. However, he didn't feel that the behaviorist theories were adequate to explain human learning. Rather, Gagné should be considered one of the early cognitive psychologists. His important research was centered on how the events in the instructional environment facilitate learning. In all, Gagné was mainly responsible for the development of five instructional theories: (1) the five domains of learning; (2) events of instruction; (3) conditions of learning; (4) role of the media; and (5) integrated goals. These theories are briefly described below.

Five Domains of Learning

Three researchers (Krathwohl, Bloom, and Masia 1964) had defined taxonomies of learning in the cognitive and affective domains. Alternatively, Gagné and Briggs defined five domains: verbal information, intellectual skills, motor skills, attitudes, and cognitive strategies. Within the intellectual skills domain, they described five kinds of learning outcomes: problem-solving, rule-using, defined concepts, concrete concepts, and discriminations (see Figure 1).

The intellectual skills domain interested Gagné most, as he saw it as a hierarchy of skills. In other words, to learn to solve problems it was first necessary to learn rules and procedures; to learn to apply rules and procedures it was first necessary to learn concepts; and to learn concepts it was necessary to be able to make discriminations. From this hierarchy, Gagné built a theory of prerequisite skills. The theory is that in order to learn higher-order skills, lower-order skills must be attained first. In order to describe the hierarchy of skills, say, for a particular rule, one would ask the question, "What concepts must the student first acquire to be able to apply this rule?" Similar questions would be asked between other adjacent types of learning.

The intellectual skills hierarchy implies that learning is cumulative and that certain prerequisite skills are necessary for learning related higher-order skills. The most obvious example is that long division can't be mastered before multiplication, because multiplication is used in the long division process. It doesn't mean the concept of division can't be learned before multiplication, but the process itself can't be learned.

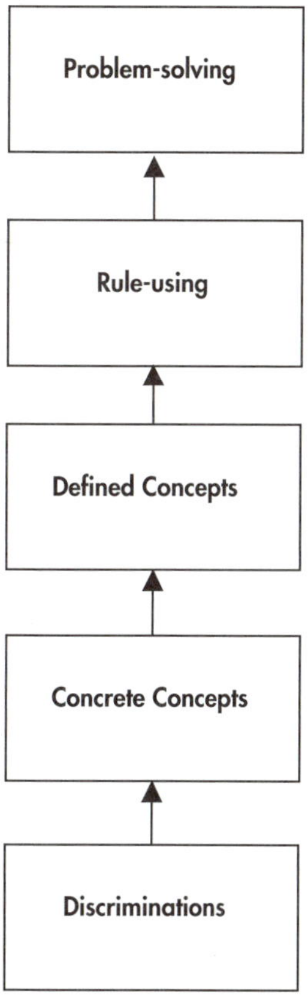

FIGURE 1: Intellectual Skills Hierarchy

SOURCE: Created by the author.

The intellectual skills domain represents most of the important skills learned in school. Problem-solving skills are at the highest level and involve what Gagné referred to as productive, or generative, learning. Gagné felt that the product of problem-solving was the generation of a new rule or procedure—a synthesis of other rules and related concepts that were never used together or learned together in the particular way they were applied.

Verbal information involves names, labels, and facts. Verbal information is generally learned by reception; it isn't deduced or inferred. For example, an insect has six legs. This is a fact. It is a definition, or a declarative statement. It is also an attribute of the concept of insect. However, being able to recall the verbal statement does not mean that an individual would necessarily be able to apply it or use it in a knowledgeable way. A lot of what is taught in school is information. Information is best learned and remembered when it can be placed into a context, or schema.

Motor skills are behaviors that involve coordinated muscular movements. Gagné believes that motor skills have an executive component that is initially a mediator of motor performance. This executive component can be learned as verbal information, such as the seven steps for parallel parking a car. However, with practice the verbal mediator drops out and the motor skill becomes automatized, or performed without thinking about it. In order to unlearn incorrect motor performance, sometimes a coach will drop back to the verbal components of the performance to correct a step in the coordinated performance.

Gagné classifies strategies related to learning as cognitive strategies. An example of a cognitive strategy would be underlining text as you read a book. This is an focusing strategy, to reduce the text to fewer important pieces of information to be remembered. There are also formal mnemonic strategies, such as making acronyms for memorizing lists of items. And there are processing strategies, such as Survey, Question, Read, Recite, and Review, a strategy for comprehending text. Sometimes cognitive strategies look like metacognitive strategies. Metacognition is cognition about learning and is often directed toward monitoring on-task behavior, understanding, and so on.

Attitudes, according to Gagné, are choice behaviors. These seem to be mostly social issues such as volunteerism, honesty, bravery, eating habits, and the like. Attitudes are often mediated by values, opinions, and beliefs. Gagné suggests that in order to teach attitudes it is helpful to have a human or anthropomorphic model, respected by the learner, that is shown engaging in the desired behaviors (such as refusing a cigarette or drugs). Attitudes can never be directly measured because the obtrusiveness of measuring them may cause a change in the voluntary nature of the behavior.

Events of Instruction

Gagné believes that information goes through a series of processes before it is stored in long-term memory (Gagné, Briggs, and Wager 1992) (see Figure 2). These processes involve:

1. reception of stimuli by receptors
2. registration of information by sensory registers
3. selective perception for storage in short-term memory (STM)
4. rehearsal to maintain information in STM
5. semantic encoding for storage in long-term memory (LTM)
6. retrieval from LTM to STM
7. response generation by effectors
8. performance in the learner's environment
9. control of processes by executive strategies

Events external to the learner can influence the processes of learning, especially processes 3 through 6. The external events as they relate to the information processing steps are:

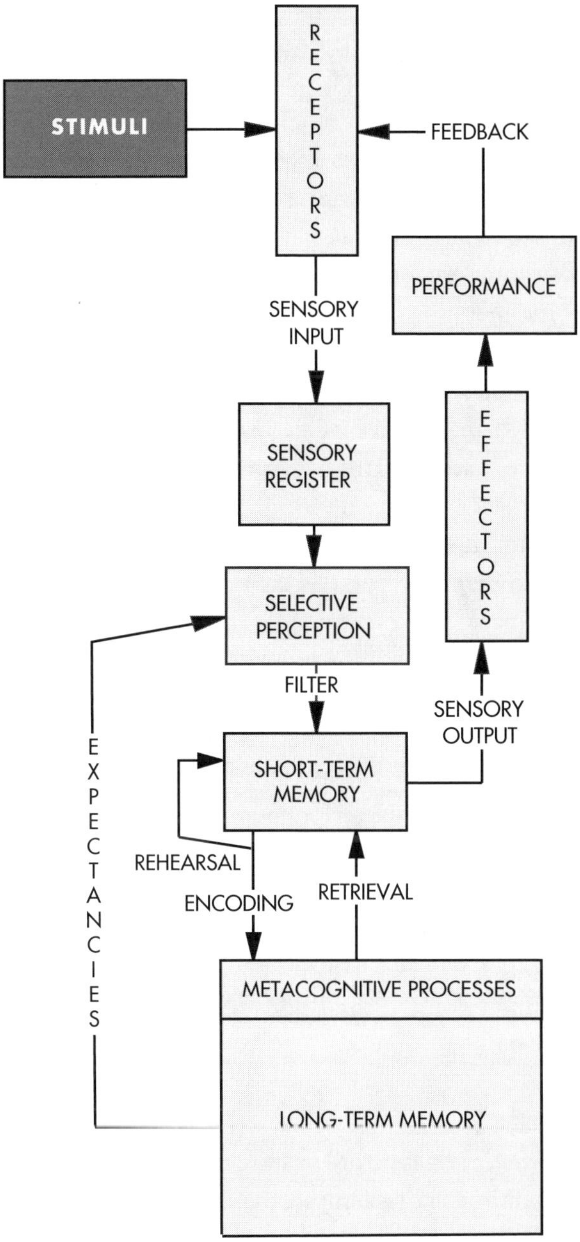

FIGURE 2: An Elaborated Model of Learning and Performance Underlying Gagné's Conditions of Learning Model

SOURCE: Created by the author.

1. Gain attention to ensure reception. Techniques for gaining attention can involve physical or mental activities. The idea is to focus the learner on the instructional environment.
2. Present the objective to establish expectancy in the learner and start the filtering process. Expectations affect selective percep-

tion. If the student knows the lesson is about classifying types of abnormal behavior, she can focus on the attributes for such behavior.

3. Reminding the learner of related material she has already stored in LTM. Recalling prerequisite concepts helps the learner encode and store the new concepts.
4. Presenting the material in a clear and focused way to aid selective perception. Organizing the material in a clear and well sequenced manner aids in making connections in memory.
5. Guidance of learning to aid semantic encoding. Guidance may include examples, elaborations, or additional information.
6. Eliciting performance to involve response generation and development of cues. Eliciting performance serves many functions, but probably most important is allowing the detection of misunderstanding.
7. Providing confirmative or corrective feedback about performance. Feedback can confirm correctness or remediate errors.
8. Assessing performance. Assessing performance certifies a level of understanding and performance, indicating whether further learning is needed.
9. Providing opportunities for use in context to aid retention and transfer. Learning often deteriorates quickly if it is not exercised.

These nine events of instruction can be used as building blocks for developing instructional materials; such events have been called learning activities (Dick and Reiser 1996). However, what goes into the events will be different depending upon the conditions of learning for a particular kind of learning outcome.

Conditions of Learning

According to Gagné, each type of learning outcome has different internal and external conditions. For example, the internal condition for learning a rule or procedure is the recall of prerequisite concepts into working memory. The external event is some type of learning activity that requires the recall of prerequisites, and the internal condition is that the proper concepts have already been learned and therefore are to be recalled. The condition for learning problem-solving skills is that the learner must have a reasonable set of rules, principles, or procedures that can be recalled in considering solutions to the problem. The external environment presents the problem, but the way it is presented can make it easier or more difficult to solve, because it can contain cues that allow the learner to more readily recall the appropriate rules, bring them into working memory, and synthesize new rules to solve the problem.

The Role of Technology and the Media

Gagné has always been interested in the role of technology in learning. From his early experience with designing training simulators to his later experiences in the classroom, he tried to draw connections between media and technology to events of learning. Gagné worked on many media selection models based on the notion that the purpose of media was to create the learning environment, and to do so they had to be able to provide the stimuli called for by the conditions of learning at the appropriate time according to the events of instruction. With another researcher (Reiser and Gagné 1983), he created a flowchart that allowed an instructional designer to select media based on a number of logistical and psychological considerations. The user would follow the flowchart, answering "yes" or "no" to questions presented at each node. Eventually the user would come to a list of candidate media that had been whittled down by the questioning process to those that could provide the needed stimuli. The authors then assumed that the user could select among the suggested media as a matter of convenience or availability, since theoretically they would all be equally effective. Gagné viewed computers as a potentially powerful teaching tool because of their ability to recognize classes of errors and to give corrective feedback.

Integrated Goals and Enterprise Schema

Toward the end of his academic career, Gagné developed the notion of integrated goals (Gagné and Merrill 1990). Integrated goals involve multiple learning outcomes, but most notably they also involve an individual's tacit knowledge about how to achieve goals. He called such tacit knowledge an enterprise schema. An enterprise schema is all the goal-related knowledge and skills pertaining to an enterprise, that is, a purposeful effort. The concept of enterprise schema leaves room for speculation and research on how schemas affect learner motivation and time management, as well as how individuals view a learning task. In fact, enterprise schemas might be associated with the concept of the proximal zone of development, in that they might indicate what types of scaffolding would be appropriate for learners, depending upon their goal schema.

Gagné was president of the American Psychological Association's divisions on military psychology and educational psychology, as well as president of the American Educational Research Association. He also served as consulting editor to several professional journals, including the *Journal of Educational Psychology, Instructional Science, Human Learning,* and the *Journal of Instructional Development.* His honors included the AERA–Phi Delta Kappa Award for distinguished educational research (1972), the E. L. Thorndike Award in educational psychology (1974), and election to the National Academy of Education (1974). Gagné retired from Florida State

University in 1991 and moved to a retirement community in Signal Mountain, Tennessee. Gagné died peacefully in his sleep on April 28, 2002.

Walter Wager

See also Cognitive Psychology; Instructional Design; Instructional Objectives; Research on Media and Learning

References

Ely, Donald P. 1995. *The Field of Educational Technology: Update 1995—A Dozen Frequently Asked Questions. ERIC Digest.* Syracuse, NY: ERIC Clearinghouse on Information and Technology, No. ED387117

Gagné, Robert M. 1965. *The Conditions of Learning.* New York: Holt, Rinehart, and Winston.

———. 1989. *Studies of Learning: Fifty Years of Research.* Tallahassee: Learning Systems Institute, Florida State University.

Gagné, Robert M., and Leslie J. Briggs. 1974. *Principles of Instructional Design.* New York: Holt, Rinehart, and Winston.

Gagné, Robert M., and Marcy P. Driscoll. 1988. *Essentials of Learning for Instruction,* 2nd ed. Englewood Cliffs, NJ: Prentice-Hall.

Gagné, Robert M., and Karen L. Medsker. 1996. *The Conditions of Learning: Training Applications.* Fort Worth, TX: Harcourt Brace College Publishers.

Gagné, Robert M., and M. David Merrill. 1990. "Integrative Goals for Instructional Design." *Educational Technology Research and Development* 38(1): 23–30.

Gagné, Robert M., Leslie J. Briggs, and Walter W. Wager. 1992. *Principles of Instructional Design,* 4th ed. Fort Worth, TX: Harcourt Brace.

Krathwohl, David R., Benjamin S. Bloom, and Bertram B. Masia. 1964. *Taxonomy of Educational Objectives, Handbook 2: Affective Domain.* New York: McKay.

Reiser, Robert R., coauthor. 1996. *Instructional Planning: A Guide for Teachers,* 2nd ed. Boston: Allyn and Bacon.

Reiser, Robert R., and Robert M. Gagné. 1983. *Selecting Media for Instruction.* Englewood Cliffs, NJ: Educational Technology Publications.

H

Handheld Technologies

Handheld computers (also known as handhelds, handheld devices, and palmtops) are small multifunctional computing devices. They range in size from those small enough to fit into one hand and use a stylus to those with keyboards that are a fraction of the size of a typical laptop computer. Although relatively new to instructional settings, handhelds hold promise for education because of their versatility and low cost. Handheld devices may make it possible for every student and teacher to have access to a computing device at all times.

One of the first handheld computers marketed commercially was the Apple® Newton®, which was introduced in 1993 and discontinued in 1998. The chairman of Apple at the time, John Sculley, accurately predicted handhelds would become ubiquitous tools. The Newton used a proprietary operating system (OS). Unfortunately, the Newton was a bit ahead of its time and did not find enough of a market to survive.

In 1996 U.S. Robotics introduced the first Palm OS device, the PalmPilot™. (The generic use of the term "PalmPilot" for all handheld computers, though popular, is incorrect.) The PalmPilot helped people organize information, and this handheld was quickly embraced by the business world.

In 1996 Microsoft introduced the Windows® CE operating system, based on the popular Windows for desktops. Casio, NEC, Compaq, Philips, and Hewlett Packard all have produced handhelds running under the Windows CE operating system. In 2000, Microsoft introduced its Pocket PC 2002, its latest version of the Windows CE–based operating system. (See Table 1.)

TABLE 1: Handheld Manufacturers by Operating System and Advantages of Each

	Operating System	
	Palm OS	**Pocket PC OS**
Handheld computer manufacturers	Palm, Handspring, Sony, AlphaSmart, Handera, Acer	Compaq, Toshiba, Casio, Hewlett-Packard, Dell, Acer
Advantages	Simple to use Lower initial cost More software available Longer battery life	More multimedia and software built in More processing power Solution is Microsoft-compatible

SOURCE: Created by the author.

The successful early handheld devices, intended primarily for business or personal use, were called personal digital assistants (PDAs). They were designed to hold and organize address and phone lists, calendars, and other personal information.

The devices have evolved to become faster and more powerful. While they still organize calendars and contact lists, they can also act as word processors, spreadsheets, databases, e-book readers, graphing calculators, Internet browsers, digital cameras, scrapbooks, graphic organizers, global positioning systems, digital audio recorders, digital music players, telephones, and game devices. Thousands of software programs are available for handheld computers.

Today, a handheld device can be purchased for less than $100. They are portable and can be carried in backpacks, pockets, and purses. Handhelds use batteries and do not require an external power source while in use. Handhelds can be used in the classroom, at home, and out-of-doors on the playground or a field trip. Properly configured, handhelds can provide wireless access to the Internet and e-mail. They are relatively easy to use. Users can work together and share information by "beaming," or using infrared technology to transfer data from one handheld to another. Educational software spanning all content areas and grade levels is available for handhelds, and administrative and management tools for teachers are also readily available.

One of the most promising aspects of handhelds is the ability to provide equitable access to computers to all students. Handhelds make it possible to achieve 1:1 student-to-computer ratios, providing students with computing tools anytime, anywhere. By comparison, in 2001 the ratio of students to computers in the United States reached an all-time low of one computer to nine students. Many of those computers are in lab settings, and as a result many students have only a few minutes of access to a computer at school each week. It is difficult to expect technology to produce gains in student achievement with such limited access, so the use of handhelds is gaining much attention. While every student cannot be provided with a $1,000 desktop computer, it is not outside the realm of possibility to imagine a time when every student will have his own handheld computer.

The challenges of using handhelds in instructional settings include the lack of compatibility between operating systems. In addition, the small devices can be misplaced or stolen. Because it is a small device, the screen is small and can be difficult to read. Little is known about the prolonged use of the handheld computer's possible impact on student health. Concerns include eyestrain and carpal tunnel syndrome. Handheld technologies and applications are evolving rapidly, so obsolescence will be a continuing problem, as in desktop technology. Additional challenges include the ongoing needs for professional development and school-based technology support.

Handhelds come with a hardware cradle that connects to the desktop computer through a universal serial bus or serial connection. A push of a button begins the synchronization process. Data is updated from the handheld to the desktop and from the desktop to the handheld (bidirectional synching). Users may also specify unidirectional data flow, from the desktop to the handheld or from the handheld to the desktop, for specific applications.

Just as with larger desktop and laptop computers, handhelds have a variety of ways to input and output data. Both Palm OS and Pocket PC devices have handwriting recognition features, using a stylus on the face of the device. Handwriting recognition schemes are not difficult to learn and provide the most convenient method for data entry at this time. Desktop computers may be used to input data, and then the handheld must be synchronized (or synched) to the desktop. Other data input techniques include receiving an infrared beam from another device, using an onscreen keyboard, and using an external keyboard. External keyboards range in size from thumb-type mini to full-size keyboards. Some of the current handheld devices will accommodate the addition of memory modules such as flash cards and memory sticks to access data. Data may also be input with probes, modems, network cards, digital cameras, and other peripherals.

The primary output device is the handheld's own screen, which can vary in size, resolution, and readability. Data may be beamed to another infrared device, including another handheld or a printer. Flash memory modules and memory sticks may also be used to store data, or data can be sent to the desktop computer during the synching process.

The ability to communicate wirelessly is becoming a popular feature of handhelds. Beyond infrared, which has line-of-sight and distance issues, other wireless capabilities exist, including cellular (using a similar service to cell phone technology). Another variety of wireless service lets users connect to a local area network. Wireless technologies hold particular promise for education in the area of assessment. With wireless assessment tools, teachers can get real-time feedback on student mastery of content.

Many software companies are providing handheld versions of popular software, and other companies are providing solutions solely for handhelds. Applications include word processors, databases, spreadsheets, presentation software, drawing programs, and graphic organizers. These can be used to create multimedia reports, charts, graphs, lesson plans, lists, e-books, outlines, journals, timelines, notes, and illustrations. Peripheral devices will allow handheld users to gather data (probeware), take photographs (digital camera), record music or voice (digital recorder), make a presentation (VGA card), determine and map locations (global positioning system), and access the World Wide Web and send e-mail (modem or network card).

Commercial software developed specifically for teachers includes gradebook programs, attendance modules and student information systems, curriculum standards databases, rubric generators, assessment tools, concept mapping tools, lesson plan generators, quiz generators, teaching tips, and idea organizers. Students will find software developed especially for their use, including web browsers, e-books and readers, reference tools, graphing calculators, graphing tools, music editors, visual scrapbooks, educational games, content review tools, concept mapping tools, and other software to help manage and organize information. New applications are being developed rapidly.

Under a grant from the National Science Foundation, the Center for Highly Interactive Computing in Education at the University of Michigan (www.handheld.hice-dev.org) has developed a number of free software applications for education for Palm OS handhelds. They include a word processor, a photo scrapbook, a visual organizer, an animation and drawing tool, a flashcard application, a virus simulator, and a handheld-based quiz and worksheet generator. The Concord Consortium (www.concord.org), a nonprofit educational research and development organization, focuses on the use of probeware for handheld computers and has an online library of free software.

Donna Baumbach
Karen Fasimpaur

See also E-texts and Readers; Probeware; Virtual Library; Wireless Networks

References

Curtis, Michael, et al. 2003. *Palm Handheld Computers: A Complete Resource for Classroom Teachers.* Eugene, OR: International Society for Technology in Education Publications.

Fasimpaur, Karen. 2003. *101 Great Educational Uses for Your Handheld Computer: A Comprehensive Guide for Using Handhelds in Education for Administration, Teaching, and Learning.* Long Beach, CA: K12 Handhelds.

Southeast Initiatives Regional Technology in Education Consortium. 2002. *SEIR*TEC Newswire* 5(2). Greensboro, NC: SERVE, Inc. Available online at www.seirtec.org.

WestEd Regional Technology in Education Consortium. 2002. *TechScape Voices* 1(2). Sacramento, CA: WestEd RTEC.

Human-Computer Interaction (HCI)

Human-computer interaction is the study of people using computers. Although the principles involved are applicable to more general systems, in practice the goal is to make computational and networked technologies easier to use for people, improving *usability.* The reasons for this include making people more efficient in accomplishing their tasks, reducing errors, and increasing the subjective experience.

Historically, computer time was costly and rare, so the burden of communication was put on the user; people found themselves translating problems into language the machine understood. As computer power increased and prices dropped, the emphasis shifted to the point where now we can put the burden of translation on the computer. Pragmatically, HCI has involved the design of computer *interfaces*—the representation through which the computer communicates to the user and the mechanisms provided for the user to communicate to the computer.

HCI involves a scientific component, including a body of theory about how people accomplish tasks mediated by tools such as computers, as well as an applied component indicating how to design usable systems. As an interdisciplinary field, involving technology and people, many different approaches are relevant. A minimal list includes social, cognitive, and organizational psychology, sociology, linguistics, artificial intelligence, software engineering, ergonomics, and mathematics as relevant disciplines. More recently, anthropology, activity theory, and semiotics are among the areas that have contributed to analysis. The underlying notion is to understand how people act in the world, alone and together, and how tools or cognitive artifacts, specifically computers, can mediate those actions.

The concern is broader than just an individual at a computer; it includes other computational and networked devices such as handheld computers, interactive kiosks, wearable computers, augmented workspaces, and the like. In addition, it concerns the context in which computing is performed, the other tools used as part of a task, collaboration and communication between individuals, and the organizational contexts in which the task is situated and the design processes occur.

HCI and Educational Technology

There are similarities with instructional systems design, which combines research and theory on how people learn with an applied approach to the design of educational technology. The main conceptual difference is that with HCI the user is assumed to know the task to accomplish, whereas with educational technology the learner is inherently involved in something new. The boundaries blur when the systems at issue are performance support systems. There has been cross-fertilization as well; for example, the HCI field first popularized the notion of user-centered design in moving to a design perspective focusing on the end user, motivating ISD to move the focus to learner-centered design.

Principles

The science side of HCI is concerned with frameworks that give researchers leverage in understanding how people act. As an interdisciplinary field, there is a rich set of such frameworks that have been brought to bear in understanding the ways users interact with a system. There has been a synergistic strength from the diversity of approaches that have been the familiar tools from each field.

Involved is the study of input devices like mice, trackballs, keyboards, voice, and the like, as well as output devices such as screens, printers, and speech generators. Perhaps more important, the issues involve how to map the ways people understand those devices to operate and how to use the output representational capabilities to present a coherent *system model,* or representation, of how the system works. Thus, we have seen graphical user interfaces that use metaphors (such as the desktop) to convey a system model and provide mappings between common input devices and different tasks (such as using a mouse to control one-dimensional scroll bars).

One of the most useful has been the seven stages of action model of Donald Norman (see Figure 1). This model captures the cycles of a user's interaction with the system. Users start with high-level goals that they need to accomplish. They need to transform these goals into specific actions that the system understands. This forms a gulf between the user and the system that the interface bridges. A similar gulf exists between the output of the system and the information that users need to determine whether their goals are met and how to accomplish the next task.

From this diagram we see that users, to achieve their goals, must specify their current intention, determine an action to accomplish the intention, and execute some physical motion that implements the action. Once the system has responded, the user must perceive the response, interpret the response with respect to the current system, and evaluate the response in relation to the overall goal.

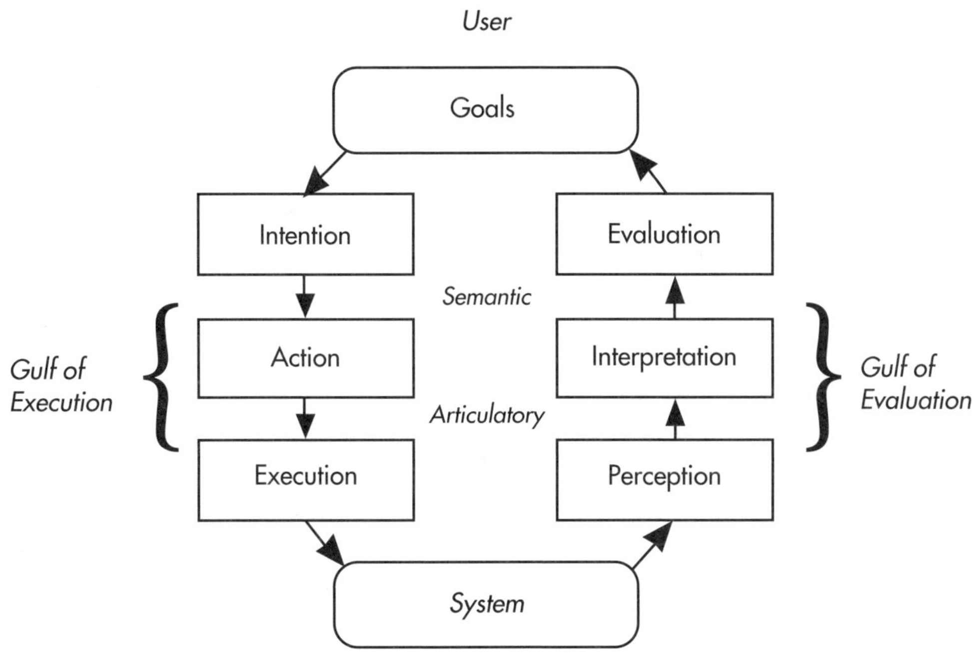

FIGURE 1: Norman's Seven Stages of Action Model

SOURCE: Adapted from D. Norman (1990), *The Design of Everyday Things* (New York: Basic Books).

More concretely, imagine that a user has an overall goal to clean up a current file directory. At one point this may translate to an intention to remove an old file. The user must then decide how files are removed in the current system to generate an action. Let us say that in this system the user must use the DELETE command. Then the user must figure out how to execute a DELETE command, specifying the command and the file. Does the user type "delete" or "del," hit a DELETE key or key combination, or pick the delete command from a menu? Does the user point to the file, name it, or use a wildcard abbreviation? Which of these maps most closely to the user's preferred level of thinking about goals? (Note that the correct answer, generally, is all of the above. Redundant interaction options allow the user to be flexible and choose which options are currently most convenient.)

Once the user has executed a command, the system can respond. That response may be that the file disappears from a visible list, or the system prompts the user that the file has been deleted, or, simply, a new prompt is available. The user must then decide whether the response indicates successful completion of the task. At each of these points, errors can occur. This potential for error is a guiding motivation for user-centered design.

Using the terminology from the diagram, we wish to minimize the distance between the system and the user, bridging the gulfs of execution and evaluation. One way is to train the user to be familiar with how a particular system operates. This puts the burden on the user. The other way is to design the system to work at a level close to the user's goals. This is the ultimate purpose of user-centered design and the way to make technology transparent to the user.

The focus has broadened to recognize that these elemental interactions are not independent of the multitude of contexts that surround them. The nature of the organization and cultures that surround the practice fundamentally affect the characteristics of the interaction. Recent work has focused on understanding these contexts and how interactions are situated in the broader world.

Elements of Design

HCI extends beyond understanding interactions to specifying a design process that can achieve a system that matches user goals. Ideally, one pass through a process would result in a final design. However, the real world does not submit to such finality. One of the results of understanding people is the realization that people are not completely deterministic: They make mistakes, they get caught in old patterns, they can't remember all the necessary information, they can't foresee all possibilities, implementations often don't work the first time, and so on. To truly apply what we know about people, we must not only design for common foibles but also use a design process that accounts for those same foibles on behalf of designers.

Original design methods (e.g., software engineering) called for a straightforward progression through the design process. The components of the general design process include:

- *analysis:* define the task
- *specification:* design the solution
- *implementation:* develop the solution
- *evaluation:* test the solution

However, this method has been recognized to produce software that requires revision and improvement due to the foibles of users (and designers) identified above. The standard approach fails to work because people are not capable of specifying what it is they want the first time through. Working from a definition of the task to be performed and applying the relevant principles are not enough to ensure good design.

Furthermore, design principles can and do conflict. Two such principles are (1) design for consistency, and (2) match the way the user thinks about the task. Both seem obvious and useful, but they can come into conflict in even seemingly simple devices. There will be no one right answer for everybody; the constraints will always result in making tradeoffs. And since users can be inconsistent, arguments for consistency can conflict with arguments for considering the user.

Process Guidelines

The approach that has been advocated to remedy this problem is user-centered design. Three major strategies that characterize the user-centered approach are that the design process should be *iterative, formative,* and *situated.* "Iterative" means that the design process should cycle through stages of the design process, with evaluation followed by reanalysis, redesign, and so on. Iteration is tied to outcomes that when achieved indicate success. "Formative" means that the results of evaluation feed back into the design process. "Situated" means that evaluation needs to be performed in realistic settings, based in the real activities of the task, and not designed for hypothetical and abstract situations. Representative user participation is required to test these prototypes. Data must be collected on these intermediate tests to illuminate further design. Each stage of the design process needs to be modified to reflect the user-centered approach.

Analysis

As in ISD and design in general, the first part of designing a system is determining the need. The initial phase of data collection includes surveying the potential users to determine the range and then picking representative users to participate in design. Design can be viewed as a process of *multiple constraint-satisfaction.* Imagine that there is a space of potential designs. Each constraint on the design helps to narrow down the solution space of possible designs. It may come to be the case that the design space is effectively constrained out of existence. In this case, some constraints may need to be relaxed. Typically, however, the greater the number of constraints, the easier the design process is. Thus, effective information-gathering is critical.

Although such a systematic approach does not guarantee a brilliant design, it is more likely to produce a good design. The exhaustive nature of the analysis stage suggests, in addition to an initial conceptual analysis, considering how the task would be accomplished without support, how it is currently being accomplished, and what solutions others have proposed. To assist with the specifics of the information-gathering exercise, checklists should be consulted. One of the limitations of designers is their ability to recall all the relevant criteria. The use of external memory support provides a greater likelihood that all issues have been examined.

Design

The processes of design are fairly well understood. Approaches such as divergent thinking (using techniques like brainstorming) followed by convergence are as common in interface design as in other design tasks. Although there is still a role for art in the design of an interface, increasingly systematicity is depended on to yield a high probability of a successful outcome. Again, the limitations of designers are to focus too early on evaluating designs, and support for extending the divergent phase tends to yield benefits.

A valuable empirical lesson has come from HCI approaches to design. The approach of design rationale—the deliberate recording of assumptions, arguments, and ultimate decisions for the stages of the design process—has proved valuable in extended design processes. This record prevents revisiting arguments due to a lack of recollection, supports new team members in understanding the project, and supports ongoing development in future updates and maintenance.

Understanding the relationship between the conceptual model and the representation is important. The emphasis is to make a distinction between the *information architecture*—the underlying structure of the system—and the interface—the representation of that structure to the user. Design processes recognizing this place the design of the architecture before the design of the interface.

Implementation

An important element is prototyping to answer questions that user analysis cannot answer. Design rules of thumb (or heuristics) are increasingly viewed as insufficient due to their ad hoc nature and potential conflicts. The focus has passed to quick implementation and testing of alternatives to help resolve design questions. The emphasis is on low investment in technology for prototypes, using thought experiments, storyboards, and paper prototypes early and often along the path to successful design. Designers tend to find it hard to abandon work invested in prototypes, so minimizing the investment supports greater willingness to experiment with alternate conceptions.

Evaluation

One of the major areas is determining how to evaluate the results of the design process in ways that are effective and efficient. Full evaluation would require large amounts of user testing. Pragmatically, that is often prohibitively expensive. Heuristic approaches have been investigated to combine expert review with limited user testing to good effect. Most approaches are eclectic and must be customized to the particular characteristics of the design problem.

Methods used include inserting code in the prototype or running a separate application to trace user activities with the system. This can be combined with videotaping user activities, audiotaping voice protocols (or records of user thoughts), or observation by experimenters. Subjective measures include questionnaires and interviews with users. Expert reviews have a place in the methods for usability analysis as well.

The measures used should align with predetermined goals for evaluating system suitability. Typical measures include time per task or tasks per time, time taken to acquire proficiency, number and type of errors per task or time unit, and subjective satisfaction with the software. The issues of when and where to emphasize qualitative versus quantitative data increasingly appear to have situational and contextual answers. A combination approach is often recommended.

Cost-Justifying Usability

One common problem in improving usability is justifying the cost and time to perform these iterations. There are numerous opportunities for benefits, however. Savings can be attributed to decreased time to learn, lower support costs, fewer errors, more productivity, and greater satisfaction, which can and should be tied to design criteria. Obstacles come from trying to integrate usability practices into existing development processes. Recent work has attempted to identify ways in which to address organizational and process barriers to improving usability.

Clark Quinn

See also Analysis; Electronic Performance Support System; Instructional Design; Performance Support; Rapid-Prototyping; Usability

References

Baecker, Ronald, et al., eds. 1995. *Readings in Human Computer Interaction,* 2nd ed. San Francisco: Morgan Kaufmann.

Helander, Martin, Thomas Landauer, and Prasad Prabhu, eds. 1998. *Handbook of Human Computer Interaction,* 2nd ed. Amsterdam: North-Holland.

Norman, Donald A. 1990. *The Design of Everyday Things.* New York: Doubleday [orig. publ. as *The Psychology of Everyday Things (1988),* New York: Basic Books].

———. 1993. *Things That Make Us Smart.* New York: Addison-Wesley.

Hypertext

The term "hypertext" refers to blocks of texts creatively joined together by a series of links, or paths. The concept is extremely important to the field of educational technology, as it forms the basis of interaction on the World Wide Web. When users click on webpage links—a word or set of words connected to other pages and known as hyperlinks—they are transported to a new block of text (generally a new webpage on the same or even a different website). These hyperlinks usually take the form of underlined words in web browsers. Hypertext is a central tenant of hypertext markup language (HTML), one of the primary programming languages for the World Wide Web. Hypertext is often used interchangeably with the term "hypermedia," which actually refers to a system that is more than just text-based (textual). Thus, in hypermedia, text and images can be linked to other blocks of text, images, videos, sound sources, and even other programs. Some researchers and scholars differentiate hypertext from hypermedia in order to focus on the concept of hypertext as a literary term and concept. With great fluidity in the definition of the concept of hypertext, some have argued that there is no practical agreement on a real or true hypertext. Researchers and scholars agree only that it is a mode of interacting with text, in both electronic and nonelectronic formats. It is a definition, like the concept, that continually evolves.

Although consensus has not been reached on a pure definition, it is important to note that hypertext refers to more than simply a functional, *electronic* linking of texts. Theodor H. Nelson first coined the term in the 1960s in the development of his computer-based Xanadu™ system, but cognitive psychologists and communication theorists argue that we thought and spoke in hypertext long before it materialized into computer technologies. Carolyn Guyer, representing this perspective, states that human beings are associative in their thinking processes, and so text and other forms of language have always actually been hypertext. This history of hypertext existing a priori to technology is one hypothesized reason the hypertextual format of technologies such as HyperCard, HyperStudio, and the World Wide Web have been so quickly adopted. Proponents argue that these tools do not present a new schema or way of thinking; rather, they fit into what we as humans already do without computer technologies.

Vannevar Bush (1945), in a famous article entitled "As We May Think," was one of the first to consider an *electronic* hypertext. He asked readers to consider the future use of a device known as a memex, a mechanized device used to store large amounts of personal information such as books and records. Being mechanized afforded the memex user opportunities to quickly and flexibly search and access stored data. This "memory supplement" was designed to store and retrieve information based on the weblike structure of the mind. Bush's system even came complete with links. This associative indexing (or coding) would allow any item to immediately and automatically be related to other items.

Roland Barthes was another major theorist who discussed this new textuality in detail. He described text blocks (lexia) composed of words or images that were electronically linked by multiple paths or trails. This open-ended system, one that was perpetually unfinished, created textuality with limitless and interactive networks of text galaxies. More important, it had no beginning and no end. Its limits were constrained only by the infiniteness of language.

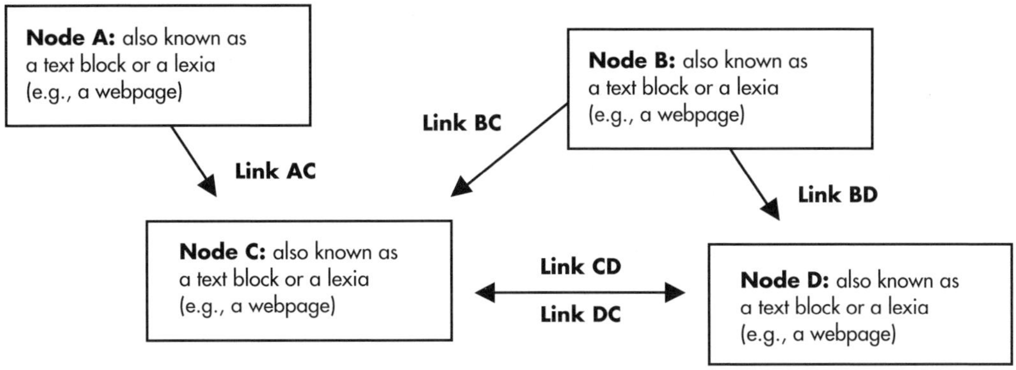

FIGURE 1: The Use of Nodes and Links in the Concept of Hypertext

SOURCE: Created by the author.

Drawing on both Barthes and Bush, Theodor Nelson (Bush's student) coined the term "hypertext" while designing an electronic form of nonsequential writing. In Nelson's mode of publication, readers were able to choose different pathways through text blocks. A text block, also known as a node or a lexia, was joined electronically to other text blocks, creating multiple branches of text and thus numerous choices for the reader (see Figure 1).

Notice in Figure 1 the links between nodes. This illustration also demonstrates the difference between hypertext and hypermedia. Hypermedia refers to nodes and links that are more than textual. Thus, in hypertext a text block contains text links to other text nodes; in hypermedia text blocks could contain text and graphic links to other text nodes as well as to sound, video, graphics, and even other programs.

The birth of the term in the 1960s spawned more than twenty years of research and development on large-scale hypertext systems. These systems were designed to realize the potential in the writings of Bush, Barthes, Nelson, and others. Innovations included nonelectronic writings (such as *Choose Your Own Adventure* books and Nabokov's *Pale Fire)* and electronic inventions (such as Doug Englebart's Augment/NLS hypertext system and Nelson's Xanadu, as well as more globally in technologies such as bulletin boards and news groups).

Perhaps the concept of hypertext gained its greatest fame, though, in the late 1980s and early 1990s when it became more concrete, available, and personal to the novice user. In 1987 Apple began packaging HyperCard with every computer purchase. HyperCard is a hypertext authoring system that resembles Rolodex cards or library card catalogs. With HyperCard, users can create various sets of cards that contain text and graphics. However, authors can also put buttons on cards that trigger events like playing sounds or linking to different cards. The simplicity and ease of use made the concept of hypertext more concrete to computer users and nonusers alike.

Another important event in the history of personal use of hypertext was the creation of the World Wide Web in 1991. Tim Berners-Lee is credited with the creation of the World Wide Web while he was working at the Conseil Européenne pour la Recherche Nucleaire (CERN), in Geneva, Switzerland. Drawing on the concept of hypertext and Nelson's Xanadu system, Berners-Lee's proposal for the World Wide Web included technologies for writing, distributing, and reading webpages. The proposal called for webpages to be written using hypertext markup language. Essentially a system of text attributes and tags, HTML allows authors to format text and create hyperlinks to other webpages. HTML is similar to standard generalized markup language (SGML), developed in 1986 by the International Organization for Standards. Whereas SGML provides the rules for tagging elements, HTML provides the definitions and interpretations of those tags. HTML is important to the concept of hypertext because it provides standards for the creation of nodes and the links between them.

The other two components in the original CERN proposal were Hypertext Transfer Protocol (http) and the web browser. Http is the protocol that defines how messages are formatted and transmitted. The web browser is the application that allows webpages to be displayed. Mosaic was the first web browser (the first that had a graphical user interface); current web browsers include Netscape Communicator and Internet Explorer. Recent surveys estimate that more than 513 million users worldwide experience the concept of hypertext via use of the World Wide Web.

The concept of hypertext, regardless if the hypertext is electronic, nonelectronic, systemic, or personal, is important to the field of educational technology for various reasons. First, it is the underlying concept behind numerous educational software titles (i.e., HyperCard and HyperStudio) and, on a larger scale, the World Wide Web. However, it is also important because it gives learners opportunities to guide their own learning. This type of learning—one based on theories that espouse the individual as the creator of knowledge and meaning-making—can be tested empirically as students navigate their own paths to information. Although hypertext research in education is a relatively new area of study, some early research has demonstrated evidence of the possible importance of hypertext-based educational tools.

First, the use of hypertext improves knowledge transfer. It highlights critical relationships between various components. Second, hypertext allows individuals to move freely within programs or within text. That feature, in turn, provides opportunities for interaction and subsequently higher scores and performances on posttest measures. Third, hypertext-based systems allow for adaptive learning. Users create personal strategies for accessing information, which increases the chances of retrieval and application. Finally, hypertext lets users navigate and choose their own paths through information and text. As highlighted above, this puts the individual as the creator of knowledge and affords opportunities for serendipitous learning. Future research in this area will include gathering and analyzing empirical evidence comparing hypertexts with more structured texts.

Richard E. Ferdig

See also Web-Based Instruction

References

Bush, Vannevar. 1945. "As We May Think." *Atlantic Monthly* 176(1): 641–649. Available online at www.theatlantic.com/unbound/flashbks/computer/bushf.htm.

Landow, George P. 1997. *Hypertext 2.0: The Convergence of Contemporary Critical Theory and Technology,* 2nd ed. Baltimore: Johns Hopkins University Press.

Information Literacy

Information literacy is the ability to locate, evaluate, and use information in an effective and appropriate manner. Information literacy has gained significance as a means for addressing the problem of information overload, prompting educators to stress its importance as a skill necessary to support lifelong learning. An information-literate person is able to evaluate information from a variety of sources. Although the term has a far-reaching impact on many disciplines, it is most closely aligned with library science. One of the leading organizations advocating the importance of information literacy is the Association of College and Research Libraries (ACRL), a subdivision of the American Library Association. The ACRL describes an information-literate individual as one who is able to:

- Determine the extent of information needed;
- Access the needed information effectively and efficiently;
- Evaluate information and its sources critically;
- Incorporate selected information into one's knowledge base;
- Use information effectively to accomplish a specific purpose; and
- Understand the economic, legal, and social issues surrounding the use of information, as well as access and use information ethically and legally (Association of College and Research Libraries 2000, 3).

Students should be able to describe and understand how scholars and researchers use information and keep currently informed, how practicing professionals use information and keep currently informed, how the use of information can improve the quality of scholars' and professionals' work, the commodity nature of information (who generates, controls, and uses information, in particular the role that the government plays), and the costs of misinformation (Olsen 1989).

Four Main Features of Information Literacy

In order to use information effectively, the user must first recognize the need for information. The information-literate person should be able to define and articulate his need for information, identify a variety of types and formats of potential sources of information, and reevaluate the nature and extent of the information needed (Association of College and Research Libraries 2000).

Second, an information-literate person must locate appropriate information sources. Locating information includes knowing how to select and search databases, constructing and implementing an effective search strategy, knowing what resources are available, retrieving information online or in person using a variety of methods, and refining the search strategy as necessary (Association of College and Research Libraries 2000).

Third, one must evaluate retrieved materials. Evaluating information requires determining the authorship of a publication, determining the publication date, reviewing references, and determining whether it is from a peer-reviewed source, an important criteria for determining quality. Additional criteria include being able to distinguish popular from scholarly treatment of a subject, distinguishing between primary and secondary sources, defining various standard formats for storage of information (CD-ROM, print, online, etc.), and identifying appropriate information resources and references within the discipline (encyclopedias, directories, indexes, etc.) (Olsen 1989).

Finally, the ability to organize and manipulate the information retrieved must be demonstrated (Olsen and Coons 1989). Relevant to this is choosing a communication medium, applications, and tools. The information-literate person, individually or as a member of a group, should be able to use information effectively to accomplish a specific purpose (Association of College and Research Libraries 2000). This may include using a bibliographic file management package to organize downloaded citations and personal files of references, conducting a needs assessment, using electronic spreadsheets to reformat and analyze numeric data, using a word-processing package to format papers and construct bibliographies, and writing correct bibliographic citations. In short, one must apply new and prior information to the planning and creation of a specific end product, as well as communicate the end result effectively to others in various formats (e.g., face-to-face, presentation software, transmission over the Internet, videotape).

The Challenge of Information Overload

The introduction of the personal computer as a gateway to the World Wide Web of information resources has heightened the importance of information literacy. This increased access to information requires a greater degree of skill to effectively find and use information, as far more information is available today than there has ever been in the past. The problem is being able to wade through all of this information to find the specific pieces needed. To become information, raw data must be processed into a usable form. In turn, once information is applied toward further understanding, it may be considered knowledge. This differentiation between data, information, and knowledge is key to understanding information literacy as a developmental process. But sheer abundance of information is not enough (Association of College and Research Libraries 2000). The ability to find relevant data is what makes information valuable as knowledge.

Increasingly, information comes to individuals in unfiltered formats, raising questions about its authenticity, validity, and reliability. In addition, information is available through multiple media, including graphical, aural, and textual formats, which poses new challenges for individuals in evaluating and understanding it. For example, using search engines to find websites, as well as evaluating those sites, often has a direct impact on the quality of information located. The evaluation of websites requires specific considerations. It is important to look at the currency of the information (when was the webpage last revised or updated?); the content (is it organized, accurate, and pertinent, and is the same information available in another format?); the authorship (who is the author and where does the site originate from?); and ease of navigation (is it clear and readable, with a clean layout and easily identifiable links, and do the images and other multimedia support the information?). Consequently, the ability to work with search engines is an important aspect of information literacy. This is not as simple as plugging in a keyword search but involves understanding how the search engine works, as some search engines are better for certain topics. Search engines vary in their collection methods, quality control, and how they index and rank results.

Information Literacy in Education

The ACRL supports student-centered learning environments where inquiry is the norm, problem-solving becomes the focus, and thinking critically is the dominant process (Association of College and Research Libraries 2000). Information literacy competencies are crucial to supporting this learning environment. These abilities are also the basis for lifelong learning and enable people to become self-directed learners and to take control over their own learning. Increasingly, librarians and faculty are developing collaborative relationships, as a team effort is needed to create new courses and curricula within the framework of information literacy (Stoffle 1998).

Another important example of how information literacy serves the education process is in the evaluation of research studies, a task that reaches across all disciplines. When evaluating retrieved information such as research articles, one must consider many of the following:

Purpose
What are the goals of this study?
What is the problem or issues at hand?
What assumptions, if any, are important to note?
What are its limitations?
What practices will it influence?
Why do the authors want to conduct it?
Why should we care about the results?
Why is the study worth doing?

Conceptual Context
What is going on?
What theories, conceptual frameworks, literature, and experience will guide the study?
What studies have been done in the past?
Have the important terms been defined?

Research Questions
What is trying to be understood?
What is not known about the phenomena being studied?
What questions will the research attempt to answer?

Methods
What will actually be done?
Has the methodology been described in detail and is it complete?
How will observations be made?
What approaches and techniques will be used?
Why was the research site selected?
How will interviews be convened?

Validity
How might the study be wrong?
What are the plausible explanations?
What is the researcher's relationship to the study group?
Why should the results be believed?

Findings

What was actually discovered or determined?
Where should future research focus?
Is the report clearly written?
Is the report logically organized?
Does the tone of the report display an unbiased, impartial scientific attitude?
Are the conclusions significant?
Are the conclusions relevant to the problem?
Are the conclusions described clearly?

Although technology skills must be learned in order to use and apply higher information-literacy skills, "information literacy" is a broader term than "computer literacy." This has stirred confusion over what is covered under information literacy. Information literacy deals with deeper concepts: understanding the role and power of information in society—its use and misuse; being able to handle the varieties of information formats; understanding the systems used in organizing information; and being able to generate information and manipulate it using electronic processes. Apart from these differences, some argue that computer literacy should be a prerequisite to information literacy (Horton 1983). Others argue that separate computer and information literacy programs only confuse students and teachers (Johnson and Eisenberg 1996). Increasingly, many educators believe that computer skills should not be taught in isolation and that separate computer classes do not really help students learn to apply computer skills in meaningful ways. They prefer a blending of concepts and instruction since they do go hand-in-hand. This shift to a more integrated approach supports a broader notion of information literacy, ensuring a far greater impact on the education process, as opposed to separate and unrelated experiences in the library.

Odin Jurkowski

See also Curriculum Integration; Cyberculture and Related Studies; Netiquette; Virtual Library

References

Association of College and Research Libraries. 2000. *Information Literacy Competency Standards for Higher Education.* Chicago: American Library Association.

Horton, Forest. 1983. "Information Literacy vs. Computer Literacy." *Bulletin of the American Society for Information Science* 9: 14–16.

Johnson, Douglas, and Michael Eisenberg. 1996. "Computer Literacy and Information Literacy: A Natural Combination." *Emergency Librarian* 23: 12–16.

Olsen, Jan, and Bill Coons. 1989. "Cornell University's Information Literacy Program." In G. Mensching and T. Mensching (eds.), *Coping with Informa-*

tion Illiteracy: Bibliographic Instruction for the Information Age.* Ann Arbor, MI: Pierian Press.

Stoffle, Carla. 1998. "Literacy 101 for the Digital Age." *American Libraries* 29: 46–48.

Instructional Communications

The systematic approach to communications theory dates back to the late 1940s when Claude Shannon and Warren Weaver, scientists working on ways to improve electronic communications at Bell Technical Laboratories, introduced the communications cycle. This cycle consists of a sender/encoder, a signal, and a receiver/decoder. Noise is an intervening (and disruptive) element, and feedback is the returning message from the receiver to the sender. (See Figure 1.)

Today many are revisiting this traditional concept in the context of instructional communications. It is a very useful tool because so many instructional communications are now electronic and involve the concepts of time (synchronous and asynchronous) and distance. There are two ways to define "instructional communications": as a systematic series of outgoing and incoming signals designed to expand a learner's field of experience (Heinich et al. 1996), and as a field of study that informs educators of all disciplines about the communication skills necessary to function competently in the classroom (Simonds 2001). These approaches are different but not mutually exclusive, and understanding one helps inform the other.

Instructional communications is also a specific academic discipline that seeks to investigate and apply the principles of human communication when the goal is to share and refine knowledge and skills. The need for more rigorous study in the field of instructional communications is being driven, in large part, by the growing impact of technology on our traditional instructional practices. Today's trend toward technology-supported instruction, and the accompanying increased costs, are driving the need to evaluate both efficiency and effectiveness to justify those increased costs. It is forcing us to take a critical look at our teaching practices. Several authors (see, e.g., Willis 1992; Palloff and Pratt 1999; Heinich et al. 1996) agree that one of the best attributes of distance education is that it forces faculty members to organize their content. This is not a cynical comment but rather a statement of fact.

Instructional Communications: A Series of Actions

Instructional communications can be viewed as messages between a sender and a receiver where the intent is to transmit information and knowledge in order to improve comprehension and/or performance. These actions are usually considered in sets, or cycles. Instructional com-

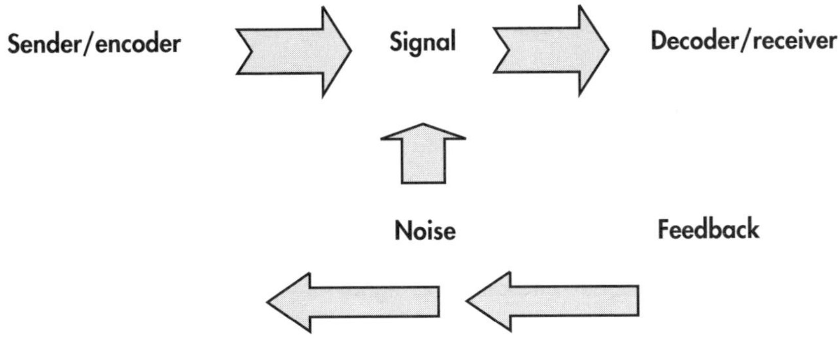

FIGURE 1: The Basic Communications Model

SOURCE: Created by the author.

munications cycles differ from standard communications cycles because the elements of guidance and assessment are introduced. For example, if a user were to go to the Internet to locate a map of Mexico and found and printed the map, she would have completed a communications cycle. If the user went to the Internet and found a virtual tour of Mexico, it would certainly be more engaging than just a map, and there is a good chance she would gain new knowledge. This transaction, however, would still not have met the criteria for an instructional communication. If the user went to the Internet and found a tutorial on the geography of Mexico, learned new facts, tested knowledge, and revisited the facts missed, she would have completed an instructional communications cycle. Not only would she have acquired new knowledge; she also would be secure in that knowledge because she had received guidance and assessment to confirm its accuracy.

When we apply this real-life example to the Shannon-Weaver model described above, however, we see it is not an ideal fit. The original model was designed to help in electronic and mechanical innovation, so naturally it was mechanistic in its approach. More recent versions (Schramm 1954) were able to humanize it by adding descriptors to the areas around sender and receiver such as "field of experience" to show that the communication must be put into a context. Between 1960 and 1990, this model fell somewhat out of favor because it was simplistic and tended to understate how drastic the disruptive effect of noise could be. But today we see its elegance and utility as we work to improve distance and other electronically enhanced instruction.

To apply the Shannon-Weaver model to instructional communications, we must accept the idea that the model can be duplicated and modified an infinite number of times. (See Figure 2.)

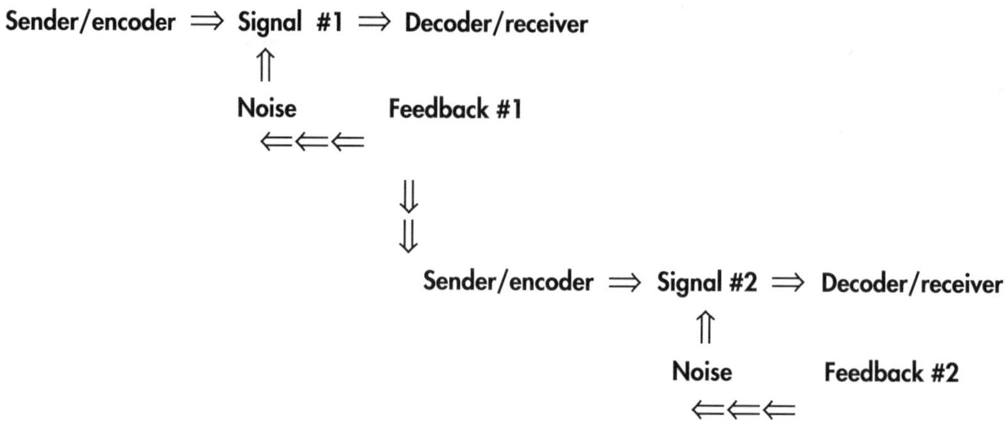

FIGURE 2: The Looped Communications Model

SOURCE: Created by the author.

Consider this analogy: If the Shannon-Weaver model represents a loop in a thread, instructional communications represents a series of connected yet varied loops. An entire lesson would become a woven fabric. Every loop or feedback cycle connects to the past, corrects any misconceptions, picks up new information, and discards what no longer needs to be carried. The feedback informs the next cycle. The new information is aimed at producing the final feedback from receivers, confirming that they have met the target level of success (comprehension and/or performance) at the end of the series.

W. Schramm's communication model builds on the Shannon-Weaver model by adding the humanistic concept of "field of experience." (Cognitivists often refer to this as the "schema" of the sender and receiver.) Experience, or the schema, colors the effect of the signal on both sides of the communication. An instructor sends a signal, and receivers process it by building on what they already know. If what receivers know is correct and they have formed the correct relationship between old and new information, the feedback should confirm that receivers are ready to receive a new concept. If receivers have made connections that are incorrect, the next message must be designed to correct these inaccuracies. The success of this application relies on frequent, thoughtful feedback.

It is human nature to assume that the field of experience of our learners is at least fairly similar to our own. Therefore, it is helpful to take purposeful steps to confirm or deny the assumption early in the communications process. If feedback from the opening communications cycles verifies common understanding of key terms and concepts, there is a solid foundation for instruction. If there are differences in the fields of experiences that will alter meaning, the sender and receiver must reach common ground before more meaningful instructional communications can occur.

The more recent constructivist approach to understanding learning presents the idea that receivers not only process information within their field of experience but also have the benefit of all previous signals and feedbacks. They are constructing meaning based not only on decoding the new signals but also on past experiences and the experience of developing and sending the feedback. Constructivists believe that the total is more than the sum of the parts because the receivers have constructed broader and richer meaning. When the model includes more than one receiver, the feedback can be developed and delivered by a group. Each member of the group benefits not only from the incoming signal but also from the additional signals, including the constructed meanings, of other group members. The feedback has become more complex; therefore the next signal will be more complex. This model supports richer exchange but presents more opportunity for error.

Although the theory of instructional communications as a series of actions sounds fairly straightforward, it is often a real challenge to develop efficient and effective instructional communications. The solution to meeting that challenge may lie in the series of assumptions and decisions that the sender (i.e., instructor) makes and in which order those decisions are made. Here is where we can put the concept of instructional communications as a field of study to very good use.

Instructional Communications as a Discipline

According to one researcher, "Much of the current theory and research in education emphasizes the important role that communication plays in a learning process. Unfortunately, that research is rarely informed by mature theories about communication" (Darling 1992, 205).

Part of the difficulty is that in the academic world, "communication" and "education" are typically seen as two distinct disciplines when, in fact, they grow from the same roots. They are more alike than they are different, and they are becoming more alike every day. Visualize, for a moment, a television tuned to CNN where the screen includes several active windows of information. Now picture a computer screen, again with several active areas. Finally, visualize a computer screen for a distance learning course. It is more than just similar technology that is driving these shared designs. Everyone, regardless of discipline or profession, is applying what current research has told us is the most effective design in the use of graphic arts, type design, spacing, color, and motion.

Today's educators are often called upon to be both instructors and communications designers. And since most educators already have enough on their plates dealing with increased workloads and new instructional methods, knowing when to seek proficient technical support is probably a key. When that technical support comes from those who have applied communications research findings to their work, the chances of success are greatly increased.

Technology has made instructional communications more competitive. Learners are becoming increasingly more critical and selective (Brock 1990). They are now expecting far more than basic media presentations to capture and hold their attention. Learners today are more demanding because they have seen what technology can do. If they can receive a movie preview at their favorite website, why can't they also see a video clip of their distant instructor making the major points in a lecture or lesson?

Tools of Instructional Communications

As instructional communications relies more and more on technology, education administrators have become more critical and selective in acquiring new technologies. Not only is there the initial cost of purchase; there are also the costs of staff training, installation, and maintenance. In the same way that technology has made us evaluate our teaching methods, we must now also evaluate our selection of media to deliver instructional content.

Not only must the message be well designed; the signal path must also be appropriate for the message. Should this be a synchronous or an asynchronous communications cycle? Does the convenience of asynchronous delivery and feedback outweigh the instructional benefit of immediate (synchronous) feedback? Is text the best vehicle for delivery, or are audio and video also needed? How much scarce and valuable communication time should be spent on feedback and assessment, and how much responsibility should be placed on the learner to meet the comprehension and/or performance goals? The answers to these critical questions underpin instructional design.

When instructional communications is delivered in highly structured electronic formats, such as the learning platforms used in distance and web-enhanced instruction, many design decisions must be made in advance. It is not practical to alter formats or presentation schemes in the middle of a course. Fortunate instructors can build on their own experience and the experience of others. Because distance course materials are often used more than once, decisions about the message and the signal path are made not with feedback from the current receiver but with feedback from those who previously had the experience.

When the course is launched, part of the initial field of experience communication must focus on the management of the learning platform, or signal path. Students must become familiar not only with the instructional content and style but also with the design of the course site. Where are course materials located, and how are they accessed? Are all materials located at the site, or are students expected to research and contribute additional materials? What feedback is expected from learners and in what format? A thorough understanding of instructional communications theory is critical in making good design decisions.

Challenges to the Instructional Communications Cycle

One of the most serious challenges to the instructional communications cycle is "noise." In the original Shannon-Weaver transmission model of communication, noise was literally that: extraneous electronic sound that interfered with the clear reception of the signal. Today we use the term to cover a wider range of possible interference. In a traditional classroom, noise can be something as innocuous as the looming homecoming game that has distracted learners. In distance education, noise can be actual or virtual. Actual noise can include technical problems with websites and learning platforms, connectivity, transmission speed, and timing-out. Virtual noise is subtler but can be equally disruptive. A lack of 24/7 technical support; issues with locating online library resources; problems with students participating equally in chat rooms and online assignments—all have the same effect of distorting communications.

In traditional classrooms, instructors receive direct, immediate feedback from multiple sources, including body language, eye contact, spoken language, and written work. This provides intentional and unintentional feedback, both of which help the instructor realize the degree of noise in the environment. Instructors can quickly identify potential noise by a process of observation and elimination. In distance education, the issue of noise is more complex because the instructor typically has only intentional feedback to identify a problem. The instructor may not know that a link to a resource is broken unless a student tells her. The instructor may not know a quiz has gone offline prematurely unless she receives a frantic e-mail asking her to repost it. As for attention span, unless the instructor is using interactive television or is in a synchronous chat room, it becomes the learner's and not the instructor's problem.

The instructional communications cycle for distance education and web-enhanced instruction is more complex than face-to-face learning, but we would not be where we are today with growing ranks of distance students if it were not for a clear understanding of basic communications models. Good teaching is good teaching. The technologies of instructional communications do not make teaching better or worse. What they do is magnify the instruction. Instruction becomes more accessible, more visible, more tangible, and therefore more open to scrutiny. We're no longer singing in the shower but have been moved onto a stage. The new environment is more challenging but also more rewarding—and more fun.

Karen Hughes Miller

See also Communication Theory; Computer-Mediated Communication; Distance Education; Message Design; Schramm, Wilbur

References

Brock, Dee. 1990. "Using Technology to Deliver Education." *Bulletin of the American Society for Information Science* (August–September).

Darling, Ann. L. 1992. "Instructional Communication Research: A Master Syllabus for a Graduate Seminar." *Communications Education* 41: 204–214.

Heinich, Robert, et al. 1996. *Instructional Media and the Technologies of Instruction.* Englewood Cliffs, NJ: Prentice-Hall.

Palloff, Rena M., and Keith Pratt. 1999. *Building Learning Communities in Cyberspace: Effective Strategies for the Online Classroom.* San Francisco: Jossey-Bass.

Schramm, Wilbur. 1954. "Procedures and Effects of Mass Communication." In Nelson B. Henry (ed.), *Mass Media and Education* (pp. 113–130). Chicago: University of Chicago Press.

Simonds, Cheri J. 2001. "Reflecting on the Relationship between Instructional Communications Theory and Teaching Practices." *Communication Studies* 54 (4): 260–265.

Willis, Barry. 1992. "Instructional Development in Distance Education." *Educational Resources Information Center Digest.* ERIC Document Reproduction Service No. ED351007.

Instructional Design (ID)

The discipline of instructional design offers a series of complementary design practices that enable the identification of needs and the design of best solutions. These solutions are most often instructional in nature, but they can also be informational or activity-oriented—focusing on improved processes, tools, and conditions for human activity. ID is based on a systematic methodology for assessing needs and developing goals, analyzing the content domain and target population, and designing, developing, and evaluating solutions.

ID Practices

A hallmark of the instructional design discipline is the attention that it places upon the needs of the target population. Through the practice of *needs assessment,* the designer solicits evidence of need from members of the target population and from those having a stake in their success. The format of needs assessments can take many forms, including interviews, onsite observations, focus groups, surveys, and analysis of resources such as documentation or standards; depending on the situation, it can be complex and time-intensive or flexible and implemented throughout the design process. Learning and performance problems are made evident by examining differences between desired and actual performance. The physical and organizational contexts are also carefully examined to determine influential factors and to align possible solutions with the mission of the organization. An analysis of the needs data leads to one of the final products of needs assessment: the goal statement, indicating what members of the target population will be able to do upon completion of the instruction (or other initiative). Another outcome is an indication of whether this goal can be met by instruction (or practice, or feedback on performance, etc.) or by some other intervention (anything from performance support, like checklists or users' guides, to environmental modifications, such as task lighting or allocation of personnel or resources).

Clients often seek to have an instructional design created after deciding that they have an instructional need. They sometimes resist the conduct of a needs assessment, stating that they already know what intervention is needed. Needs assessment, however, has the potential for uncovering very different needs than initially anticipated; lack of performance is not always due to lack of knowledge or skill, and in this case an instructional solution will not result in achievement of the goal. When the client's assumed need *is* correct, needs assessment still enables a much better understanding of the need and the environment in which it is present.

In *instructional analysis* (also known as front-end analysis), an instructional goal is broken down into component parts, or tasks, in order to identify the knowledge, skills, and information that must be learned, as well as the type of learning each task represents (intellectual, psychomotor, or metacognitive skills; verbal information; and/or attitudes). Characteristics of the target population and environment are also considered to identify any characteristics that will affect the learning process. Specific learning objectives are then defined that specify the desired performance and, at times, the conditions and criteria for performance. In order to determine the effectiveness of the instruction, assessment methods are also defined at this point; these directly reference the performance, conditions, and criteria indicated in the objectives. The result is a clearly articulated set of specifications for the instruction that needs to be produced, as well as a methodology for assessing the effectiveness of that instruction for achieving the goal and objectives.

Design is the heart of the discipline. Here, instructional strategies are devised to help the target population achieve each one of the objectives. Frameworks such as the Nine Events of Instruction (Gagné 1974, 1977) are useful tools that guide the designer in:

- Capturing the learners' interest and attention
- Informing learners of the learning objectives
- Encouraging learners to recall relevant prior learning
- Providing distinctive stimuli
- Guiding the learner
- Providing practice opportunities

- Providing feedback on performance
- Assessing learner performance
- Encouraging retention of what is learned for transfer to novel situations

The design serves as the blueprint for the *instructional development* practice. Designers first search for existing instructional materials that address the instructional goal and objectives and that employ the desired strategies. If appropriate materials are found, they may be used as is or adapted to better meet the instructional goal and objectives. If new materials are designed, the instructional designer will select the most appropriate medium (if not already dictated by client or learning environment) via models (see, e.g., Reiser and Gagné 1982, 1983). The desired instructional materials are then produced and immediately subjected to evaluation.

The design and development practice of *formative evaluation* helps to determine the effectiveness of the instructional design. Rough materials developed in response to the design are tried out by increasingly larger groups of learners, with revisions to the materials between each evaluation. Although evaluation of materials in this formative stage is critically important, ideally evaluation takes place throughout the practices of instructional design.

An emergent trend in ID practice that encourages early evaluation is *rapid prototyping,* in which designs are evaluated beginning with instructional strategies or with materials in very early, very rough stages. Such prototypical drafts of the design or of its key features are evaluated by members of the target population. The design is subsequently revised and implemented a bit more fully with each version. The benefits of this approach are obvious: The designer arrives at the end of the instructional development phase with a set of materials that are already well on their way to meeting learner needs. Rapid prototyping also brings instructional designers and materials developers together early in the process; this early collaboration can increase the likelihood of appropriate media selection, strategies design, and development.

Summative evaluation, considered to be part of the instructional design process, takes place once the instructional materials have been completed and are in use. Effectiveness data are collected to help others decide whether to adopt the materials, or to determine whether preexisting instruction is meeting learner and stakeholder needs. In this way, it can lead directly to a needs assessment, and the ID cycle is begun anew.

Evolution of Instructional Design

The tools of instructional design practice are built upon the foundations that form the profession's specialized knowledge, including systematic design, learning theory, communication, and technology. The foundations of these diverse disciplines can be traced as far back as the Elder Sophists, who developed tutorial systems that built upon expository lectures and group discussions (c. 500–380 B.C.), yet the relationship between them and the design of instruction began to evolve around the turn of the twentieth century. At this time, an extraordinary number of scientific discoveries emerged, and technological advances in communications, energy, and transportation changed the way people viewed the world. Systems thinking and the scientific method were embraced as methods for managing this new world, and the science of learning came to the forefront of public awareness.

Calling for a science linking theory to educational practice (Dewey 1916, 1944), educational psychologists integrated and expanded upon the theories of early learning and systems thinkers such as Aristotle, Comenious, Pestalozzi, and Herbart. Edward Thorndike's theory of connectionism, for instance, prescribed carefully refined and sequenced subject matter. It resulted in extensive laboratory research into instructional methods during the 1920s and 1930s that prescribed linear procedures for attaining new skills. This laboratory research also led to the formulation of several theories that would later influence the instructional design process. Franklin Bobbit argued that educational goals should come from an analysis of the skills necessary to succeed in society, and thus the idea of job (task) analysis was born. Ralph Tyler introduced user evaluation in 1933 when he argued that behavioral objectives and assessment could be used to evaluate and revise instruction until the instruction becomes effective in helping learners reach an appropriate level of performance (Pinar et al. 1995).

J. Dewey's most influential methods had their greatest impact on the instructional design process with the advent of constructivist-based design in the 1980s. They were the introduction of the progressive movement, individualized exploratory learning supported by instructors rather than direct instruction, and *reflective judgment,* a five-stage approach to the scientific method that placed emphasis on the learner's experience and on problem-solving as an iterative process (Saettler 1990). Instructional design's earliest ties to technology began during the 1920s with the birth of the audio-visual education movement. Technological innovations provided increasingly efficient means for communicating in the 1920s and 1930s and their usefulness in education was explored; film became popular as a medium for delivering information, and in the 1930s radio-based instruction was implemented. New markets for technology, including public schools, museums, and clubs, increased the need for audiovisual materials. Traveling educational museums, for instance, needed portable exhibits such as stereographs, slides, films, prints, and other curricular aids. Commercial enterprises saw a market for educational media and focused on the development of innovative technologies and their potential for instructional products (Reiser 1987).

Prior to World War II, much of the learning that took place was via the apprenticeship model: Experts were responsible for training adult workers by sharing what they knew and demonstrating their skills. Apprentices would learn these skills under their supervision and develop the skill required for a lifelong career. The onset of World War II, however, resulted in unprecedented demands in the United States for efficient training materials and instructional methods designed for mass distribution. Thousands of military personnel and industrial workers needed to be trained quickly and efficiently to perform and manage thousands of tasks, whether on the front lines or in the factories (Saettler 1990). Teams of instructional developers were created to meet this challenge, consisting of instructional designers, educational psychologists, subject-matter experts, and producers. Instructional design evolved as a systematic process that integrated design and development, implementation, evaluation, and revision (Dick 1993)

Introduced in the 1940s, general systems theory (GST) is considered one of the primary influences in the development of instructional design practice in the 1950s. In the work of one biologist (Bertalanffy 1969), GST was quite similar in principle to earlier systems analysis, yet it formalized the study of systems analysis as a discipline. It was a science exploring environments, their component parts, and the influences of input and output of components from other systems. Using the basic components of a living organism as a metaphor for systems analysis, GST offered a framework for instructional design by systematizing a means for analyzing tasks or problems and their effects on people or events. Grounded in the assumption that learning should be developed methodically and have measurable outcomes (Seels and Glasgow 1990), the systems view described the world as a hierarchy of interdependent components. When analyzed, they defined patterns that, in turn, could help humans understand the complexities of the world. Systems models describe these patterns and enabled users to interpret situations by guiding decisions and actions (Rowland 1999).

One of the earliest learning theories to influence instructional design was Bloom's taxonomy (Bloom 1956), a classification system defining intellectual behavior as cognitive, affective (attitudinal), and psychomotor (physical) domains. Bloom and his team of educational psychologists explored the stages of intellectual development within the cognitive domain. They identified six levels of cognition that ranged from simple recall, or recognition of facts, to the higher-order abstract, or complex, levels of cognition, evaluation, and comparative contrast. They advocated mastery learning as a way to develop cognition (Bloom 1956) and suggested that instructional media could facilitate this process by offering a variety of instructional methods that met learner needs. This was one of the earliest methods directing designers to match content with instructional media and the learner.

Robert Gagné (1965, 1974, 1985) developed methods for creating behavioral objectives with a learning taxonomy that classified six different types of learning outcomes. Closely resembling Bloom's taxonomy, Gagné's taxonomy began to develop these methods when he designed instruction for the Air Force in the late 1950s.

All of this work informed the development of ID models. Prominent among them is the classic systematic design of instruction, a model advanced by Walter Dick and Lou Carey (1978, 1985, 1990, 1996, and 2001), which remains the predominant model used to teach the ID process. Core elements include goal analysis, learner analysis, objectives, design, development, evaluation, and revision. The most notable addition to this ID model is the needs assessment (Kaufman 1972). By the 1980s, more than sixty ID models had been identified (Andrews and Goodson 1980), accompanied by research and debates over which model offered the best method for designing instruction. Most of these ID models contain the ID practices described above.

Applicability of the ID Approach

The instructional design discipline is sometimes criticized as being slow and inflexible. Although ID can be time-consuming, it can also be performed relatively quickly and its practices flexibly applied. For example, in some situations only a needs assessment may be required, whereas in others design may be coupled with evaluation in an iterative prototyping phase. With increasing experience, designers are able to determine how to adapt ID practices in efficient and creative ways that are nonetheless effective.

Teachers often wonder if ID methods can be useful to them. After all, they already teach and guide their students and manage the classroom learning environment. Do they really need to adopt ID methods? The answer is frequently yes, because "teachers in today's restructuring schools may find themselves increasingly involved in design activities" (Morrison, Ross, and Kemp 2001, 13), particularly at schools attempting to meet national initiatives for educational reform. These initiatives frequently call for student-centered, activity-oriented approaches to solving real problems, methods that call on the teacher to design and implement appropriate instructional materials and methods (Morrison, Ross, and Kemp 2001). The need for a systematic approach to ID is even more important if teachers plan to share their instructional materials and methods with others; the ID process helps to ensure that the resulting materials are effective and can stand on their own without the teacher-designer being present to adapt and revise them en route. And when the content area is new and/or the student population is unfamiliar, ID is particularly valuable in providing structured processes for exploring content domains and learner needs.

The importance of good instructional design methods is important not only for the education of our children but also for the training and retraining of adults. The U.S. Census Bureau estimated that investment in corporate training, which was virtually nonexistent prior to World War II, grew to more than $36 billion in 1993. It is continuing to rise at 3–6 percent each year (Bremmer 1993). Organizations need workers who have the technology and cognitive skills that will enable them to solve problems and customize products and services in a timely fashion. Employees are finding themselves in positions where their tasks and responsibilities are in a constant state of flux. Technologies render some disciplines obsolete while spawning demands for new skills, or even careers, overnight. It is not uncommon that individuals must learn new skills or change career tracks. Learning is therefore becoming a lifelong undertaking, with the need for retraining being a constant. Instructional design will be a valuable tool in this process, facilitating success across learning environments.

Mable B. Kinzie
Marti F. Julian
William C. Davis

See also ADDIE Model; Analysis; Association for Educational Communications and Technology; Bloom, Benjamin S.; Evaluation; Gagné, Robert Mills; Instructional Technology; Performance Support; Rapid Prototyping

References

Andrews, Dee H., and Ludwicka A. Goodson. 1980. "A Comparative Analysis of Models of Instructional Design." In G. J. Anglin (ed.), *Instructional Technology: Past, Present, and Future,* 2nd ed. (pp. 161–182) (reprinted from *Journal of Instructional Development*). Englewood, CO: Libraries Unlimited.

Bertalanffy, Ludwig V. 1969. *General System Theory: Foundations, Development, Applications.* New York: Braziller.

Bloom, Benjamin. 1956. *Taxonomy of Educational Objectives, Handbook 1,* 1st ed. New York: Longmans, Green.

Bremer, Jennifer. 1993. *Trading Up.* Report on a symposium to provide state and private sector input into the National Export Strategy. U.S. Dept. of Commerce National Technical Information Service.

Dewey, John. 1916 [1944, 1994]. *Democracy and Education.* ILT Digital Classics. Available online at www.ilt.columbia.edu/academic/texts/dewey/d_e/contents.html.

Dick, Walter. 1993. "Enhanced ISD: A Response to Changing Environments for Learning and Performance." *Educational Technology* 33(2): 12–16.

Dick, Walter, and L. Carey. 1978. *The Systematic Design of Instruction.* Glenview, IL: Scott, Foresman.

———.1985. *The Systematic Design of Instruction,* 2nd ed. Glenview, IL: Scott, Foresman.

———.1990. *The Systematic Design of Instruction,* 3rd ed. Glenview, IL: Scott, Foresman.

———.1996. *The Systematic Design of Instruction,* 4th ed. Glenview, IL: Scott, Foresman.

Dick, W., L. Carey, and J. O. Carey. 2001. *The Systematic Design of Instruction,* 5th ed. New York: Addison, Wesley, Longman.

Gagne, R. M. 1965. *The Conditions of Learning.* New York: Holt, Rinehart, and Winston.

———. 1977. *The Conditions of Learning,* 3rd ed. New York: Holt, Rinehart, and Winston.

———. 1985. *The Conditions of Learning,* 4th ed. New York: CBS College Publishing, Holt, Rinehart, and Winston.

Gagné, Robert M., and Leslie J. Briggs. 1974. *Principles of Instructional Design.* New York: Holt, Rinehart, and Winston

Kaufman, Roger. 1972. *Educational System Planning.* Englewood Cliffs, NJ: Prentice-Hall.

Morrison, Gary R., S. M. Ross, and J. E. Kemp. 2001. *Designing Effective Instruction,* 3rd ed. New York: Wiley.

Pinar, W., et al., eds. 1995. *Understanding Curriculum.* New York: Peter Lang.

Reiser, Robert A. 1987. "Instructional Technology: A History." In R. M. Gagné (ed.), *Instructional Technology: Foundations.* Hillsdale, NJ: Erlbaum.

Reiser, Robert A., and R. M. Gagné. 1982. "Characteristics of Media Selection Models." *Review of Educational Research* 52(4): 499–512.

———. 1983. *Selecting Media for Instruction.* Englewood Cliffs, NJ: Educational Technology Publications.

Rowland, G. 1999. *A Tripartite Seed: The Future Creating Capacity of Designing, Learning, and Systems.* Cresskill, NJ: Hampton Press.

Saettler, Paul. 1990. *The Evolution of American Educational Technology.* Englewood, CO: Libraries Unlimited.

Seels, Barbara, and Zita Glasgow. 1990. *Exercises in Instructional Design.* Columbus, OH: Merrill.

Instructional Objectives

Instructional objectives are clear statements that describe what a learner will be able to do or know as a result of instruction. Although there are many theories and models for developing instructional objectives, each with slight variations, objectives typically include references to the behavior, product, or performance that is expected, the criteria outlining successful performance, and the conditions under which the behavior is to be performed.

In theory, objectives are guidelines for curriculum developers, instructors, and learners. An explicit statement of the expected outcomes for instruction helps guide the development, delivery, and assessment of instruction while also helping learners set appropriate expectations and measure their progress. Because objectives are so useful, most models of instructional design or curriculum development include the creation of objectives as a critical component. In fact, the development of objectives is required for many types of projects, and virtually all curricularists and instructional designers consider sound objectives as indispensable to good practice in the design and development of technology-aided instruction.

Benjamin Bloom, along with a team of educational psychologists and university examiners, published what has become the standard model of educational objectives beginning in the mid-1950s. Bloom's taxonomy, as it has become known around the world, consists of three domains: cognitive, concerned with matters of knowledge and thinking; psychomotor, concerned with physical skills; and affective, concerned with attitudes, values, and feelings. Most instruction, however, is centered on cognitive tasks.

Bloom's model consists of a hierarchical ranking of skills. At each level, particular types of learning occur, and lower-level skills are prerequisites for higher-level ones. Though most instruction often takes place in the lower levels only (knowledge and comprehension), students seem to retain and transfer knowledge better when they engage in learning at the higher levels (application, analysis, synthesis, and evaluation). Implicit in the construction of the taxonomy is the alignment of instructional materials with test questions, such that each instructional objective has a corresponding test item.

Within a decade after Bloom's taxonomy for the cognitive domain was published, objectives had become ubiquitous in educational circles; they became especially popular among instructional systems design theorists, whose focus is training outside the context of traditional schooling. Robert Mager's model, for example, further clarified the construction of objectives and prescribed that they explicitly include the three basic elements: what the learner is to do, what criteria are to be used to measure success, and under what conditions it is to be done. Robert Gagné's model created a slightly different taxonomy and described a five-component objective, which included more explicit items: a description of the environmental characteristics under which the learner will operate; a "learned-capability verb" that relates the objective directly to Gagné's taxonomy; a description of the content or product of the task; an action verb that describes the task; and a list of any tools, constraints, or special conditions that may be required for the performance of the task.

In the instructional design process, the development of objectives is typically intertwined with the process of task analysis. By deconstructing

TABLE 1: The Looped Communications Model

Taxonomy Level	Appropriate Action Verbs
Knowledge	list, state, name, define, recall, recognize, identify
Comprehension	distinguish, compare, contrast, describe, discuss
Application	write, apply, create, use, choose, operate, construct, select
Analysis	diagram, assess, infer, explain, analyze, categorize
Synthesis	develop, design, synthesize, create, compose, generate, plan
Evaluation	decide, test, measure, rank, justify, judge

SOURCE: Created by the author.

the component steps that make up a particular task, instructional designers can be more certain of the individual skills that a learner must know; those skills usually translate into objectives. Because objectives written in this way are so specific, they almost always correspond directly to testing and assessment items.

When designing objectives, it often helps to start by defining the action that the learner should be able to perform. Action verbs are key, as they can describe the learner's behavior concretely and specifically. A selection of common action verbs for the cognitive domain, and the cognitive category of Bloom's taxonomy that they represent, are listed in Table 1.

Almost inseparable from the action is the behavior or product that the learner will develop. When describing the outcome, be as specific as possible about the behavior or product. For instance, the following statement of behavior, though clear, is not as specific as it could be: "On an exam, the student will be able to *name the parts of speech.*" Therefore, better choice for an instructional objective would be the following: "On an exam, the student will be able to *name the eight major parts of speech in the English language.*" When defining the outcome, it is important to focus on what the learner will do to demonstrate mastery of the task rather than focusing on the instruction or the subject matter.

Once the product is determined, consider how to assess learners' performance of the task. Again, be as specific as possible and think only in terms of what criteria actually is to be considered when determining successful performance. Usually, performance criteria are measures of things like time, accuracy, and quality. For instance, the following statement has a relatively vague criterion for successful performance: "The student will be able to write a research report *as discussed in class.*" A more measurable objective might have the following statement, which describes in more detail what level of quality is expected in the finished product: "The student will be able to write a research report *of at least six pages, which contains the five basic parts of a research report, as listed on page 5 of the course pack.*"

Finally, performance objectives should list any conditions under which the behavior or task should be performed. This statement should include particular tools that the learner can use, or perhaps that the learner will not be able to use, when demonstrating mastery of the skill. Think of these as the definition of the performance environment, as in the following example: "*Without the aid of a calculator,* the student will be able to accurately complete four long division problems in two minutes." The condition limits the objective to a particular context so that it can be more accurately taught and learned.

As tasks become more cognitively complex, it becomes much more difficult to write objectives that meet the requirements outlined by the models of Gagné, Mager, and others. For instance, tasks at Bloom's levels of synthesis and evaluation may be difficult to state in terms of observable student behavior. On tasks like these, for which assessment may be relatively subjective or difficult to quantify, be sure to specify as much information as possible about the desired outcome; doing so provides a maximum amount of guidance to developers as well as learners. Vague objectives are meaningless and generally not helpful.

When developing objectives for technology-based learning projects, it is important to remember that objectives should be clear, concise, and focused on what learners should be able to do after they've completed the instruction. Avoid the temptation to create loosely defined objectives that don't clearly specify outcomes, and try not to focus objectives on the subject matter or the technology itself. Regardless of the medium of instruction, well-planned and well-written objectives can help developers create focused and effective learning environments.

Gabriel Reedy

See also Bloom, Benjamin S.; Bloom's Taxonomy; Gagné, Robert Mills; Instructional Design

References

Bloom, Benjamin S., ed. 1956. *Taxonomy of Educational Objectives: The Classification of Educational Goals, Handbook 1: Cognitive Domain.* New York: Longmans, Green, and Company.

Dick, Walter, and Lou M. Carey. 1996. *The Systematic Design of Instruction.* New York: HarperCollins.

Gagné, Robert M. 1985. *The Conditions of Learning.* New York: Holt, Rinehart, and Winston.

Mager, Robert F. 1984. *Preparing Instructional Objectives.* Belmont, CA: David S. Lake Publishers.

Instructional Technology

In popular usage, the term "instructional technology" refers to the use of communications media—hardware and software—to help people learn. At different times, different media have been central to educators' conception of instructional technology. The term first came into widespread use in the early 1960s; James D. Finn, a professor at the University of Southern California, was probably the most vigorous proponent. At that time, the focus was on audiovisual media such as radio, television, film, slides, filmstrips, and audio recordings. By the middle of the 1980s the focus had shifted dramatically to computers.

Within the field of practice of instructional technology, the term from its inception connoted a process or way of thinking about instruction. The idea of technology as a process was supported by economists such as John Kenneth Galbraith, who referred to technology not as things but as "the systematic application of scientific or other organized knowledge to practical tasks" (Galbraith 1967). This notion received influential sanction in 1970 when the Commission on Instructional Technology (1970, 21) issued a report defining instructional technology as "a systematic way of designing, carrying out, and evaluating the total process of learning and teaching." This definition also highlighted the growing commitment to a systems approach to solving problems. The systems approach emerged in the 1960s as a "soft" application of the scientific procedures of systems analysis that developed around the use of computers for military operations research during World War II. This orderly, logical process gave rigor to the rather artistic approaches that previously had characterized film and TV production and the integration of media in teaching.

By 1977 the largest professional association in this field, the Association for Educational Communications and Technology (AECT), adopted an official definition that began by referring to a complex, integrated process for analyzing learning problems and for creating, testing, and implementing solutions to those problems. This definition focused on processes and the systems approach. The most recent definition adopted by AECT refers to "the theory and practice of design, development, utilization, management, and evaluation of processes and resources for learning" (Seels and Richey 1994). By this time, the emphasis is firmly settled on intellectual processes.

There are several difficulties in defining this term. First, there is the popular conception versus the conception held within the field. Although the process notion took firm root within the field of practice, the popular meaning continued to revolve around things, particularly computers. Second, even within the field, the process notion is embraced more by those with a theoretical bent than those engaged in daily use of the tools of the field. So the officially accepted definition, which dwells on processes, is rather disconnected from most users' concrete experiences.

Third, the engineering sorts of attributes encompassed in the process notion are at odds with the values of teachers, professors, and other educators, who are the principal users of instructional technology. The concept of a systematic approach implies to many educators a paint-by-numbers approach that does not resonate with their own view of how instruction is created. Also, the concept of efficiency, which is at the heart of the meaning of technology, evokes the emotive connotation of putting time and money before humane considerations. However, it is difficult to avoid the issue of efficiency. The main claim of instructional technologists is that they are able to achieve effective learning outcomes with less investment in time and resources than through other means. The centrality of this commitment to economy was expressed well in a nineteenth-century book on railroad construction:

> It would be well if engineering were less generally thought of, and even defined, as the art of constructing. In a certain important sense it is rather the art of not constructing: or, to define it rudely, but not inaptly, it is the art of doing well with one dollar, which any bungler can do with two, after a fashion. (quoted in Petroski 1992, 213)

Finally, attempts at defining instructional technology may fail to distinguish between instructional technology as a theory and as a field of practice. Viewed as a theory, instructional technology can be seen as the proposition that some ways of creating and presenting instruction are more efficacious than others. As a field of practice, instructional technology is what people do when they are applying the theory: analyze problems, plan solutions, create materials, use the materials, evaluate the results, and so on.

It is possible to construct a definition that deals with these difficulties explicitly, uniting theory and practice, encompassing both processes and things, recognizing both the artistic and scientific elements, and asserting efficiency claims in concert with humane ones. Thus instructional technology can be viewed as the art and science of designing, producing, and using—with economy and elegance—solutions to instructional problems; these solutions may combine verbal or audiovisual media and may be experienced with or without human mediation and may take the form of lessons, courses, or whole systems that facilitate learning efficiently, effectively, and humanely.

Michael Molenda

See also Association for Educational Communications and Technology; Computer-Assisted Instruction; Instructional Design; Research on Media and Learning

References

Anglin, Gary. 1995. *Instructional Technology: Past, Present, and Future.* Englewood, CO: Libraries Unlimited.

Commission on Instructional Technology. 1970. *To Improve Learning: A Report to the President and Congress of the United States.* Washington, DC: U.S. Government Printing Office.

Fuller, Jim, and Jeanne Farrington. 1999. *From Training to Performance Improvement: Navigating the Transition.* San Francisco: Jossey-Bass.

Galbraith, John Kenneth. 1967. *The New Industrial State.* Boston: Houghton Mifflin.

Heinich, Robert, et al. 1999. *Instructional Media and Technology for Learning.* Columbus, OH: Prentice-Hall.

International Society for Performance Improvement (www.ispi.org).

———. *Performance Improvement.* Washington, DC: International Society for Performance Improvement.

Performance Improvement Quarterly. Tallahassee: Learning Systems Institute, Florida State University, in cooperation with the National Society for Performance and Instruction.

Petroski, Henry. 1992. *The Evolution of Useful Things.* New York: Alfred A. Knopf.

Saettler, Paul. 1990. *The Evolution of American Educational Technology.* Englewood, CO: Libraries Unlimited.

Seels, Barbara B., and Rita C. Richey. 1994. *Instructional Technology: The Definition and Domains of the Field.* Washington, DC: Association for Educational Communications and Technology.

Stolovitch Harold, Richard Clark, and Steven Condly. 2002. *Incentives, Motivation, and Workplace Performance: Research and Best Practice.* Washington, DC: ISPI/Site Foundation.

Intelligent Tutoring Systems (ITS)

Intelligent tutoring systems are computerized educational systems that incorporate artificial intelligence methods. The original motivation for the development of computerized tutors was the observation that one-to-one human tutoring is much more effective than classroom teaching. Artificial intelligence methods provide reasoning comparable to human experts and are used in ITS development to make decisions in the role of a teacher. Although human tutoring for all students would be ideal, it is not always feasible, and many researchers believe that individualized computer teaching is an improvement over standard lecture methods.

More than forty years of research on intelligent tutors has produced a variety of systems and contributed to the basic understanding of human learning. Systems have been developed to support all educational levels, from grade school through college, as well as industrial education and military training. Subjects range from mathematics, physics, languages, speech, and computer science to customer support, factory equipment maintenance, and medical applications such as surgical training and cardiac care.

Intelligent tutoring methods have been built into computerized forms of standardized tests, homework systems, and complex simulations and embedded in training systems in factories and ships. These systems can operate on personal computers or may be distributed widely over the Internet for group training in large organizations.

An intelligent tutoring system may appear like any other educational system in actual usage. The student interface might be an educational game, a simulation, or a series of questions. Interaction can be graphical, textual, spoken, or involve other sensors. Any task that a computer can monitor can be the basis of an ITS. The intelligence, however, modifies the system response, problem selection, and adaptive customization.

The distinction between intelligent and other systems is ambiguous; to be considered *intelligent* a tutoring system requires computer algorithms that make significant instructional decisions. The knowledge upon which these decisions are based can be obtained from expert teachers, experts in the subject matter, psychologists, or other experts in cognitive science. Some knowledge may be derived from records of past student performance using machine learning techniques or statistical analysis. Applied research methods may be used to obtain specific knowledge required for the development of an educational system.

The capabilities of an ITS can be divided into domain knowledge, pedagogical knowledge, and student modeling. The domain knowledge of a system is its ability to do expert work, such as the ability of a physics tutor to solve physics problems, or the ability of an architectural tutor to design buildings. The pedagogical knowledge of a system is the teaching ability (at least to the degree that it can be considered separately). Manipulating information about a student's needs and abilities is considered to be student modeling.

Sophisticated pedagogical knowledge improves the effectiveness of the educational approach. Some systems can vary their presentations from discovery modes to an example-driven approach or a Socratic style. Other systems always follow a single style of teaching. A sophisticated pedagogical model allows the system to reason intelligently about its teaching approach. Alternatively, a system with a fixed teaching approach may be said to encode substantial pedagogical knowledge if it expresses the approach of one or more expert human teachers or if the system has been used extensively and refined empirically. In either case, the system operates on the basis of substantial knowledge of how students really learn the material.

The student model in a tutoring system is a description of an individual student, including the student's abilities, preferences, and history. Automatic evaluation of a student's ability is generally based upon some analysis of the topics that can be learned in the domain. An elementary school mathematics tutor might evaluate the student's knowledge of fractions, decimals, and percentages, whereas a high school geometry tutor would consider lines, points, angles, and so forth. Other forms of student modeling can be based upon fitting a student into predefined categories, called stereotypes or profiles, or adapted from psychological testing techniques.

The important aspect of a student model is that it gives an educational system the ability to adapt instruction to suit the needs of individual students. The primary adaptive mechanism of an intelligent tutor is in the selection of problems and lessons. The goal for automatic problem selection is to keep the student learning material that is challenging but not overwhelming. It is obvious that learning cannot be effective when material is too easy or too hard; it is much less obvious how to automatically select material of an appropriate difficulty level.

Adaptive mechanisms that control factors other than problem selection have been developed but are generally less useful. Specialized adaptivity to ameliorate various handicaps are of great benefit but do not particularly involve tutoring mechanisms. The idea that students have different learning styles is often considered as a basis for student modeling, but few learning styles (other than visual versus textual) have been clearly defined. It has not been established that reasoning about learning styles contributes to better student learning in most systems.

The knowledge captured by an intelligent system can be an explicit, flexible, and modifiable part of the educational software, or the knowledge may be implicitly stored in the design or media content of the system. Neither an explicit nor an implicit manner of using knowledge is better; the reasons for any specific approach are complex. However, it is one aspect of a system that must be considered in selecting one for any specific need.

Often, the terms "knowledge systems" and "intelligence systems" are used interchangeably, but it is helpful to distinguish them. "Intelligence" refers to problem-solving based upon the explicit manipulation of representations, whereas "knowledge" is a measure of how much expertise is encoded in the system. Powerful intelligent reasoning makes a system more flexible in its responses.

Knowledge can be encoded by giving the system more reasoning ability (intelligence) or by providing more and better media for the system. A system with a large library of fixed problems created by a capable author might have more knowledge and greater teaching ability than a system with sophisticated problem-solving algorithms providing a high degree of intelligence. From an educational point of view, it is important that tutoring systems respond appropriately; the mechanism generating this response is only a means to an end.

Systems can be classified as intelligence-based, knowledge-based, or media-based, and there may be a historical progression in that order. An intelligence-based system uses flexible algorithms to compute responses directly from first principles, whereas a knowledge-based system responds using stored solutions to problems.

As developers gain experience with intelligent software, the best response to many choices becomes clear, and reasoning from first principles can be replaced by knowledge recorded in a database. As a system evolves so that more actions are controlled by knowledge, rather than reasoning, the term "knowledge-based" becomes more appropriate than the term "intelligence-based." As the technology to meet educational needs becomes more specialized, the need for flexibility is reduced, handcrafted media (e.g., text, graphics, and animation) can replace explicitly encoded knowledge as well as reasoning from first principles. A system that is fully scripted to follow essentially predetermined paths should be called "media-based."

Education and learning are influenced by many complex factors, some of which cannot be controlled or even known. It is easy to overlook the fact that a student's private thoughts are of overwhelming significance and completely unknown to a computer system (as well as to human teachers). At best, tutors (human or computerized) can make educated guesses about a student's private thoughts (motivations, attention, goals, prior knowledge, and personal theories); as a result, there is a fundamental limit in the degree to which tutoring can be customized for individual needs.

Moreover, systems that automatically change their approach in some fundamental way are very difficult to implement, understand, or deploy. Engineering requirements increase dramatically as a system becomes more complex. The quality of media (pictures, text, sound) is important for effective educational systems and must be developed separately for every approach used by a system. Real-time generation of graphic computer media might someday allow greater flexibility but is not currently feasible for most development projects.

Taken together, the great difficulty involved in creating extremely flexible systems and the unknowable aspect of private student thoughts provide justification for the current mainstream approach to intelligent tutoring, which is to build systems with substantial amounts of carefully crafted media combined with limited intelligent mechanisms.

Research on intelligent tutors has been in progress since the early 1960s and has resulted in contributions of significance far beyond the actual software developed. Early efforts to develop intelligent tutors greatly contributed to knowledge about education, learning, psychology, and computer science. The discovery that children make systematic rather than random mistakes in subtraction was based upon analysis performed during the development of an intelligent tutoring system. Researchers also claim that a primary impetus behind the invention of personal computing systems was the concept that computers could be used for teaching children. Tutoring applications were the original inspiration behind the development of some of the graphic technology now commonly used for computer animation in popular movies.

The long-term prospect for intelligent tutoring systems is a matter of speculation, but a global integration of educational knowledge into flexible teaching systems could develop quickly and have extraordinary consequences.

Chris Eliot

See also Computer-Assisted Instruction; Knowledge Management; Learning Styles; Performance Support

References

Forbus, Kenneth D., and Paul J. Feltovich. 2001. *Smart Machines in Education: The Coming Revolution in Educational Technology.* Menlo Park, CA: AAAI/MIT Press.

Proceedings of AI-ED 2001/2003 (www.cs.usyd.edu.au/~aied).

Proceedings of User Modeling 2003 (forthcoming) (www2.sis.pitt.edu/~um2003/index.html).

Interactive Television

Interactive television is the transmission of televised content ranging from a unidirectional telebroadcasting format with audio feedback to fully interactive videoconferencing via satellite, telephone line, or computer network. It has been a popular medium of distance education, particularly when supplemented by other technologies such as telephone, e-mail, or conventional mail used to transmit feedback. A good instructional design will exploit the advantages of interactive television while at the same time trying to work around the disadvantages. There are at least five advantages to interactive television (Claassen 1994):

1. *Instantaneousness:* a geographically dispersed population can be reached immediately, irrespective of location;

2. *Simultaneousness:* information provided to all remote areas at the same time, regardless of their distance from the source;
3. *Unfiltered information transfer:* All receivers get the same message;
4. *Accessibility:* the ability, particularly thanks to satellite technology, to reach anywhere within its footprint, irrespective of distances or geographic obstacles; and
5. *Affordability:* economies of scale (the larger the target population, the smaller the unit cost).

Although the power of conventional television is widely recognized, one of the frequently cited shortcomings of the medium is its lack of interactivity. Furthermore, although schools may have programs broadcast to them, it can be difficult to fit the broadcast schedule into the school's timetable. Nevertheless, television's strong visual emphasis makes it a much-considered medium to support (and sometimes even carry) education. A proposed solution to many of these problems is sought in interactive television. Thus interactive television's value as an educational technology relies on the level of interactivity and the modes of interactivity that are determined by the instructional design of the communication features. The level of the interaction is defined by the interdependence of participants. The mode of interactivity that takes place is the number of participants, the time frame, and the level of technological sophistication.

Level of interactivity may be determined by the degree to which parties are dependent on one another. There is a useful taxonomy of communicative interdependence, of which interactivity is the highest level (Berlo 1960). From lowest to highest, these four levels are:

1. *Definitional-physical interdependence:* without a sender there is no receiver, and vice versa;
2. *Action-reaction interdependence:* the existence of feedback confirms the sender's success and prescribes anticipated responses to future messages; if the audience supplies laughter as feedback, for instance, future messages will be tailored to evoke more laughter in the same way;
3. *Interdependence of expectations:* requires the receiver to infer meaning based on what the communicating party is doing;
4. *Interaction:* "When two people interact, they put themselves into each other's shoes, try to perceive the world as the other perceives, try to predict how the other will respond" (Berlo 1960, 120).

Most typically, television broadcasts support interactivity on the level of definitional-physical interdependence. Programming content is broadcast with a particular view in mind. When television is regarded as a conversation between the broadcaster and the audience, higher levels of interactivity may be achieved. Situational comedies, for example, provide canned laughter to simulate action-reaction interdependence. Television game shows may take this a step farther via a question-and-answer format. The presenter asks the question and studio audience or home audience try to answer before the contestant does. In this way, one can simulate interdependence of expectations in one-way broadcast television.

In addition to the level of interaction, the mode of interaction is an important aspect of interactive television. The mode of interaction reflects how many people participate interactively, in what time frame, and by which devices. These aspects are known as participation ratio, synchronicity, and symmetry. The level of interaction is directly affected by each of these variables.

Participation ratio is the number of participants and the way in which they participate. This can vary from one-to-one, one-to-many, and many-to-many. A single pupil at a remote site tutored by an individual mentor is participating in a one-to-one setting. Alternatively, thousands of matriculants watching a broadcast, and then phoning in their questions, are participating in a many-to-one setting. Four or five individuals linked to a team of several experts suggest a many-to-many setting. Typically, as participation ratio rises, interactivity drops. In a one-to-one situation, interaction can be potentially very high, whereas it is very difficult to reach the same level of interaction in a 1:800 or 1:8,000 situation.

Synchronicity is the time frame in which the interaction takes place. Synchronous interactivity is that which takes place simultaneously (i.e., both participants are present at the same time). The most obvious example of synchronous interaction is face-to face contact. Live television broadcasts are synchronous but not interactive. Synchronous interactive television is usually facilitated via an open telephone link. The students react either verbally or via digitally networked technologies. Asynchronous interactivity is that which occurs when the participants are not present at the same time. This is most often achieved with written feedback. A television viewer sends in a written request and the presenter replies on air, which the viewer sees on videotape at a later time. The clearest benefit of asynchronicity is that it makes scheduling easier. One of the most frequent downfalls of interactive television lies in the fact that it is very difficult to get all participants to be available at the same time. Often participants convert synchronous communications to asynchronous ones by videotaping, but in doing so they lose the interactivity. In the same way that the level of interactivity decreases as the participation ratio rises, it also decreases as the participation becomes asynchronous.

Symmetry is the combination of technologies mediating the interaction. Symmetrical interaction takes place when both parties use the same medium. In videoconferencing, for example, each participant sits in front of a camera and watches a screen that displays one or all of the other parties. Symmetry is related to bandwidth and cost. Although it might be infinitely more preferable to have bidirectional video, this will cost a great deal more. For cost-saving purposes, therefore, most systems are asymmetrical; video is sent in one direction and feedback usually occurs through audio, fax, or e-mail. This can have educational consequences; for example, if the instructor does not see the learners, he is unable to receive valuable feedback cues such as body language and class response.

It can be seen then that interactive television can range from a low level of interactivity (such as definitional-physical interactivity) to a very high level of interactivity (such as interdependence of expectations), but that true interaction (Berlo 1960) is seldom achieved. Moreover, economic and technological constraints, as well as human nature and scheduling problems, lead to a tendency toward high participation ratios, a preference for asynchronous communication, and the use of asymmetrical mediation, all of which lower the level of interactivity.

It has been proposed that interactivity rests on immediacy of response, nonsequential access of information, adaptability, feedback, options, bidirectional communication, and appropriate grain size (Borsook and Higginbotham-Wheat 1991). Each of these is important for the effective educational use of interactive television.

Immediacy of response accounts for the preference of the telephone above written correspondence. In a synchronous time frame with a small participation ratio, such as a telephone conversation between two people, the response is immediate. In a synchronous time frame with a large participation ratio, such as a live broadcast with phone feedback, the response is delayed, at least until question time and until it is the questioner's turn to be answered. In an asynchronous time frame, an immediacy of response among many participants can be achieved by electronic answering systems or even digital keypads that allow users to punch in numerical responses and have them evaluated immediately.

Nonsequential access of information means that synchronously one can interrupt the speaker and prompt a deviation from the original sequential preparation of the talk. This is what distinguishes live television from videotape. Synchronous nonsequential access of information is extremely difficult to achieve through high-participation interactive television, which means that instructional designers should look for other media instead. Asynchronous nonsequential high-participation access to information will be achieved only when video on demand is technologically feasible given bandwidth constraints (Gates 1995).

Adaptability both of what is said and how it is said is hard to achieve in an asymmetrically mediated interactive situation, given the lack of visual cues between presenter and viewer. The adaptation of the presenter may therefore be highly dependent on the effectiveness of a remote-site facilitator.

Feedback is the foundation of interactivity. "Feedback allows interactive systems to personalize and adapt instruction" (Borsook and Higginbotham-Wheat 1991, 12). The problem with interactive television, though, is that so little feedback can be obtained, particularly if there is a high participation ratio. As indicated previously, digital keypad technology allows feedback to be sent to a computer that can process the feedback and echo it to the presenter in a graphic format. That is to say, in a well-designed system every remote learner may be equipped with a keypad. The learner's response to a question is in the form of selecting a multiple-choice answer and then entering it. This information is then relayed to a computer that lets the presenter know how many users selected each distracter. The presenter can then *adapt* the presentation in accordance with the processed feedback

Options describe the ability of the receiver to select among several choices. One limit of high-participation-ratio interactive television is that if a certain option does not exist on the keypad, for example, then the learner has to select that which is next closest. Open questioning is very difficult to achieve.

Bidirectional communication means that the speaker and listener have to exchange their positions from time to time; otherwise there is no interactivity. Again this is difficult in large-scale interactive broadcasts and might have to be simulated by the methodology.

Grain size is the length of a presentation sequence before input is required. The larger the grain size, the lower the interactivity. The remote-site facilitator should be trained to give feedback in terms of grain size.

In spite of certain limitations, it can still be argued that interactive television needs to be explored as an educational alternative, provided the instructional designer exploits the advantages of interactive television and designs around its disadvantages. Thus it is worth noting two primary disadvantages of interactive television.

A lack of visual feedback is one of the filters that may obscure the interactive communication process. Another may be a lack of knowledgeable facilitators at the remote site. If the instructor/presenter is unable to see what is going on at every desk and cannot obtain a feel for the audience, there will be few clues that viewers are not comprehending the televised content. One solution is to ensure that a trained facilitator be available at remote sites who can inform the presenter if, for example, the pace is too fast or too slow.

A second concern should be the availability of instructional resources (i.e., prepared lessons that can be presented) and secondary resources (printed material to which the instructor may refer learners). Although the image, voice, and surroundings of the presenter may be broadcast, it is impossible to broadcast a library or an object needed to practice a skill.

Interactivity is likely to only be meaningful if it is based on activity. In deciding what learners should do, one might consider the factors that contribute to eliciting responses from television viewers (Fisch and McCann 1993). These are:

1. *Appeal:* refers to the degree in which the instructional design appeals to the target audience. S. Fisch and S. McCann (1993) describe, for example, how a game-show metaphor was used to appeal to a school-age target audience of a mathematics broadcast. Although the broadcast was noninteractive, interaction was simulated both by the presence of a studio audience and by the fact that participation was elicited from viewers.
2. *Clearly defined problems:* Clear problem definition depends on a number of factors. One should conceptualize the problem, then formulate a broad problem followed by subproblems. Fisch and McCann describe a mystery metaphor to engage the audience in problem-solving activities. Noninteractive television has many examples of such mystery games. An example mystery metaphor would be a treasure hunt in which the audience has to guess where a certain object is hidden while a television presenter searches for it. The audience can see what the searcher sees but can provide only audio input. The problem is clearly defined, though: Solve the clue and find the hidden object. The learners should have to achieve mastery of certain content before they can do so.
3. *Think time:* difficult to achieve during broadcast time, as a pause in the program will negatively influence tempo. A better way of allowing think time is by giving homework problems to which the solutions are broadcast in later sessions.
4. *Room for educated guesses:* Educated guessing means that the audience is allowed to "generate a solution even when they are not completely sure of their answer" (Fisch and McCann 1993, 107). This is done by giving clues, much in the same way as a detective story feeds the audience with clues as to whodunit. In interactive television the

presenter could pose a problem and ask the learners to discuss the problem among themselves and try to guess at a solution with the information at their disposal. At a later stage (after some think time) various groups could be asked for their solutions and the correct one could be given.

R. Clarke (2001) argues that it is the method and not the medium that influences the quality of what we learn. This suggests that it is not the technology of interactive television that matters but the way in which instruction is designed to take advantage of its strengths, such as visual impact and many-to-many broadcasting, while also seeking maximum opportunity for high levels of interaction and participation.

Johannes C. Cronjé

See also Clark, Richard E.; Digital Video; Distance Education; Instructional Design; Schramm, Wilbur; Television and Learning

References

Berlo, David. 1960. *The Process of Communication: An Introduction to Theory and Practice.* New York: Holt, Rinehart, and Winston.

Borsook, Terry K., and Nancy Higginbotham-Wheat. 1991. "Interactivity: What Is It and What Can It Do for Computer-Based Instruction?" *Educational Technology* (31)10: 11–17.

Claassen, P. 1994. "Address" to University of Pretoria interest group on interactive television October 18.

Clark, Richard E., ed. 2001. *Learning from Media: Arguments, Analysis, and Evidence.* Greenwich, CT: Information Age Publishing.

Fisch, S. M., and S. K. McCann. 1993. "Making Broadcast Television Participative." *Educational Technology Research and Development* 41(3): 103–109.

Gates, William H. 1995. *The Road Ahead.* London: Viking.

International Society for Performance Improvement (ISPI)

Founded in 1962, the International Society for Performance Improvement is the leading association dedicated to improving productivity and performance in the workplace. ISPI represents more than 10,000 members throughout the United States, Canada, and forty other countries. ISPI's mission is to develop and recognize the proficiency of members and advocate the use of human performance technology (HPT). ISPI works toward achieving this mission by sponsoring an annual conference and exposition, conducting other educational events such as institutes and workshops, publishing books and periodicals, and supporting research. Improving organizational capability; increasing the ability to compete; and finding, hiring, and developing talent are issues that top the agendas of every business leader. When organizational leaders face the challenges of improving the performance and productivity of their teams and their people, it is increasingly likely that they will find themselves seeking support and advice from a member of ISPI.

ISPI has a long history beginning in 1962, when its first members formed the Programmed Learning Society during a dinner on January 29, 1962, at the officers' club of Randolph Air Force Base in San Antonio, Texas. The original purpose of the society was to collect, develop, and diffuse information about programmed instruction. In February of that year the first issue of the *Programmed Learning Society Newsletter* was published. In March the society adopted a new name—the National Society for Programmed Instruction (NSPI)—and adopted its first constitution.

Word of this new organization traveled quickly, and just a year later NSPI received inquiries about forming chapters in France and Germany. In March 1963 the organization held its first convention, and more than 650 people attended; that year, the NSPI publications began to bloom with the merging of the Institute of International Research and Development's magazine with NSPI's newsletter. The new publication, *NSPI Journal,* was one of the first to devote itself entirely to the area of performance improvement.

Throughout the late 1960s and early 1970s, NSPI continued to develop as an organization dedicated to research and practice. It was soon clear that with the size and growth of the organization professional management was needed, and in 1971 NSPI hired its first executive director.

A groundbreaking moment arrived in March 1972. Although training and instruction served as the initial focus of NSPI, the organization published a journal with much greater breadth. *Improving Human Performance* became a research quarterly. To match this philosophical shift, the organization changed its name once more in 1973. It was now the National Society for Performance and Instruction.

Over the next ten years, NSPI continued to gain greater recognition and win numerous awards for its programs and publications. In 1983 NSPI added corporate patrons as a new membership category. Its first patron member was DuoTech, Inc., from San Diego, California. Continuing its corporate outreach, four years later ISPI gained its first advocate: IBM.

In 1992 the human performance technology model was developed. That same year, the *Handbook of Human Performance Technology,* edited by Harold Stolovitch and Erica Keeps, was copublished with Jossey-Bass. The *Handbook* served as a unifying doctrine for the profession of performance technology while providing readers with the fundamental knowledge necessary to improve performance in the workplace.

In recognition of its international presence, NSPI again considered a name change and in 1995 became the International Society for Performance Improvement. With members living and working in more than thirty countries, this change more accurately reflected the membership while acknowledging the varied interventions offered by performance technologists.

Since then ISPI has continued to serve as the leading organization in the field of human and organizational performance. Members of ISPI include performance technologists, organizational consultants, academic leaders, training directors, human resources managers, instructional technologists, and human factors practitioners. They work in a variety of settings, including business, government, health services, banking, academia, and the armed forces. ISPI's vision is that members have the proficiency and insight to customize human performance technology to meet the needs and goals of their organizations and clients, so that the members are recognized as valued assets.

Human performance technology is a systematic approach to the selection, analysis, design, development, implementation, and evaluation of programs to cost-effectively influence human behavior and accomplishment. It is a systematic combination of three fundamental processes: performance analysis, cause analysis, and intervention selection and can be applied to individuals, small groups, and large organizations. HPT encompasses a wide range of interventions that are drawn from many disciplines, including behavioral psychology, instructional systems design, organizational development, and human resources management. It stresses a rigorous analysis of current and desired levels of performance, identification of the causes for existing performance gaps, a wide range of interventions with which to improve performance, guidance of the change management process, and evaluation of the results. ISPI members recognize four principles core to the practice of HPT:

1. *HPT focuses on outcomes:* Focusing on outcomes, or results, allows for questioning, confirming, and reconfirming that people share the same vision and goals; that the job procedures support productivity, efficiency, and quality; and that people have the knowledge, skills, and motivation they require to perform at the level expected within their organizations. The initial question asked by human performance technologists is generally, "Where is there an opportunity or a performance gap, a difference between the present and the desired levels of performance?" The outcomes, or results, of an intervention will be measured to determine whether or not performance has improved. Sometimes it is necessary to challenge the assumed answer to a problem or the preconceived outcomes of an intervention and instead focus on the accomplishment or business need that is the client's true priority.
2. *HPT takes a systems view:* Because organizations are complex systems, a change in one area is likely to affect operations in another. As

a result, competent human performance practitioners take a systems view in their approach to problem-solving. It is important to distinguish a systems approach from a process model. A process contains inputs and outputs with feedback loops. A system implies an interconnected complex of functionally related components. The effectiveness of each unit depends on how it fits into the whole, and the effectiveness of the whole depends on the way each unit functions. A systems approach considers the larger environment that impacts processes and other work. The environment includes inputs; more important, it includes pressures, expectations, constraints, and consequences.

3. *HPT adds value:* This is an assessment that clients will be asked to make. Clients should be offered a process that will help them fully understand the implications of their choices, set appropriate measures, identify barriers and tradeoffs, and take control. While HPT requires a focus on intermediate goals (such as improving quality, customer retention, and cost reduction), its success is measured in improvements in desired business outcomes (such as revenues, profitability, and market share). Alignment of individual performance to intermediate and business outcomes is critical to the HPT methodology. Measurement of results at both of these levels serves two important purposes: communicating the importance of what is being done, and assessing the amount of performance improvement.
4. *HPT establishes partnerships:* Performance improvement professionals work in partnership with clients and other specialists. A collaborative effort involves relevant stakeholders in the decisionmaking process and working with specialists in their areas of expertise. Working collaboratively means sharing decisions about goals, process steps, and implementation strategies. Partnerships are created from listening closely to clients and colleagues, with trust and respect for each other's knowledge and expertise.

To address HPT issues, ISPI members make use of six systematic disciplines familiar to human performance improvement practitioners:

1. *Systematically assess needs and opportunities:* Analysis occurs in the beginning of the project. Needs or opportunity analysis is about examining the current situation at any level or levels (societal, organizational, process, or workgroup) to identify the external and internal pressures affecting it. This process determines the deficiencies or performance gaps that require remedy. The output of this phase is a statement describing the current state, the projected future state, and the rationale or business case for action or nonaction.

2. *Systematically analyze the work and workplace to identify causes and factors that limit performance:* Cause analysis is conducted to determine why a gap in performance or expectations exists. Some causes are often very clear, such as a lack of required skills among newly hired employees. This step in the systematic process determines what performance factors should be addressed to improve performance. The output is a statement of why performance is not happening or will not happen without some type of support. Job task analysis includes the identification of the important tasks that employees must perform as well as the knowledge, skills, and abilities to perform them. The output is performance objectives that describe the desired performance, delineate the conditions under which the performance is done, and identify the criteria for successful performance.
3. *Systematically design solutions:* Design is about identifying the key attributes of a solution. The output is a communication that describes the features, attributes, and elements of a solution and the resources required to actualize it.
4. *Systematically develop the solution and its elements:* Development can be done by an individual or by a team. The output is a product, process, system, or technology. Examples include performance support tools and equipment, measurement and feedback systems, a new or reengineered process, the redesign of a workspace, a change in compensation or rewards, election and placement of employees, or a training program.
5. *Systematically implement the solution:* Implementation is the deployment of the solution and management of the change required to sustain it. The outputs are changes in or adoption of the behaviors that are believed to produce the anticipated results or benefits. This standard is about helping clients adopt new behaviors or use new or different tools.
6. *Systematically evaluate the process and the results:* Changes in performance and learning systems are significant events and represent substantial investments of time, energy, and money. Evaluation is necessary to determine the impact a solution has had on the business. Likely measurement targets include the efficiency and effectiveness of what was done, how it was done, and the degree to which the solution produced the desired results so that the cost incurred and the benefits gained can be compared.

There is considerable debate over when to conduct an evaluation. As noted above, it is essential to determining if the desired results have been achieved. Many practitioners believe that evaluation is necessary after each phase of the process. There would be general consensus that some type of review is necessary. In the early phases of a project, formative evaluation would be used to assess the performance analysis, cause analysis, intervention selection and design, and intervention and change phases. Then evaluation focuses on the immediate response of employees and their ability and willingness to perform the desired behaviors. The final evaluations are centered on improvement of business outcomes (such as quality, productivity, sales, customer retention, profitability, and market share), as well as determining return on investment for the intervention.

All ISPI members operate under a set of professional standards intended to promote ethical practice in the profession. The society has recently developed and implemented a professional certification process not unlike those for accountants and financial planners. Certification applicants sign a statement agreeing to conduct themselves in a manner that upholds six core values on which the code is based. These values include:

1. *Add value:* The field of HPT would soon die if its members were incapable of adding value. ISPI expects its members to conduct themselves, and manage their projects and their results, in ways that add value for their clients, their customers, and the global environment.
2. *Use validated practices:* There is craft, and there is science. Craft is nonreplicable. There is no doubt that many sciences began as crafts—the medical field, for example. Although there are many beliefs about how to make people healthy, the majority of the population relies on remedies that have been rigorously tested and approved. In the field of human performance improvement, practitioners are also expected to use and promote only those practices, strategies, and standards that have been validated. First, do no harm.
3. *Collaborate:* ISPI members work in partnership with clients and users, functioning as trustworthy strategic partners.
4. *Continuously improve:* ISPI members are often viewed as organizational physicians. They diagnose, prescribe, and then reassess to ensure the systems are improving. As in the field of health care, the profession is constantly changing, and new advancements in the field allow organizations and their people to improve better, faster, and at less cost. ISPI members must remain abreast of these advancements if they are to serve their communities of interest.
5. *Uphold confidentiality:* Because of the nature of their work, performance consultants frequently have access to confidential material related to the performance of individuals and organizations. It is always necessary to maintain client confidentiality, not allowing for any conflict of interest that would benefit themselves or others.

6. *Integrity.* ISPI members are expected to be honest and truthful in their representations to clients, colleagues, and others with whom they may come in contact while practicing performance technology.

Certification

Consumers, customers, and clients are entitled to information that better enables them to discriminate effective from ineffective performance improvement products, services, and practices. Practitioners are entitled to a set of standards that better enables them to judge their own work and manage their professional development. ISPI believes that certification:

- Encourages practitioners to pursue further professional education and development;
- Improves practitioners' career opportunities through professional contacts; and
- Results in greater recognition by colleagues and employers because certified practitioners have demonstrated their expertise in performance improvement.

ISPI supports the professional development of its members by offering a wide range of reading material on subjects related to the field of HPT. Available resources include:

- *ISPI bookstore:* ISPI offers a wide range of professional books, manuals, handbooks, and other publications through its bookstore. The bookstore operates onsite during conferences, institutes, and workshops and virtually via the ISPI website.
- *Performance Improvement Journal:* A monthly publication sent to all international members; publishes articles related to concepts, research, and the practical application of HPT methods.
- *PerformanceXpress:* The monthly newsletter of ISPI, published entirely online. *PerformanceXpress* has gained a huge following. Articles tend to be 300–500 words so readers have bare-bones facts about recent developments in the profession and on a wide range of solution ideas.
- *Performance Improvement Quarterly:* A peer-reviewed journal created to stimulate professional discussion in the field and to advance the discipline of HPT through publishing scholarly works. A table of contents for all issues published since 1998 is available via the web.
- *ISPI Buyer's Guide:* This guide is distributed to members in order to help them find products and services relevant to perfor-

mance improvement, training, instructional design, and organizational development.

- *Membership directory:* The directory provides current contact information for all international members. Local chapters also produce a directory for local members. The international directory is available via the web. Through it, members can locate others, renew their memberships, and keep ISPI up to date on address and profession changes. The online directory is updated twice a month.
- *Advertising opportunities:* Because ISPI represents a core part of the organizational improvement community, many companies that offer related products and services seek to advertise in ISPI publications. A comprehensive listing of ISPI advertising opportunities is available via the ISPI website or by calling the headquarters office.

One of the main factors that differentiates ISPI from other similar professional organizations is its commitment to research. ISPI has within its ranks a solid contingent of professionals with a long history of research in the many fields related to the development of human and organizational improvement. While many of these individuals have primary homes within academia, there are numerous others from the business community.

Each year, ISPI sponsors research projects that can add to the body of knowledge associated with HPT. Supporting these grants is a dedicated special fund. Each year, a specially designated committee reviews applications for grants. The board of directors reviews their recommendations, and grants of varying size are distributed. Grant recipients are asked to provide reports upon conclusion of their work, to publish a related article in one of the many ISPI publications, and to present their findings at the annual conference.

As part of its mission, ISPI seeks to recognize outstanding achievement. The ISPI Awards of Excellence program is designed to showcase the people, products, innovations, and organizations that represent excellence in the field of instructional and human performance technology. Nominations are accepted annually, typically in October, and awards are presented at the conference the following year. In addition to its achievement awards, ISPI presents three special honorary awards that recognize outstanding individuals and organizations for their significant contributions to human performance technology and to the society itself.

ISPI conducts many events each year for the purposes of educating members about developments in the profession, providing opportunities for collaboration and camaraderie, supporting the need for professional development, and recognizing achievements in the field of HPT.

- *Conference and exposition:* This is ISPI's flagship event. Held in the spring of each year, the conference draws nearly 2,000 members from around the globe, providing the opportunity for them to share their expertise and learn from others. Over four days, attendees have opportunities to select from more than 200 concurrent sessions, keynote speakers, and masters from related fields.
- *Institutes:* These three-day events are conducted as public seminars and, upon request, can be customized for in-house delivery for businesses and other organizations. There are two main themes to the institutes, and the ISPI staff can help organizations select the one that is best for their needs.
- *Principles and practices:* This program offers the fundamentals of HPT from experts in the field. These first-of-their kind events provide participants a forum in which to work with a faculty of human performance improvement experts and colleagues to learn more about the theory, process, and practice of HPT.
- *Making the transition to performance improvement:* This event is designed to provide the knowledge, skills, and resources necessary for training, human resources, and performance professionals to make a successful transition to a human performance improvement organization. It is built around case studies.

On occasion, institutes are also held virtually. This innovation helps those who may wish to attend but cannot travel. The virtual institutes are fully interactive, require multiple collaborative exercises, and have led to the development of many postclass professional relationships.

ISPI Leadership

The day-to-day leadership of ISPI resides with an executive director who manages a small staff out of offices in suburban Washington, D.C., and San Francisco. The staff manages a strategic plan developed in conjunction with an elected board. The board is led by a president who serves for a one-year term, preceded by a year as president-elect. In addition, there are five board members, each of whom serves for a two-year term. The board meets quarterly at sites across the United States.

Supporting various projects, programs, and initiatives are numerous task forces and committees. Leaders of these teams are determined by the board and serve one-year terms that are renewable at the discretion of the board. The committee/task force structure is a key method by which ISPI develops its leadership talent pool.

Member Opportunities and Benefits

Professionals in the field of HPT gain numerous benefits from their ISPI memberships, and there are many ways to become a member. Individuals may join as regular international members, as students, or as retirees. These members receive the membership directory, *PI Journal,* and discounts to conferences and other offerings. In some cases, people choose to join a local chapter. Although this does not bring with it the international benefits, it is often a good way to increase one's ability to meet others in their local environs who share the same professional interests.

Organizations have multiple ways they can support, and be supported by, ISPI. Membership options include patron, sustaining, and advocate. Each level brings various benefits, from multiple individual memberships for employees to special recognition at conferences and institutes. For organizations with numerous employees interested in ISPI, an organizational membership is often the most cost-effective.

Jim Hill

See also Electronic Performance Support System; Evaluation; Just-in-Time Training; Knowledge Management; Performance Support

International Society for Technology in Education (ISTE)

Since its inception, the International Society for Technology in Education has been dedicated to helping educators use technology effectively. ISTE is a nonprofit professional membership society with a worldwide membership of leaders and potential leaders from K-12 and teacher education institutions. The organization's mission is to provide leadership and service to improve teaching and learning by advancing the effective use of educational technologies. It meets its mission through knowledge generation, professional development, and advocacy.

ISTE grew out of a study committee of the Oregon Mathematics Education Council, which was formed in 1971. The organization started as a statewide technology group, known in the 1970s as the Oregon Council for Computer Education (OCCE). Dave Moursund, who later served as executive officer of ISTE for nine years, was a member of that first committee. This group was interested in teacher training and disseminating information about using computers in education. These tasks were accomplished with an annual meeting and a monthly newsletter.

In May 1974 Moursund published the first issue of *The Oregon Computing Teacher,* a periodical targeted at computer-using educators, especially at the K-12 levels and in teacher education. It became *The Computing Teacher* in 1979, when materials and advertisers from *Calculators*

Computer Magazine were transferred to OCCE. At this point, there were 320 members from thirty-two states and six countries. OCCE created the International Council for Computer Education (ICCE), with the OCCE executive committee providing ten of the eleven persons on the board of directors of ICCE. The title of *The Computing Teacher* was changed to *Leading and Learning with Technology,* effective with the May 1995 issue.

ISTE publications designed to support computer-using educators began to emerge in the early 1980s. These ranged from *Evaluator's Guide for Microcomputer-Based Instructional Packages* to a policy statement on network and multiple-machine software. The *Evaluator's Guide* was developed to help educators evaluate educational software. The policy statement outlined the responsibilities of educators and hardware and software vendors for ensuring copyright compliance. In 1987 the ICCE committee on ethics and equity issued the Code of Ethical Conduct for Computer-Using Educators. This code covered nine main areas: curriculum issues, issues relating to computer access, privacy/confidentiality issues, teacher-related issues, student issues, the community, school organizational issues, software issues, and hardware issues.

In 1989 ICCE merged with the International Association for Computers in Education to create a new organization, the International Society for Technology in Education. Its function was to carry on the publishing business and to provide an international organization for those people who had been OCCE members primarily for its publications. In addition to publications, ISTE has provided professional development opportunities for educators since 1979 through its support of the National Educational Computing Conference, which that year was called the Conference on Computers in Undergraduate Curricula. The conference has grown from offering thirty-two sessions to offering more than 350 sessions in 2002.

ISTE has offered professional development opportunities for educators and administrators for more than twenty years through the National Educational Computing Conference. The organization has supported other educational technology conferences as well, such as the Florida Educational Technology Conference and the Northwest Council for Computers in Education's annual conference. ISTE has shared best practices and current research through its journals and other publications.

ISTE is probably best known for its work on technology standards. The first technology standards were developed by the accreditation committee and were adopted in 1994. These standards outlined what teachers should know about technology. In the 1990s, ISTE formed a partnership with educators at all levels, as well as government, businesses, and private foundations, to develop technology standards, guidelines, and tools through the National Educational Technology Standards (NETS) project. To date, they have developed technology standards for students, teachers, and administrators. The NETS standards have been adopted by schools, school districts, colleges, and other educational agencies.

The new ISTE is a result of a merger between the National Educational Computing Association and ISTE. The merger was completed in June 2002. Board members from both organizations felt a merger would strengthen their efforts to support improved teaching, learning, and administration through the appropriate use of technology.

The mission-driven programs that ISTE offers include: National Educational Technology Standards for students, teachers, and administrators; professional development services; the National Educational Computing Conference (NECC); and research and evaluation. ISTE's leadership team provides standards-based training for teachers, technology and curriculum coordinators, media specialists, administrators, and policymakers. NECC provides educators an opportunity to benefit from hands-on workshops, lectures, interactive sessions, and the largest educational technology exhibit in the world. ISTE's research team helps educators and administrators make informed decisions for technology planning and program design using the latest research findings and proven best practices.

For more information, visit www.iste.org or www.neccsite.org

Sally Brewer

See also Association for the Advancement of Computing in Education; Society for Information Technology and Teacher Education; Teachers: Preparation and Training; Technology in K-12 Schools

References

International Council for Computers in Education. 1983. *Evaluator's Guide for Microcomputer-Based Instructional Packages.* Eugene, OR: International Council for Computers in Education. Eric Document Reproduction Services Number ED 240 699.

———. 1983. "ICCE Policy Statement on Network and Multiple Machine Software." *Computing Teacher* 11(2): 18–22.

———. 1987. "Code of Ethical Conduct for Computer-Using Educators: An ICCE Policy Statement." *Computing Teacher* 14(5): 51–53.

ISTE press release. 2002.

Internet Safety

Internet safety entails the guidelines and procedures implemented in public and private settings to protect the well-being of Internet users. Specific Internet safety measures include laws prohibiting illicit activity online and protecting children's privacy; acceptable use policies that stipulate rules for children to access the Internet in schools; family guidelines and parental supervision; instruction in Internet literacy; and implementation of browser access controls and software to filter, block, and monitor children's access to certain sites. As an educational tool, the Internet offers access to extensive resources and information; however, the speed and openness of this media necessitates awareness and intervention to promote safety and well-being.

Although many presume that a child who is linked to the Internet is safe under the protective refuge of his school or home, the potential risks online are comparable to the dangers noted in any expansive environment. The risks include exposure to hate, violence, misinformation, consumer exploitation, and sexual predators. Nonetheless, the Internet provides an extraordinary opportunity for enriching teaching and learning. Internet safety encompasses those initiatives that mediate the online experiences that are disadvantageous to a child's physical, cognitive, and socio-emotional functioning.

Online Risks

The recognition of threats to children in cyberspace is an important first step in developing constructive solutions and a plan of action that fosters protective and productive learning experiences. Without Internet safety, children's vulnerability to the dark side of computer technology is exacerbated by the relative degree of online anonymity coupled with the lack of system-imposed restraints. Although the details of the risks are discomforting, it is important to keep in perspective that awareness and Internet safety measures can facilitate children's safe surfing of those aspects of the Internet that are educational and secure from danger. Moreover, Internet safety promotes the socially relevant use of digital media, including mobile phones, online games, instant messaging, and other emerging online technology.

Crimes against Children

The Internet has provided a new medium for the victimization of children. Children are often naive and trusting of others and simultaneously are in need of attention and affection. This combination of traits contributes to children becoming an easy target to be enticed into illicit interactions with predators.

Sexual exploitation in cyberspace encompasses a continuum of perils to children, including solicitations by predators who lure young victims into face-to-face meetings to molest the child, and engagement with perverse individuals who send lewd messages, engage them into online sex, or supply them with pornography. Reliable figures about the prevalence of the physical molestation of minors by predators in cyberspace are unavailable. However, the Internet creates a context where children are accessible to predators who can easily locate victims and where they can validate their interests by receiving support and advice for aberrant interactions with children through online newsgroups.

These offenders may be pedophiles who prey on children in cyberspace; however, adolescents are also at risk for engaging in situational offenses as they search online for pornography and sexual opportunities. The vast and loose-knit network masks identities and provides a new context where curious and rebellious minors can be seduced and manipulated into indirect victimization through the transfer of sexually explicit information and direct exploitation.

The online predator is skilled at collecting information from children, searching profiles for vulnerable targets, and acquiring personal information on a specific child. Information available online can be used to engage in an online friendship, which is the initiation of the grooming process. Trust is established between the predator and the child through the sharing of information, the use of false identities, sending of gifts and pictures, and eventual desensitization to pornographic content. Subsequently, a meeting between the predator and the child may be arranged.

Infringements on Privacy

The interactivity of the World Wide Web can compromise children's safety when they reveal information about themselves to others. As a child surfs among websites, he leaves a trail that can be used by website operators to make improvements in content or to target children with advertisements. The collection of information from children has been a common practice in cyberspace. Chat rooms, bulletin boards, games, contests, and other online forums have facilitated the disclosure of personal information to strangers. Passive and active information collection has resulted in detailed accounts and profiles of young people online. This widespread accumulation of personally identifiable information about children undermines their privacy. Children are at risk of having their safety compromised when this information is accessible to others interested in online and offline contact. In the interest of protecting children's privacy online while also safeguarding First Amendment and privacy values, federal legislation was passed and implemented through the Children's Online Privacy Protection Act (COPPA). Further protective action has been facilitated through awareness efforts to promote development of practical skills in data protection, personal privacy, and consumer protection of youth online.

Access to Inappropriate Information and Illicit Activity

Without much effort, a child may inadvertently or deliberately be exposed to online content that is obscene, pornographic, violent, racist, or otherwise offensive. The lack of oversight of the content in cyberspace means that misinformation is rampant, and access to contraband, including drugs, alcohol, guns, and gambling activities, goes unchecked. Among young people, computer misconduct can be more tempting in an environment that at least appears to be anonymous and devoid of standard rules of conduct. Without appropriate and clear identification of content and actions that are unsuitable for minors, the well-being of young people online is at risk. This reality necessitates explicit rules regarding access to content on the Internet, critical analysis skills, and implementation of reasonable consequences for violations. In educational contexts, some teachers also undertake the added precaution of prescreening all online material, including e-mail communication, as a preventative measure.

Harassment, Stalking, and Fraud

Youths may be both victims and perpetrators of harassment online. Some Internet users seek power and control by verbally attacking other vulnerable individuals online. Electronic harassment, threats, and humiliation can ensue. Online stalking can be a frightening result when online correspondents continue their harassing behaviors.

Young people who engage in online exploration may be curious about the allure of criminal acts and antisocial behavior. The intrigue and thrill of misconduct in a context that appears anonymous and devoid of consequences can result in the escalation of violations to computer crimes, including privacy infringements, computer hacking, and system intrusions.

Networks of young people may organize themselves online for purposes of collaboratively engaging in the activities of an online gang that harasses others and/or participates in illicit activity. These interactions are part of an online milieu that is conducive to deception. There are many examples of defrauders who attract victims through this inexpensive and quick means. In addition to the activity of the online gangs, children must be wary of deceptive online advertising practices, hoaxes, chain letters, pyramid schemes, and fictitious virus announcements. The challenge for users is to adequately judge the veracity of information online and to take precautions to protect themselves.

E-Commerce

Since young people are a large consumer base, e-commerce has specifically targeted adolescents and children. These age groups are especially susceptible to manipulative marketing techniques with offers of prizes, games, and products. Children's privacy is compromised through marketing and advertising practices that solicit detailed personal information and track online computer use. The development of personal profiles for young people enables marketers to design individualized messages and ads.

Internet Addiction

The merits of the Internet for young people engaged in research and educational exploration are vast; however, these benefits can be overshadowed when individuals become preoccupied with online use. Adolescents may be enticed by the allure of cyberspace and may replace homework, social activities, and other important interests with online relationships and interactive games. Subsequently, excessive Internet use may contribute to a student's failure to attend to school assignments. Poor school performance may be exacerbated by sleep deprivation when sleeping hours are eroded by time spent on the Internet. Excessive users of the Internet may avoid emotional attachments with family and friends in the community and rely on online friendships as a substitute for social interaction. Other activities and interests may decline as a child becomes further obsessed with computer activities.

Legislation

Applicable legislation has not been able to keep pace with technological developments, and the constitutionality of many legal solutions has been challenged. Nonetheless, many states have crafted statutes that address child pornography and computers. Federal and state law also has evolved to ban computers as a means of soliciting children for sexual activity, and the proposal of additional legislation may lead to future statutes designed to safeguard youth.

Recently, infringements of children's privacy online resulted in the passage of COPPA, which focuses on commercial websites and the protection of children age twelve and under. The legislation requires active parental consent prior to the collection of personal information from children under age thirteen. Websites are allowed to respond to children's requests for information via e-mail without parental involvement as long as the information is used for no other purposes than the child's specific request. COPPA signals a trend toward federal efforts to ensure privacy in the interactive medium of the Internet.

Access Control Software

The arduous task of safeguarding children in cyberspace has resulted in the development of filtering and blocking tools. These tools include access control features available through the Internet service provider, websites geared toward children's safe surfing and online interaction, and software that blocks, filters, and/or monitors objectionable content. In classroom settings, tools can integrate chunks of websites into a page accessible to students without going online.

Whereas child-friendly sites provide a contained environment that features fun and educational activities, software may guide children's online exploration with control access. These applications screen sites when a user attempts to gain access and uses researched lists to evaluate the location with a predetermined database of approved and blocked sites. Preestablished lists of blocked areas may be overridden to allow greater access or to restrict online exploration. Some software also provides features that prevent children from divulging personal information, such as their name, age, address, phone number, or school name, to online acquaintances through websites and chat rooms. Users can establish a log of all sites accessed, programs used, and words and phrases typed or received. Similarly, many Internet providers and online services offer site blocking, restrictions on incoming e-mail, and children's accounts that access specific services.

The most apparent weakness of these devices is the ability of a child to disable the tool or circumvent its control. Additional limitations occur when some illicit sites are overlooked and remain accessible or when valid sites are blocked for spurious or unknown reasons. A combination of guidelines for children's use of the Internet and adult supervision are necessary supplements to protective software. Restrictions to appropriate content are only one component of a vigilant process for protecting children.

Acceptable Use Policies

School policies frequently include guidelines designed to create boundaries and barriers that promote safety when accessing Internet resources. Structuring a safe environment involves a declaration of rules, policies, and procedures that clearly communicate that the school will not support behavior that places children at risk. The acceptable use policy typically outlines expectations for behavior and consequences for infringement of the privilege to use computer resources. The signature of a child and parent denotes an understanding and acceptance of the school policies for Internet access and use.

Building Children's Defenses

Since children can be easy targets for exploitation and victimization, awareness and supervision are necessary components of any Internet safety initiative. Children may access content in cyberspace that is obscene, pornographic, violent, hateful, racist, offensive, and illegal. Consequently, the active involvement of caring adults is necessary to prepare them for safe navigation. Direct observation of children online in a public space with periodic interaction and ongoing discussions of their web experiences are the foundations of Internet safety procedures. Without this discussion, young people may be unprepared to deal with the risks that they face. They may miss the warning signs or attempt ineffective solutions that can exacerbate the problem. Embarrassment and fear may prevent them from seeking assistance when problems do occur. Conversely, continual dialogue and preparation weave a net of safety.

In conjunction with early preparatory experiences that engage a child in assessing risky situations, developing appropriate coping techniques, and practicing responses to problematic situations, children can be adequately prepared for potential risks on the Internet. Avoidance techniques, deescalation skills, and protection strategies are additional safety mechanisms children need on the Internet.

Netiquette (online manners) defines acceptable conduct when engaged in an interchange with people in cyberspace. They represent guidelines for relating in a courteous and respectful manner and emphasize an awareness that computers are merely the mechanism for communicating with other individuals. The application of rules that assist young people in making informed decisions and allow them to demonstrate an ability to apply online critical thinking skills facilitates productive social participation. Moreover, it can counter the potential disengagement of young people from positive social interactions, especially when the guidelines limit time spent on the computer. Limits for children can also be set in the form of a contract and should be accompanied by open discussions about disturbing activity and online content.

Netiquette is integrally connected with global understanding, multicultural respect, diversity, and tolerance. With the advent of the World Wide Web, there is broad access to the world, but users lack cultural sensitivity that can foster collaboration in a global community. Young people are especially prone to misperceive the perspectives and opinions of others and to refrain from respectful interactions. Dialogue is important for countering misconceptions and bolstering children's perceptions of themselves and others.

When rules for appropriate conduct are combined with skills in information literacy, young people are more capable of critically evaluating information found on the Internet. The ability to discern between commercial information, advertising, propaganda, opinion, and fact prepares young people for wondrous discoveries and counters potentially frightening realities in cyberspace. It is common for the technology skills of youth to surpass their critical thinking and judgment skills. While laws and attitudes struggle to keep pace with activity online, educators, mental health professionals, and parents have the opportunity to systematically attend to issues of accountability, responsibility, tolerance, and respect.

Michael J. Berson
Ilene R. Berson

See also Acceptable Use Policies; Netiquette

References

Aftab, Perry. 2000. *The Parent's Guide to Protecting Your Children in Cyberspace.* New York: McGraw-Hill.

Arnaldo, Carlos A., ed. 2001. *Child Abuse on the Internet: Ending the Silence.* New York: Berghahn Books.

Berson, Michael J. 2000. "The Computer Can't See You Blush." *Kappa Delta Pi Record* 36(4): 158–162.

Berson, Irene R., Michael J. Berson, and John Ferron. 2002. "Emerging Risks of Violence in the Digital Age: Lessons for Educators from an Online Study of Adolescent Girls in the United States." *Journal of School Violence* 1(2): 51–72.

Gurak, Laura J. 2001. *Cyberliteracy: Navigating the Internet with Awareness.* New Haven, CT: Yale University Press.

McKenna, Katelyn Y. A., and John A. Bargh, eds. 2002. "Consequences of the Internet for Self and Society: Is Social Life Being Transformed?" *Journal of Social Issues* 58(1).

National Center for Missing and Exploited Children (www.missingkids.com).

National Research Council. 2001. *Nontechnical Strategies to Reduce Children's Exposure to Inappropriate Material on the Internet.* Washington, DC: National Academy Press.

Willard, Nancy. 2002. *Computer Ethics, Etiquette, and Safety for the 21st Century Student.* Eugene, OR: International Society for Technology in Education.